9th All Russian Petrographic Conference with International Participation: Petrology of Magnetic and Metamorphic Complexes 2017

IOP Conference Series: Earth and Environmental Science
Volume 110

Tomsk, Russia
28 November - 2 December 2017

Editors:

R.E. Ernst
M.I. Kuzmin
V.V. Vrublevskii
G. Gutierrez-Alonso
N.N. Kruk
A.A. Tsygankov
P.A. Tishin
I.F. Gertner
T.S. Krasnova
A.G. Vladimirov
I.V. Vologdina
A.I. Chernyshov

ISBN: 978-1-5108-5530-4
ISSN: 1755-1307

Printed from e-media with permission by:

Curran Associates, Inc.
57 Morehouse Lane
Red Hook, NY 12571

Some format issues inherent in the e-media version may also appear in this print version.

Printed by Curran Associates, Inc. (2018)

For permission requests, please contact the Institute of Physics
at the address below.

Institute of Physics
Dirac House, Temple Back
Bristol BS1 6BE UK

Phone: 44 1 17 929 7481
Fax: 44 1 17 920 0979

techtracking@iop.org

Additional copies of this publication are available from:

Curran Associates, Inc.
57 Morehouse Lane
Red Hook, NY 12571 USA
Phone: 845-758-0400
Fax: 845-758-2633
Email: curran@proceedings.com
Web: www.proceedings.com

Table of contents

Volume 110

9th All Russian petrographic conference with international participation: Petrology of magmatic and metamorphic complexes

28 November to 2 December 2017, Tomsk, Russian Federation

Accepted papers received: 18 December 2017
Published online: 3 January 2018

Preface

Papers

Preface

These Conference Proceedings contain the selected papers of the 9th All Russian petrographic conference with international participation "**Petrology of magmatic and metamorphic complexes**" which was held from 28 November to 2 December 2017 in Tomsk. The conference was organized by the Faculty of Geology and Geography of the National Research Tomsk State University. It was supported by Russian Foundation for Basic Research and the Government of the Russian Federation (grants No. 17-05-20572, 14.Y26.31.0012). The Program Committee included the members of the Russian Academy of Sciences N.L. Dobretsov, V.A. Koroteev, M.I. Kuzmin, V.V. Reverdatto, V.V. Yarmolyuk, Professor of the Carleton University (Canada) R.E. Ernst, Professor of the University of Salamanca G. Gutierrez-Alonso, Professor of the University of Wyoming K. Chamberlain and other famous scientists.

The presented materials combine the results of a structural and compositional study of magmatic and metamorphic complexes in Europe, Asia and North America. Issues of petrology, formational analysis, geodynamics and mineralogy of crystalline complexes of the lithosphere were considered. This forum became a convenient platform for wider exchange of scientific information between specialists from higher education institutions, academic institutions and companies. Section for young researchers with the aim of developing their scientific and creative potential in the field of modern petrology was foreseen by the conference program.

We thank all participants for the initiative and active support of the forum.

Organizing Committee

Scientific editors:

Name	Institution	Address
Prof. R.E. Ernst	Carleton University Tomsk State University	1125 Colonel By Dr, Ottawa, ON K1S 5B6, Canada 36, Lenin Avenue, Tomsk, Russia
Academician M.I. Kuzmin	A.P. Vinogradov Institute of Geochemistry of the Siberian Branch of the RAS	1a, Favorskogo str., P.o. box 9, Irkutsk, Russia
Prof. V.V. Vrublevskii	Tomsk State University	36, Lenin Avenue, Tomsk, Russia
Prof. G. Gutierrez-Alonso	University of Salamanca	37008 Salamanca, Spain
Dr. N.N. Kruk	V.S. Sobolev Institute of Geology and Mineralogy of the Siberian Branch of the RAS	3, Ac. Koptyuga ave., Novosibirsk, Russia
Dr. A.A. Tsygankov	Geological Institute of the Siberian Branch of the RAS	6a, Sakhyanovoy str., Ulan-Ude, Russia
Ass. Prof. P.A. Tishin	Tomsk State University	36, Lenin Avenue, Tomsk, Russia
Ass. Prof. I.F. Gertner	Tomsk State University	36, Lenin Avenue, Tomsk, Russia
Ass. Prof. T.S. Krasnova	Tomsk State University	36, Lenin Avenue, Tomsk, Russia
Prof. A.G. Vladimirov	Novosibirsk State University V.S. Sobolev Institute of Geology and Mineralogy of the Siberian Branch of the RAS	2 Pirogova Str., Novosibirsk-90, Russia 3, Ac. Koptyuga ave., Novosibirsk, Russia
Ass. Prof. I.V. Vologdina	Tomsk State University	36, Lenin Avenue, Tomsk, Russia
Prof. A.I. Chernyshov	Tomsk State University	36, Lenin Avenue, Tomsk, Russia

e-mail: **conf-petrology@ggf.tsu.ru**

IOP Conf. Series: Earth and Environmental Science **110** (2017) 011003 doi:10.1088/1755-1315/110/1/011003

Peer review statement

All papers published in this volume of *IOP Conference Series: Earth and Environmental Science* have been peer reviewed through processes administered by the proceedings Editors. Reviews were conducted by expert referees to the professional and scienti c standards expected of a proceedings journal published by IOP Publishing.

IOP Conf. Series: Earth and Environmental Science **110** (2017) 012001 doi:10.1088/1755-1315/110/1/012001

Mineralogical and geochemical heterogeneity of the Bazhenov suite

I V Afonin, A V Hitarova

Tomsk State University, Tomsk, 634050, Russia

E-mail: heaven05@list.ru

Abstract. In this work, using one of the reference well the mineralogical and geochemical features of the Bazhenov suite of the Krasnoleninsky arch are considered. Based on the results of X-ray diffraction and X-ray fluorescence analysis, a number of mineral specific associations were established. This allows us clearlytodefine the order of the change in the chemistry of bassen expressing the change in oxidation-reduction and acid-alkaline sedimentation regimes.

1. Introduction

Bazhenov suite is one of the most considerable objects of petroleum geology in Russia. For a long time it was thought that its resources are inapplicable, but the opinion have changed since traditional petroleum sources run low. At the present moment this is the most promising object for petroleum industry of West Siberia. The purpose of this work is to define chemical characteristics of sedimentation basin and to model conditions of sedimentation with use of geochemistry methods.

Objects of the study are core samples of bazhen-abalak complex (U0 and U1 beds) rocks taken from one of oil wells of Em-Egorovskoe deposit.

Sedimentary rocks making up the U0 bed are clay-silicious ones with some pyritization and enriched with organic component. The bulk is clay-silicious (30-60%), with organic matter (10-47 %) and pyrite (up to 25 %). There are some fragments of terrigenous sediments (up to 8 %), some biogenic component made up of carbonates and phosphates (up to 10 %), glauconite grains are rarely observed (up to 1 %).

There are layers of clay-carbonate and carbonate rocks found throughout the section. Clay-carbonate (dolomite) rock consists of carbonates (up to 56 %), clay matter (up to 25 %), silicious matter (up to 5 %), organic matter (up to 15 %), pyrite (up to 5 %) with insignificant admoxture of terrigenous fragments (up to 1 %), glauconite (up to 1 %) and biogenic component (3 %). Carbonate rock consists of clasts of primarily clastic carbonate rock (85 %), so a repeated clastation took place, and of fine-grained calcite clasts (up to 10 %), which are irregularly pigmented with finely dispersed organic matter with pelitomorphic clay matter admixture (2 %).

The U1 bed consist of carbonate rocks interlayered with argillites. Carbonate rock consist of 74-93 % of carbonates and some insignificant admixtures as silicious matter (up to 2 %), organic matter (0,5-5 %), pyrite (1-4 %), terrigenous sediment fragments (0,5-3 %), sometimes galuconite (3-10 %) and clay matter (7 %) and also occasional biogenic fragments (1 %).Argillite mainly consists of clay material (41,5-60 %) and glauconite (18-40 %), then secondary carbonate (10 %), some of silicious matter (0,5%), organic matter (4-5 %), pyrite (3-5 %), terrigenous fragments (0,5-4 %).

2. Methods
This work is based on 93 determinations of petrogenic oxides contents with use of X-ray fluorescence analysis, trace elements contents with use of inductively coupled plasma mass-spectrometry and bulk mineral composition with use of X-ray diffraction analysis. The study was held at the Laboratory of Geochronology and Geodynamics of Tomsk State University. X-ray fluorescence analysis was carried out with use of OXFORD ED 2000 energy-dispersive X-ray fluorescence spectrometer («OXFORD Instruments Analytical»). ICP-MS analysis was carried out with use of Agilent 7500cx spectrometer (Agilent Technologies). X-ray diffraction analysis was carried out with use of X'PERT PRO diffractometer (PANalytical). BCR-2, SGHM-1, SG-3, SSL-1 were used as internal and corrective standards.

Following modules were calculated for lithogeochemicaltypification of the rocks: hydrolyzate modulus (HM), titanium modulus (TM), femic modulus (FM), alumosilicious modulus (AM), normalized alkalinity modulus (NAM). Also geochemical coefficients, which characterize: facial environments of sedimentation – La/Yb; sedimentation enviroments –Ce/Ce*, Eu/Eu* were calculated. Cr/Ni, Cr/Cu aretopofacies indicators [1-8].

3. Discussion
According to U0 bed bulk composition study results it was stated that the main mineral is quartz, which content varies from 7 to 98 %. Also there is pyrite (2-84 %) making up the intensely pyritized interval (sample№ 1-13) and local pyritization emergences further down the section. Sodium feldspar is less widespread: its content counts from trace levels up to 14 %. There is potassium feldspar found in two samples (15 – 3 %; 16 – traces). Carbonates presented are calcite (2-88 %), dolomite (4-76 %) and siderite (2-13 %). Calcite makes up the thick carbonatized layer11-26, dolomite makes up thin lens-like layers. There is 15 % of apatite in sample № 64, also there is barite in samples following: 1 – 38 %, 5 – 70 %; 7 – 9 %.

Clay component content varies from trace levels to 12 %. Clay minerals are kaolinite (10-100 %), hydromica (10-100 %), mixed-layered ones (7-47 %). There is no clay component found in samples № 1-7, 18, 20, 22-24, 27, 30, 32, 33, 45, 47, 48, 49, 53-55, 60, 64, 66.

UK1 bed composition is characterized by prevalence of quartz (4-82 %); quartz content decreases, as a rule, are taking place at intensive carbonatization intervals and limestone layers. Sodium feldspar content varies from 2 to 9 %. Almost all of the samples contain some amount of pyrite (2-21 %), also there is a slightly pyritized interval (samples № 75-78). Carbonates are calcite (2-4 %) and siderite (2-58 %), the latter occurs throughout the whole section.

Clay component content varies from traces to 31 %. It increases since № 76sample, so there is higher clayiness further. Clay minerals are kaolinite (7-80 %), hydromica (14-49 %), mixed-layered minerals (4-46 %), chlorite (2-12 %)

There is presence of organic component and amorphous silica in all of the samples.

On the basis of minerals distribution it was stated that there is cyclical composition of the section with a few assemblages alternation: quartz, clay-quartz, pyrite-quartz, pyrite-clay-quartz, carbonate-quartz, carbonate-clay-quartz pyrite-carbonate-quartz (figure 1). Separation of the assemblages was carried out according to minerals contents: more than 20 % for total amount of carbonates and about 10 % for pyrite and total clay minerals contents.

According to X-ray fluorescence analysis results 5 main associations of elements were defined: silicious, aluminum-silicious, calcium–silicious, iron-calcium-aluminum-silicious and iron-aluminum-silicious ones. This separation was carried out according to dispersion values (elements with dispersion value less than 5 were not taken into consideration) and median values (CaO=9%, Al_2O_3=15 %, Fe_2O_3=10 %.), silicon dioxide was counted for each association as the main component of rocks matrix (figure 2). Associations variations correspond to cyclicity of sedimentation processes, match with previously separated mineral assemblages almost completely and indicate changes of physical and chemical parameters of sedimentation basin:

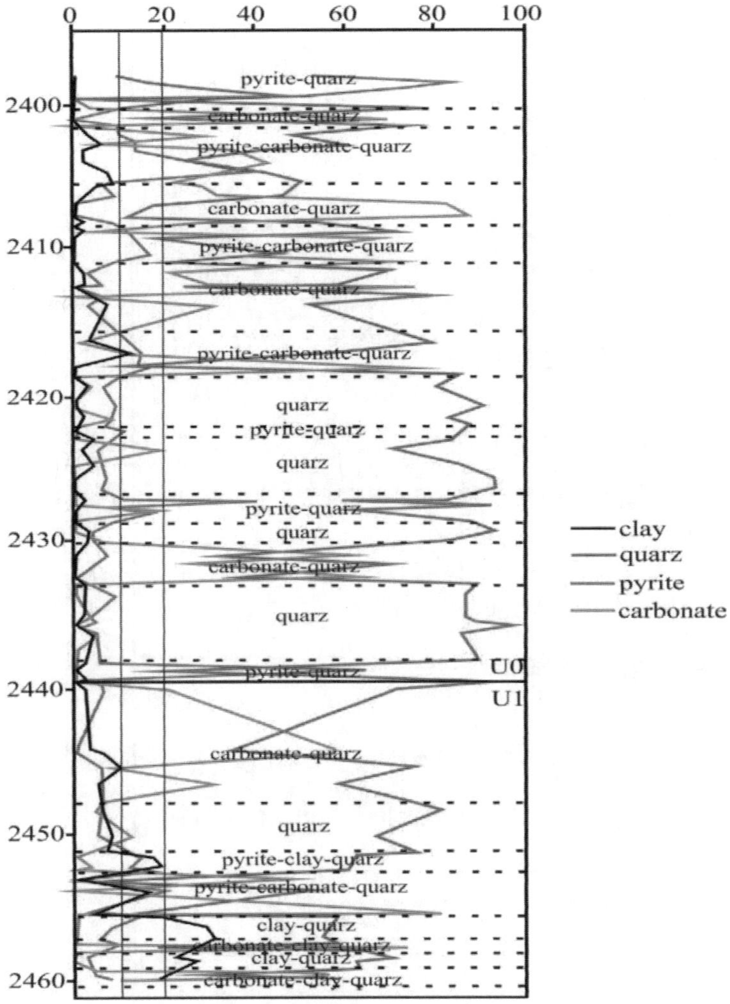

Figure 1. Mineral associations

1. Silicious association – silicon dioxide predominance among all the oxides. Match with quartz mineral assemblage. Probably formed in conditions of normal marine basin with decreased alkalinity; this caused precipitation of silica from solution and led to amorphous silica and quartz formation.

2. Aluminum-silicious association – combination of silica and alumina corresponds to bulk rock clay-silicious composition. The association is supposed to be formed in conditions of normal marine basin with calm sedimentation regime and stable physical and chemical parameters (sea coast environments are also considered).

3. Calcium-silicious association – it formed at geochemical barrier that appeared due to carbon dioxide content increase in the sea water. That, in turn, may be caused by basin depth increase or CO_2 input by sea current. Also carbonatization could take place due to biogenic causes.

4. Iron-calcium-aluminum-silicious (Iron-calcium-silicious) association – its formation may be interpreted dualisticly: on the one hand, sediment sideritization could take place, on the other hand, carbonate sediments pyritization also was possible. The formation probably took place in conditions of marine basin with reductive sedimentation regime. In case of pyritization contamination with hydrogen sulphide probably took place.

IOP Conf. Series: Earth and Environmental Science **110** (2017) 012001 doi:10.1088/1755-1315/110/1/012001

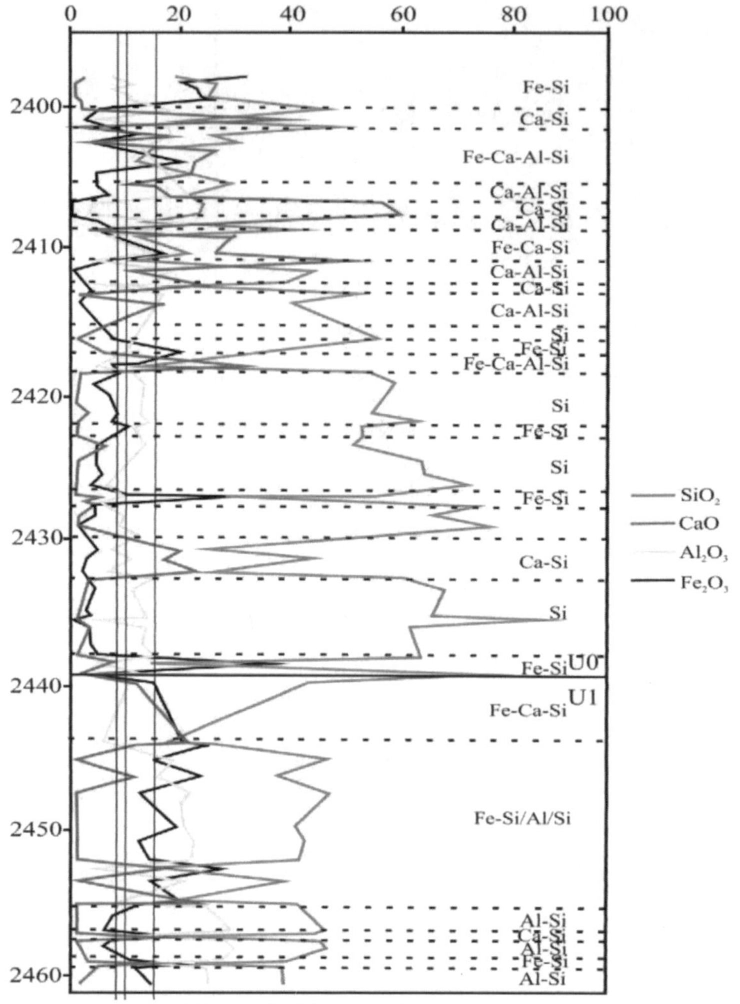

Figure 2. Elemental associations

5. Iron-aluminum-silicious (Iron-silicious) association has much in common with the one described above. Itformedduetopyritizationofclay-siliciousbulk.

There is association 5 in the top of U0 bed, in other words there are intensively pyritized rocks. There is paragenesis frequent change with carbonates prevalence in general in the middle part of the bed. The lower part consists of the first, the fifth association combination, where the latter two represent pyritization and carbonatization of silicious sediment. The U1 bed has well developed associations, in which calcium and iron take part. At that there are all of the 5 associations present within the bed.

According to classification by E. Y. Yudovich and M. P. Ketris, there are silites (HM <0,30), siallites (HM=0,31-0,55) and pseudosiallites (MgO>3 %), siferlites (HM=0,31-0,55, FM=0,75) and pseudosiferlites (MgO>3 %), hydrolyzates (HM>0,55) and pseudohydrolyzates (MgO>3 %) present in studied rocks. Meanwhile there is clear zonality in rocks types' distribution. For the upper and middle parts of the U0 bed there is siallites, hydrolyzates and their magnesial kinds alternation due to dolomite presence; the lower part mainly consists of silites with small amounts of siallites, pseudosiallites, hydrolyzates and pseudohydrolyzates. Pseudohydrolyzates are exceptionally widespread within the U1 bed, that may be caused by pyritization and sideritization of the sediment with chlorite forming.

At the same time there is general nonconvergence of clay minerals amounts and aluminum content. This is caused by underestimating of clay minerals content due to their low crystallinity and high contents of amorphous silica and kerogen, so in case with Bazhenov suite we can apply X-ray diffraction analysis results only for qualitative study but not for quantitative one.

Analysis of distribution of geochemical indicators values allows detecting cyclicity of studied beds composition. Microcycles bounds were defined by sudden symbatic indicators change, with further cycles' enlargement with close geochemical characteristics (figure 3).

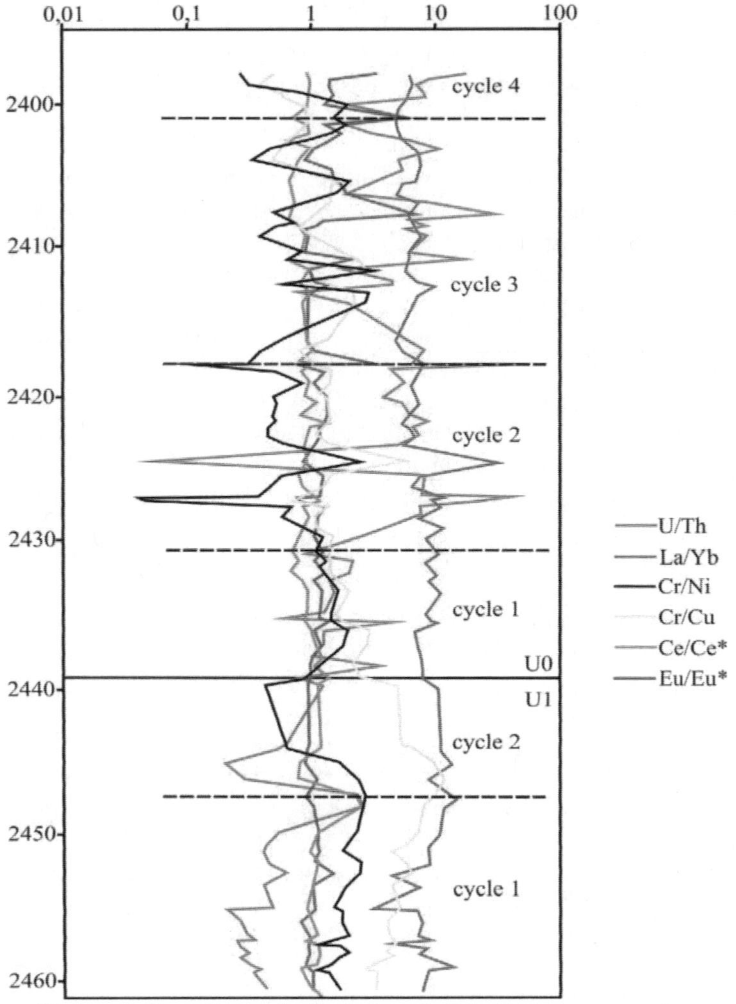

Figure 3. Distribution of geochemical indicators on section

U1 bed is composed of two cycles (figure 3). The lower one formed in conditions of marine basin with sea water and fresh water mixing areas; that is confirmed by values as Ce/Ce*=0,83-1,16 and Eu/Eu*=0,81-1,48 (the latter indicates reducing environment). Sedimentation processes took place at the same time with general regression of the sea basin that is confirmed by La/Yb, Cr/Ni, Cr/Cu values. There are similar variations of indicators' values (Ce/Ce*=0,77-1,19, Eu/Eu*=0,87-1,11) for the second cycle sediments, so it is possible to conclude that there were similar conditions of sedimentation [2, 3, 6]. The main difference is in La/Yb, Cr/Ni, Cr/Cu values decreasing, so there was a transgression.

U0 bed consists of four parts (figure 3). There are some local U/Th ratio increases up to more than 10, which are probably caused by postsedimentary transformations that led to uranium redistribution.

These values were not taken into account. Lowest cycle sediments formed in conditions of marine basin (Ce/Ce*=0,77-1,19) with reductive regime (Eu/Eu*=0,87-1,11) at the same time with increasing transgression (gradual La/Yb, Cr/Ni, Cr/Cu values decreasing form bottom to top).

Overlying sediments of the second cycle formed in similar conditions in general (Ce/Ce*=0,79-1,11, Eu/Eu*=0,76-1,37), but sedimentation area was located further from the shoreline (there is symbatic La/Yb (lanthanides chemogenic concentration increases), Cr/Ni, Cr/Cu values decrease). Third cycle sediments probably formed in conditions of transgression maximum that is confirmed by Ce/Ce*=0,66-0,95 values (marine basin conditions only), Eu/Eu*=0,93-3,36 (reductive environment only), La/Yb=4,88-7,99 (chemogenic sedimentation) and Cr/Ni, Cr/Cu indicators' consistent behaviour. Overlying sediments of the fourth cycle presumably formed in conditions of regression starting that is confirmed by Ce/Ce* and La/Yb values slightly increasing [2, 3, 6].

4. Conclusion

On the basis of obtained results it's possible to conclude that studied rocks formed in following way. U1 lower cycle sediments formed in coastal-marine environment with possibility of fresh water influence at the same time with regression of sedimentation basin; U1 upper cycle sediments formed in almost similar conditions, but there was transgression. U0 first, second and third cycles sediments formed during further transgression. At the same time there was change of physical and chemical characteristics of sedimentation basin: environment became more reductive, chemogenic sedimentation became prevalent. Fourth cycle formation took place at the same time with beginning of regression. In such a way, nonuniformity of Bazhenov suite rocks is caused by sedimentation conditions changes, which cause mineralogical heterogeneity: quartz, clay-quartz, pyrite-quartz, pyrite-clay-quartz, carbonate-quartz, carbonate-clay-quartz, pyrite-carbonate-quartz assembladges alternation. Each of these assembladges corresponds to some specific physical and chemical parameters of sedimentation.

Acknowledgements

This work was conducted as a government task of the Ministry of Education and Science of the Russian Federation, Project Number 5.2352.2017/4.6. The examinations of samples were performed on the equipment of Analytic Center of Natural Systems Geochemistry of Tomsk State University.

References

[1] Wang C Zhu W Wang Z Guicherit R 2000 Rare earth elements and other metals in atmospheric particulate matter in the western part of the Netherlands *Water, Air and Soil Pollut* **121**pp 109-118.
[2] Maslov A V 2005 Sedimentary rocks: methods of study and interpretation of the data Tutorial (Ekaterinburg: UGGU) 289 p (in Russian)
[3] Sklyarov E V 2001 Interpretation of geochemical data (Moscow:IntermetInzhiniring) 2001 288 p (in Russian)
[4] YudovichYa E Ketris M P 2000 Basis lithochemistry (St. Petersburg:Nauka) 479 p (in Russian)
[5] Ohta A KawabeI 2000Theoretical study of tetrad effect observed in REE distribution coefficients between marine Fe-Mn deposit and deep seawater, and in REE (III)-carbonate cjmplexation constants Ibid. **34** pp 455-473
[6] Balashov Yu A 1976 Geochemistry of rare earth elements(Moscow: Nauka) 267 p (in Russian)
[7] Haley B AKlinlhammer G P McManus J 2004 Rare earth elements in pore of marine sediments Ibid **68** pp 1265-1279
[8] Luo Y-R, Byrne R H 2004 Carbonate complexation of yttriym and the rare earth elements in natural waters *Geochimica et Cosmochimica Acta* **68** pp 691-699

Features structure of iron-bearing strata's of the Bakchar deposit, Western Siberia

E M Asochakova

Department Of Geology and Geography, Tomsk State University, Lenina Avenue, 36, Tomsk, 634050, Russia

E-mail: ev.asochakova@gmail.com

Abstract. The ore-bearing strata's of Bakchar deposit have complicated structural-textural heterogeneity and variable mineral composition. This deposit is one of the most promising areas of localization of sedimentary iron ore. The ore-bearing strata's are composed mainly of sandstones (sometimes with ferruginous pebbles, less often conglomerates), siltstones and clays. The ironstones are classified according to their lithology and geochemistry into three types: goethite–hydrogoethitic oolitic, glauconite-chloritic and transitional (intermediate) type iron ores. The mineral composition includes many different minerals: terrigenous, authigenic and clayey. Ironstones are characterized by elevated concentrations of many rare and valuable metals present in them as trace elements, additionally alloying (Mn, V, Cr, Ti, Zr, Mo, etc.) and harmful impurities (S, As, Cu, Pb, Zn, P). There are prerequisites for the influence of numerous factors, such as prolonged transgression of the sea, swamping of paleo-river deltas, the appearance of a tectonic fracture zone associated with active bottom tectonics and unloading of catagenetic waters, regression and natural ore enrichment due to the re-washing of slightly-iron rocks. These factors are reflected in the structure of the ore-bearing strata in which rhythmic cycles of ore sedimentation with successive changes in them are distinguished by an association of different mineral composition.

1. Introduction

Oolitic formations of iron ores are well known in many countries, and relate to sedimentary rocks of the Phanerozoic and Cenozoic age [1–5]. The iron ore strata's of Western Siberia were discovered in the 1950s and investigated at that time by [6, 7]. Since 2005 the Bakchar field again began exploration and evaluation of minerals. The average iron content of up to 37.4 % and probable reserves of tens of billions of tons, but there are setback in the technological processing of ore due to the presence of harmful impurities (mainly phosphorus and sulfur), which are present in the finely dispersed state in the form of REE phosphates (monazite, kularite) and sulfids [8–11]. The detailed lithological and geochemical structure and genesis of the Bakchar oolitic iron ore are the matter of this article.

2. Geology

Bakchar deposit is part of the West-Siberian iron ore basin, which is a broad band of Cretaceous-Paleogene coastal-marine clastic sediments with oolitic brown ironstone on the eastern edge of the Western Siberian plate. Iron ores are overlapped by Neogene-Quaternary sediments 160–200 m in width [6]. The iron-bearing strata range varies from 2 to 40 m. The iron ore horizons are traced throughout the entire area of the deposit, as well as outside its borders and are separated by the barren and weakly ferruginous rocks that overlap each other often with washouts [7]. The Bakchar field is divided into two parts – West and East. The area of the ore field in the contours of wells is about 520 km^2, the average width of 20 m.

3. Sampling and methods of analysis

Sample from iron-bearing strata of the Bakchar deposit were collected from West and East part section and investigated by a variety of mineralogical, chemical and spectroscopic techniques. The micromorphological characteristics and chemical composition of the oolitic iron ore were investigated using a VEGA II LMU field emission scanning electron microscope attached to an Energy-dispersive X-ray spectroscopy INCA Energy (OXFORD Instruments Analytical, Great Britain). The bulk mineralogical composition of ironstone was performed on powdered samples by using X'Pert Powder (PANalytical, Netherlands) powder diffractometer. Chemical analyses were performed on whole rock powders by X-ray fluorescence OXFORD ED 2000 (OXFORD Instruments Analytical, Great Britain). The whole rock trace and rare earth elements were measured by ICP-MS Agilent 7500 cx (Agilent Technologies, USA). All experimental studies were carried out at the Center for collective use "Analytical Center geochemistry of natural systems" National Research Tomsk State University.

4. Lithology and mineralogy

The ore-bearing strata of the formation form takes time an interval from the Turonian to the Eocene and lie with unconformity on the marine sandy-argillaceous sediments of Kuznetsov suite (Cenoman - Turon). The productive layers of the Bakchar deposit are confined to three stratigraphic levels – the sandy sequences of the Late Cretaceous, Paleocene and Eocene age: Narymskiy (Santon), Kolpashevskiy (Maastricht) and Bakcharkiy (Paleocene-Eocene) ore horizons [7, 10].

The ore-bearing stratum is composed of coastal-marine and marine sediments containing horizons of oolitic iron ores: sandstones of quartz-chlorite-siderite, quartz-chlorite-clayey composition, siltstones, clays, sands, gravels (Figure 1). The ores lie horizontally with a barely noticeable immersion of them to the east.

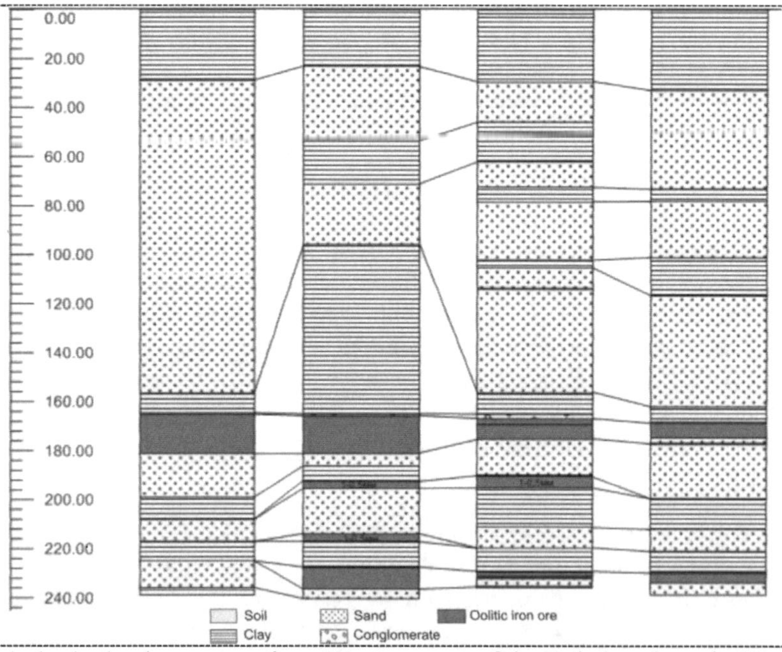

Figure 1. Lithological columnar sections Western part of the Bakchar deposit, Western Siberia.

All minerals composing ores are divided into two genetic groups: allotigenic, authigenic and polygenic (Figure 2) [8]. The allotigenic part of the ores is represented by quartz, feldspar, magnetite, ilmenite, zircon and rutile. Authigenic minerals include carbonates, clayey and hydromica minerals, glauconite, chlorite, the actual ore phases (goethite, hydrogoethite, lepidocrocite), as well as phosphates, sulfides and native silver [8]. The content of hydrogoethite in brown ores varies from 48% to 75%, in glauconite-chlorite from 19.6% to 61.2%, in transition types of ores from 22.2% to 65.1% . The sulfide association comprises microinclusions in the contact of fragmented minerals and concentres of goethite-hydrogoethite and leptochlorite composition. Pyrite is the essential sulfide mineral in oolitic iron ores. To polygenic minerals are referred kaolinite and illite, bearing signs of terrigenous and sedimentary origin.

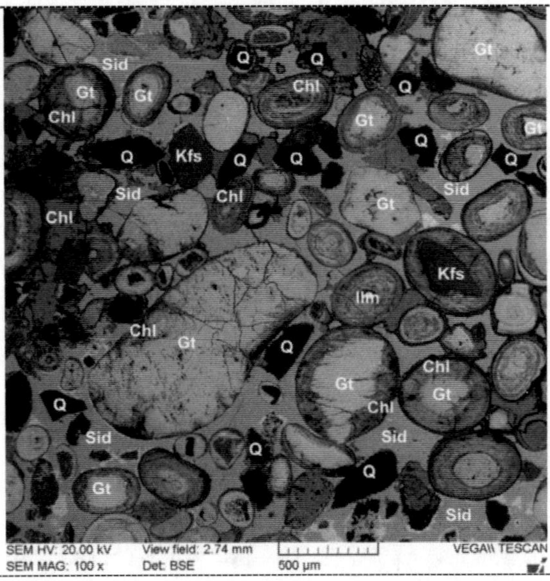

Figure 2. Picture BSE. Oolitic goethite-hydrogeotite ore. Legend: Gt = Goethite, Chl = Chlorite, Kfs = Potassium feldspar, Sid = Siderite, Q = Quartz.

5. Geochemistry

To characterize the chemical composition of oolitic iron ore were used results of X-ray fluorescence analysis (Eastern part) and exploratory testing (Western part). The total sample includes more than 250 analyses (Table 1).

Oolitic iron ores are over Clarke content of Be, Sc, Ge, Y, Mo, W, Th, P and the iron group elements Cr, Co, Ni, V and chalcophilic Zn, Pb, Sb, Ag, As [10]. Recent elements in the oolitic ores are represented by sulfides, phosphorus – REE and calcium phosphates [8, 11]. The content and composition of the REE more properly reflect the characteristics that influence the formation of the overwhelming majority of sedimentary rocks. The most important factor controlling the REE content in the sediments is a competition between the two sources: the terrigenous component and seawater. It must be borne in mind that REE are more environment characterize the specificity of accumulation, and not the influence of drift sources [8].

Previous studies of the lithochemical features of oolitic iron ores have established that the types of ores that have been isolated are due to their different chemical composition [9, 10]. According to factor analysis, the first factor (F_1) 54% determines the distribution of components and is characterized by high values of SiO_2, K_2O, MgO, Al_2O_3, and significant negative loads Fe_2O_3tot, P_2O_5, LOI. The second factor (F_2) is not significant (Figure 3). Using a factor analysis, two main factors were defined, which took more than 70 % of dispersion over.

IOP Conf. Series: Earth and Environmental Science **110** (2017) 012002 doi:10.1088/1755-1315/110/1/012002

Table 1. Representative average chemical composition of iron ore-bearing strata's, Bakchar deposit, Western Siberia

	Eastern part[a]			Western part[b]		
	1[c]	2[d]	3[e]	4[c]	5[d]	6[e]
SiO_2,Wt%	43.09	21.72	43.40	45.10	21.30	59.41
TiO_2	0.13	0.21	0.09	0.34	0.27	0.35
Al_2O_3	3.96	2.97	4.68	6.81	5.39	6.74
Fe_2O_3tot	34.45	56.07	32.89	33.39	56.15	20.82
MnO	0.05	0.03	0.05	0.10	0.22	0.06
MgO	2.67	1.51	3.12	1.17	0.66	1.15
CaO	0.61	0.42	0.70	0.48	0.30	0.51
K_2O	1.31	0.38	1.49	2.18	0.97	2.39
P_2O_5	1.21	2.17	1.15	–	–	–
LOI	9.48	12.49	9.75	–	–	–
H_2O^-	–	–	–	3.70	2.75	3.59
Total	96.96	97.97	97.32	93.27	88.00	95.02
V, ppm	850.7	1379.7	901.9	900	1400	500
Cr	295.6	304.4	264.2	–	–	–
Co	31.5	39.3	43.1	–	–	–
Ni	78.9	92	81.3	–	–	–
Cu	11.1	11	9.9	–	–	–
Zn	178.3	189.4	159.5	–	–	–
Ga	11.87	10.39	13.14	–	–	–
Ge	0.76	0.91	0.94	–	–	–
Rb	110.82	73.32	117.89	–	–	–
Sr	66.75	60.8	70.5	–	–	–
Y	38.02	44.77	38.61	–	–	–
Zr	88.98	82.84	87.52	–	–	–
Nb	8.19	7.55	8.49	–	–	–
Mo	4.5	6.45	4.49	–	–	–
Cd	0.3	0.18	0.25	–	–	–
Sn	1.46	1.25	1.38	–	–	–
Sb	6.67	8.38	6.2	–	–	–
Cs	2.13	2.18	2.88	–	–	–
Ba	230.79	181.9	219.45	–	–	–
Hf	2.54	2.61	2.84	–	–	–

[a] XRF analyses were carried out at the National Research Tomsk State University (Tomsk)
[b] exploratory testing
[c] glauconite-chloritic ore
[d] goethite-hydrogoethitic ore
[e] transitional varieties rock/ore

IOP Conf. Series: Earth and Environmental Science **110** (2017) 012002 doi:10.1088/1755-1315/110/1/012002

The components of the clastic part (SiO_2, K_2O, MgO, Al_2O_3) and the ore part (Fe_2O_3tot, P_2O_5 and L.O.I.) influenced the variation of chemical composition of the oolite iron ores significantly. When typing lithochemical ore complex all three types of ore clearly became isolated in accordance with the difference in their chemical composition, which emphasizes the cyclic structure of the section.

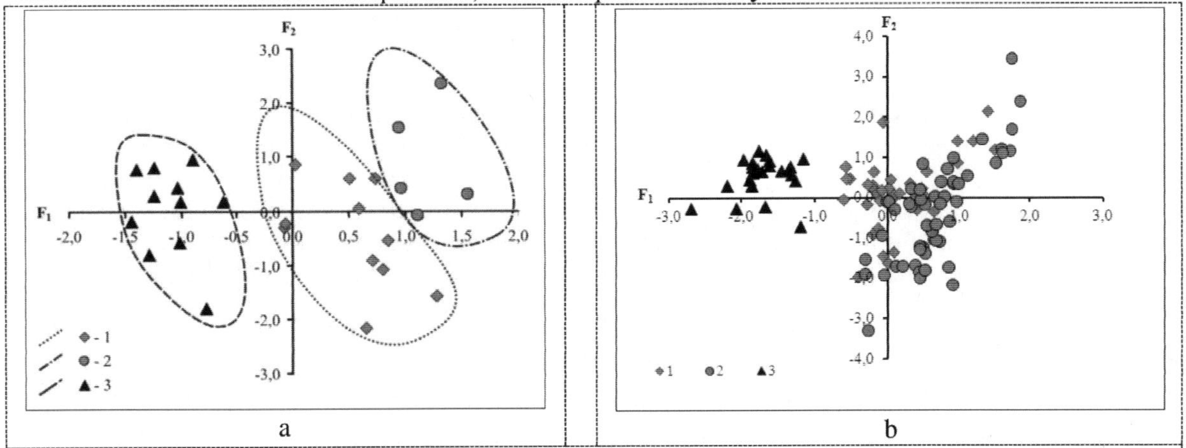

Figure 3. Factor diagram of oolitic iron ore of Eastern part (a) and Western part (b), Bakchar deposit, Western Siberia. 1 = glauconite-chloritic type ore; 2 = transitional type rock/ore; 3 = goethite-hydrogoethitic type ore.

These cycles are different, apparently, the general direction of the process of receipt of ferruginous sediments that control the speed and intensity of weathering on the paleowatersheds associated with tectonic and paleogeographic situation.

For sedimentary formations of humid climatic conditions, removal from feldspar Ca, Na and K is characteristic, which leads to an increase in the ratio of aluminum and alkalis in weathering products [3–5, 12–14]. The calculations of maturity indices of detrital material show that practically throughout the entire formation of the ore interval, the degree of maturity of the fine aluminosilicoclastics entering the basin was moderate. The magnitude of the chemical index of alteration (CIA) [12] in the ores of the entire ore formation does not exceed 70, which implies a dominance of the semi- arid-semihumid type climate throughout the Cretaceous and Early Paleocene in the paleo-watersheds (Figure 4).

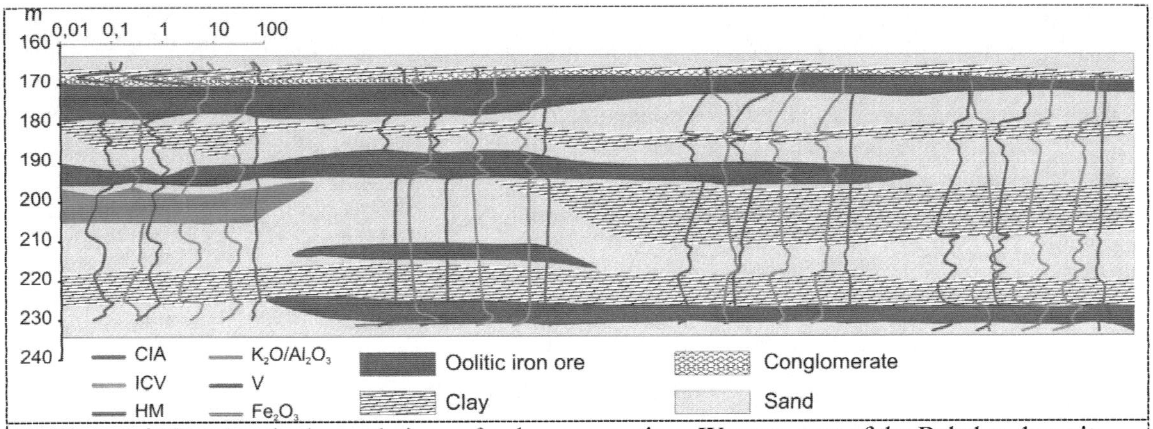

Figure 4. Geochemical correlations of columnar sections Western part of the Bakchar deposit, Western Siberia.

The index of compositional variability (ICV) [13] also reflects the degree of maturity of the fine aluminosilicoclastics entering the sedimentation area. For oolitic ores, it has values greater than 1, reflecting the presence of debris in them with a high content of non-clay silicate minerals, since more mature argillic rocks with a large number of clay minerals have lower ICV values. The values of the hydrolyzate module (HM) [14] in oolitic ores exceed 0.5, since they contain free iron oxides. The most pronounced increase in the values of HM is observed in the goethite-hydrogoethitic ores (HM to 5), in glauconite-chloritic ores it drops to 1.0. The ratio of K_2O/Al_2O_3 in oolitic ores varies from 0.09 to 0.82. The critical value is 0.30, separating according to [14] lithogenic and petrogenic formations.

Reverse correlation of CIA and K_2O / Al_2O_3 indices in the ores of the section. It indicates that the minimum values of K_2O / Al_2O_3 correspond to the epochs of the highest values of the chemical index of alteration and vice versa. The decrease in K_2O / Al_2O_3 values in the aluminasilicoclastic could be a consequence of the intensification of the processes of change in sedimentation caused either by a shift in climatic conditions to the region of more humid environments or by a change sea-level of the sedimentation basin.

6. Discussion

Within the complex of the tested ore allocated three types of sediments, the position which reflects its sectional cyclic structure of ore-bearing strata's. Clearly traced cycles, each of which consists of two ore types: brown goethite-hydrogoethitic and greenish-gray glauconite-chloritic oolitic iron ores. Transitional type ore/rocks (it may be simple sediments) fixes zone change in the conditions of sedimentation. These types have their own distinctive features, resulting in structural and textural features and variations of the mineral composition. Geochemical indexes are the elements which characterize the redox conditions and mode of formation of ore facies within the allocated cyclites [10].

According to [15] in the Cretaceous the territory of Western Siberia was a deep-sea epicontinental basin. Terrigenous material entered the marine sedimentary basin a little, and it often accumulated in the marginal parts. The biogenous sedimentation prevailed on the greater interior of the basin. The sedimentation regime was uncompensated. There was hydrogen sulfide contamination or oxygen deficiency in the waters of the basin and sediment [15]. The leading factor in the formation of the ore-bearing complex was a prolonged transgression of the sea, the maximum of which fell on the Maastricht. The appearance of stagnant depressions in the coastal part is associated with the breaking-up of the bottom of the sedimentation basin and the swamping of delta paleo-river and the formation of lacustrine-lagoon zones.

7. Conclusion

Studies have shown that the formation of oolitic iron ore is result clear staging, which is combined with the cyclical nature of mineral formation. At the heart of cycling is transgressive-regressive regime ore accumulation. Based on the authigenic composition, there are two contrasting groups of minerals in the ores, formed in reducing and oxidizing conditions. The reducing association characterizes glauconite-chloritic ores, the oxidizing composition determines the composition of brown goethite-hydrogoethite oolitic ores. These data also confirm the results of variation of geochemical parameters in the lithological section (CIA, ICV, HM, K_2O/Al_2O_3). Variations of geochemical indexes in the section are emphasized by the position and the change of different types of ores/rocks, which matched to a change in the conditions of sedimentation.

Acknowledgements

This work was conducted as a government task of the Ministry of Education and Science of the Russian Federation, Project Number 5.2352.2017/4.6. The examinations of samples were performed on the equipment of Analytic Center of Natural Systems Geochemistry of Tomsk State University.

References

[1] Salama W, El Aref M and Gaupp R 2012 Mineralogical and geochemical investigations of the Middle Eocene ironstones, El Bahariya Depression, Western Desert, Egypt *Gondwana Research* **22** pp 717–736
[2] Cornell R M and Schwertmann U 2003 The Iron Oxides: Structure, Properties, Reactions,Occurrences and Uses 2ed (JohnWiley&Sons) p 694
[3] Mücke A 2006 Chamosite, siderite and the environmental conditions of their formation in chamosite-type Phanerozoic ooidal ironstones *Ore Geology Reviews* **28** pp 235–249
[4] Sliwinski J, Gunning C, Sheridan K, Caplan B and Smith B 2010 The Upper Cretaceous Bad Heart Ooidal Ironstones: An integrated mineralogical-sedimentological-ore characterization *GeoCanada* Abstract Joint Convention
[5] McGregor F, Ramanaidou E and Wells M 2010 Phanerozoic ooidal ironstone deposits – generation of potential exploration targets *Transactions of the Institutions of Mining and Metallurgy, Section B: Applied Earth Science* Vol 119 Issue 1 pp 60–64
[6] West-Siberian iron basin 1964 ed. F N Shahov (Novosibirsk: SB RAS USSR) 447 p (in Russian)
[7] Nicolaeva I V 1967 Bakchar oolitic iron ore deposit (Novosibirsk: SB RAS USSR) 129 p (in Russian)
[8] Asochakova E M and Bukharova O V 2013 Mineral inclusions of iron ores of Bakchar deposit (Western Siberia) *Tomsk State University Journal* **369** pp 168–172 (in Russian)
[9] Asochakova E M 2013 Petrochemical typification of the oolitic iron ore from the Bakchar deposit (Western Siberia) *Mineralogical magazine* Goldschmidt Conference Abstracts p 626
[10] Asochakova E M 2015 Geochemistry of sedimentary iron ores Bakchar deposit. Western Siberia *15th International Multidisciplinary Scientific GeoConference SGEM 2015* Albene June 18–24 Book 1 Vol 1 pp 181–188
[11] Rudmin M A and Bushmanov A I 2015 Rare earth phosphates in sedimentary iron ores of Bakchar deposit (Tomsk oblast) *Modern problems of science and education* **1–1** 1826
[12] Nesbitt H W and Young G M 1982 Early Proterozoic climates and plate motions inferred from major element chemistry of lutites *Nature* **299** pp 715–717
[13] Cox R, Lowe D R and Cullers R L 1995 The influence of sediment recycling and basement composition on evolution of mudrock chemistry in the southwestern United States *Geochimica et Cosmochimica Acta* **59 14** pp 2919–2940
[14] Yudovich Y E and Ketris M P 2000 Basics of Lithochemistry (Saint-Petersburg: Nauka) 479 p (in Russian)
[15] Kontorovich A E, Ershova S V, Kazanenkov V A, Karogodin Yu N, Kontorovich V A, Lebedeva N K, Nikitenko B L, Popova N I and Shurygin B N 2014 Cretaceous paleogeography of the West Siberian sedimentary basin *Russian Geology and Geophysics* **55 5–6** 582–609

Barite-polymetallic mineralization of Zmeinogorsk ore district and some genetic aspects of its formation

K V Bestemianova and O M Grinev

Department of Geology and Geography, National research Tomsk State University, Lenin Avenue 36, Tomsk, 634050, Russia

E-mail: KsenijaVT@mail.ru, tomskgrom@yandex.ru

Abstract. Zmeinogorsky ore district is located in the northwest part of Ore Altai megatrough, which has long-lasting history of its development and complicated geological structure. Within the ore district, which is the northwest part of the devonian Zmeinogorsk-Bystrushinsky trough, ore mineralization is associated with the system of northwest border faults and cross branch faults.
There were four main stages and five phases of minerogenesis. The first stage is the stage of oregenesis beginning and quartz-chlorite-sericite wall-rock alteration rocks formation. Ore deposition and intense tectonics took place during the second stage. The third stage is the most longstanding and productive ore formation stage. There are five distinct minerogenesis phases within this stage. The fourth stage expressed in erosion development and supergene alteration of already formed ore bodies with oxidation zone formation.
Main ore minerals are pyrite, chalcopyrite, sphalerite and galena. Minor minerals are tetrahedrite, bornite, tennantite and chalcocite. Precious metals minerals are acanthite, gold, electrum, gold and silver amalgams. Barren minerals are barite, quartz, calcite, gypsum.
According to obtained data average isotopic composition of third stage sulphides is: pyrite -0,2‰, chalcopyrite 0‰, galena +0,5‰, sphalerite -1,2‰ for the first complex; chalcopyrite -1,9‰, galena -3,4‰, sphalerite -2,3‰, tetrahedrite -3,7‰ for the second complex; tennantite -12,8‰, bornite -8,9‰ for the third complex. Sulfur isotopic compoisiton variations indicate source inhomogeneity. Thus, there was dominant source change from mantle one in the beginning to crustal one in the end.
Main oregenesis stages took place in the range of temperatures between 170 and 210°C and in the mineral-forming solutions salinity range between 3 and 10 wt % NaCl equiv.

1. Introduction

Zmeinogorsky ore district is the part of Ore Altai polymetallic belt, which is one of the world's largest provinces with polymetallic deposits and lead and zinc ores reserves. There are two subformations (pyrite-polymetallic and barite-polymetallic) deposits belonging to Zmeinogorsky ore district [1]. Different type deposits formed in different geological conditions (even within the same subformation) and have mineral composition differences.

Barite-polymetallic subformation deposits bear five minable components: Cu, Pb, Zn, Au and Ag. This determines the necessity of barite-polymetallic deposits structural features and mineral composition study especially while these deposits are under dynamic production, mineral resources base is depleting and new objects finding is essential.

Barite-polymetallic subformation ores geological position, mineral composition features and genesis aspects was reviewed by the examples of three main Zmeinogorsky ore district deposits: Zmeinogorskoe, Zarechenskoe and Strizhkovskoe, which are located in the northwest part of the district.

Ore bodies contain a few different types of ores, which replace each other in the following sequence (from bottom to top): pyrite-polymetallic (among which it is possible to distinct essentially copper ones allocated within lower Paleozoic strata predominantly); polymetallic (among which it is possible to distinct essentially zinc, essentially lead or lead-zinc ones); barite–polymetallic (among which it is possible to distinct essentially barite ones). There are 45 different minerals, among which there are sulfides, sulfosalts, tellurides, oxides, oxygen acids salts, native elements and intermetallic compounds, making up the ores composition [2, 3].

2. Geological position of deposits

Zmeinogorsky ore district is the part of Ore Altai megatrough (figure 1), which belongs to Hercynian Ob-Zaysan collision area. The trough occupies boundary position between Kalba-Narym zone of Ob-Zaysan area at the southwest and Caledonian Altai at the east. Besides main structure-forming longitudianl regional faults with northwest strike there is a number of submeridional faults and a combination of faults with northeast and sublatitudinal strike [4].

Zmeinogorsky ore district is located in the north part of Zmeinogorsk-Bystrushinsky volcanogenic-terrigenous trough, which is the northern part of northeast megatrough boundary depression zone.

There are two structural levels – caledonian and hercynian – in the structure of Zmeinogorsk-Bystrushinsky trough and its ore deposits [5]. At that caledonian structural level consists of korbalikhinskaya suite dynamometamorphites while hercynian structural level consists of volcanogenic and volcanogenic-terrigenous lower, middle and upper devonian sediments of melnichnaya (D_{1-2} mn), sosnovskaya (D_2ss), zavodskaya (D_2zv) and kamenevskaya (D_{2-3} kmn) suites.

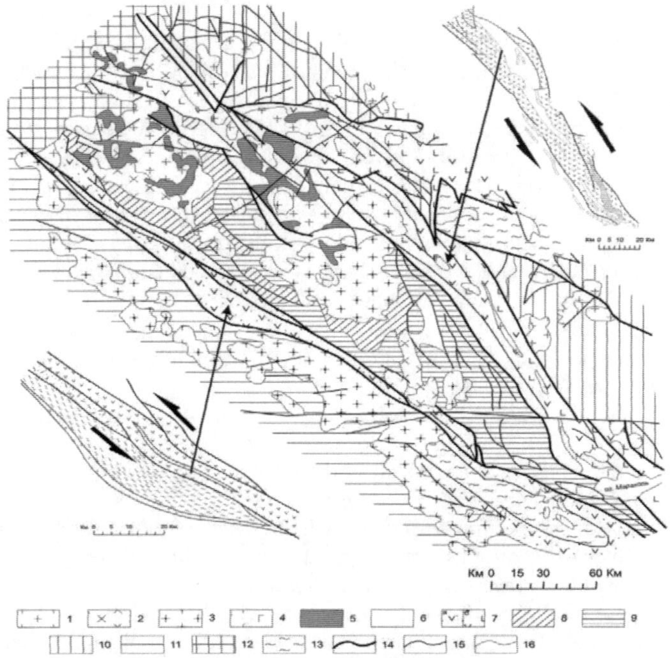

Figure. 1. Ore Altai megatrough. 1 – Devonian and predevonian zmeinogorsky complex granitoids, 2 – Carboniferous aleysky complex granitoids, 3 – early Permian kalbinsky complex granitoids, 4 – unsubdivided mafic intrusions (D, C, P), 5 – metamorphic complexes (O-S), 6 – unsubdivided strata, 7 – boundary depression zones strata a) middle and upper Devonian and b) lower, middle and upper Devonian, 8 – middle and upper Devonian bystrushinsky and beloubinsky synclinorium strata, 9 – upper Devonian and lower Carboniferous Kalba-Narym zone strata, 10 – Altai Baikalides and Caledonides, 11 – Kalba-Narym zone, 12 – Barnaul sag, 13 – dynamometamorphic rocks of basement highs (R-$Є_1$), 14 – first-order suture zones, 15 – disjunctive faults, 16 – structural-compositional complexes boundaries.

During the study it was stated that Zmeinogorsky district ore areas are located within small-scale early-middle Devonian ring-swell volcanic morphostructures. Horst-like highs of greenschist metamorphites, overlying early-middle Devonian volganogenic-terrigenous rocks and penetrating Devonian plutonic and subvolcanic bodies (e.g. dikes) take part in these morphostructures compoisiton. Ore-bearing rocks are early Devonian in general.

Main ore-bearing strata mostly are Emsian rocks of nizhnemelnichnaya subsuite where impregnated, stringer-impregnated, less often – massive ores concentrate in the lower terrigenous sedimentary stratum (argillites and aleurolites predominantly, limestones and sandstones less frequently with small tuffs layers); most part of massive ores is concentrated within bedding-plane and cross bodies at crystal tuffs horizons and terrigenous-carbonate stratum [6, 7].

Oregenesis took place during active tectonics with repeated changes from compression and to tension conditions and proper ore areas morphostructures formation. Vein-like and lens-like ore bodies are assigned to disjunctive faults, clastation and schistosity zones, cleavage and separation cracks; at that ore textures are breccia, stringer-impregnated, impregnated, massive and inequigranular ones, so structural and tectonic conditions predominate in barite-polymetallic ores distribution control. In auspicious environment when ores are forming layer by layer lithologic-stratigraphic conditions influence significantly too.

3. Methods

Data on ore and barren minerals composition and pictures of mineral associations and individuals interrelations in back-scattered electrons (BSE) mode was obtained by carrying out the electron microprobe analysis with use of scanning electron microscope Tescan Vega II LMU, which is equipped with energy-dispersive spectrometer (with Si(Li) Standart detector) INCA Energy 350.

Microelement composition was studied by inductively coupled plasma mass spectrometry (ICP–MS) with use of Agilent 7500 x spectrometer. These analyses were carried out at the collective use center «Analytic Center of Natural Systems Geochemistry» of Tomsk State University (Russia). Au and Ag contents determination was carried out by atomic absorption spectroscopy. Thermo- and cryometry of fluid inclusions were conducted with use of Linkam equipment. This analysis was carried out at the Analytical Center of multielement and isotopic research of Sobolev's Institute of Geology and Mineralogy SB RAS (Russia).

Variations of stable sulfur isotopes contents were studied at the stable isotopes laboratory of Far East Geological Institute of FEB RAS.

4. Results and discussion

There was multidtage and multiphase oregenesis at the studied objects. There are a number of mineral complexes related to four main stages and five steps [8].

During the premineral stage (the first one) wall-rock alternation (quartz-sericite, quartz-chlorite and quartz-chlorite-sericite metsomatic rocks formation) took place. Studied metasomatic rocks are early Devonian altered ones, which formed due to carbon dioxide solutions impact on poorly lithified early Devonian volcanogenic-sedimentary host rocks. Zones of their distribution are related to sublatitudinal and submeridional disjunctive faults and schistosity zones. Studied rocks have indistinct zoning: the inner part of metasomatic aureole is filled by quartz-sericite rocks while the peripheral quartz-chlorite-sericite and quartz-chlorite rocks. Mineral paragenesis of these rocks is rather plain: major minerals are quartz, chlorite, sericite and pyrite, minor ones are zircon, apatite, rutile, accessory ones are chalcopyrite, galena, sphalerite, barite, monazite and xenotime [9].

According to obtained data pyrite sulfur isotopic values in these rocks vary from -0,9‰ to +1,6‰. This indicates that there was juvenile sulfur source.

According to results of study of precious metals contents there are low gold and silver contents in altered rocks. Thus, in quartz-sericite rocks Au contents vary in 0,02–0,10 ppm range; Ag contents vary in 1,0–2,5 ppm range; in quartz-chlorite-sericite rocks there are 0,12–0,38 ppm Au contents, 0,49–2,1 ppm Ag contents; in quartz-chlorite ones there are 0,018–0,12 ppm Au contents, 0,49–0,68 ppm Ag contents. Apparently gold and silver are scattered in sulfides.

IOP Conf. Series: Earth and Environmental Science **110** (2017) 012003 doi:10.1088/1755-1315/110/1/012003

Pyrite, sphalerite, chalcopyrite and galena deposition began during the second stage. This stage took place in the end of Emsian age and the beginning of Eifelian age (D_1) and was short-time apparently. None the less intense tectonics took place. Textures are breccia and impregnated ones, structures are crush and replacement (second stage sulfides are replaced by third stage sulfides usually) ones. This stage ores make up no more than 10-15 % of ores total volume (figure 2).

The third stage of oregenesis was the most long-lasting and productive one. About 80 % of ores deposited during this stage. Syn- and postmineral tectonics and faults and jointing zones activation took place during this stage.

The first complex of the third stage includes chalcopyrite, sphalerite, galena and less often pyrite. Following textures are typical for the third stage: impregnated, stringer-impregnated, stringer and massive ones; typical structures are inequigranular, fine-, medium- and coarse-grained, emulsion, colloform, corrosion, clastic, contortion, less often replacement ones.

The second complex of the third stage includes barite, sphalerite, galena, chalcopyrite, tetrahedrite, arsenopyrite, acanthite and quartz. Barite fluid inclusions homogenization temperature varies in 190-210 °C range, salinity is 6,0-10,0 wt % NaCl equiv.

Quartz fluid inclusions homogenization temperature varies in 170-190 °C range, salinity is 3,0-4,5 wt % NaCl equiv.

The third complex of the third stage includes bornite, tennantite, chalcocite, less often – galena, chalcopyrite and barite. Minerals of this complex fill the stringers and small (up to 2 cm size) nests. There are higher gold contents (up to 60 ppm) in this complex than in the other ones.

The fourth complex of the third stage includes barren minerals mostly: there are quartz, biotite, barite, calcite, hematite and gypsum. These minerals fill the fissures, stringers and dissolution cavities.

Average chemical compositions of main sulfides included in distinguished complexes and average sulfur isotopic compositions are presented in table 1.

Table 1. Average chemical and isotopic composition of major sulfides of distinguished complexes.

Minerals	Element, wt. %											Central cases of δ34CDT, ‰
	S	Fe	Cu	Zn	As	Se	Ag	Cd	Sb	Pb	sum	
Sulfides of the first (I) complex												
Pyrite	52.8	46.8	-	-	-	-	-	-	-	-	99.6	-0.2‰
Chalcopyrite	34.26	29.7	35.1	-	-	-	-	-	-	-	99.06	0‰
Sphalerite	35.02	-	-	64.96	-	-	-	-	-	-	99.98	-1.2‰
Galena	13.86	-	-	-	-	-	-	-	-	86.06	99.92	+0.5‰
Sulfides of the second (II) complex												
Chalcopyrite	34.81	31.12	33.78	-	-	-	-	-	-	-	99.71	-1.9‰
Sphalerite	34.27	0.29	-	64.12	-	-	-	1.2	-	-	99.96	-2.3‰
Galena	13.23	0.72	-	-	-	1.05	-	-	-	85.34	100.34	-3.4‰
Tetrahedrite	24.72	1.21	38.66	7.34	4.32	-	0-17	-	23.75	-	100	-3.7‰
Sulfides of the third (III) complex												
Bornite	25.51	11.36	62.86	-	-	-	0-5	-	-	-	99.73	-8.96‰
Tennantite	30.98	0.63	40.34	7.0	15.47		0.4		5.96		100.38	-12.8‰
Chalcopyrite	35.26	30.34	34.37	-	-	-	-	-	-	-	99.7	-
Galena	13.72	-	-	-	-	-	1.43	-	-	84.17	99.32	-
Chalcocite	19.4	-	78.0	-	-	-	2.2	-	-	-	99.96	-

Postmineral stage of (the fourth one) developed significantly and expressed in partial tectonic and hypergene alteration of the ore bodies with oxidation zones formation (Zmeinogorskoe and Strizhkovskoe deposits).

Besides major components (Cu, Pb, Zn) there are also (ppm): Sc (0,06-0,7), Ti (3,0-205,0), V (0,5-10,0), Cr (0,1-4,5), Co (1,2-22,0), Ni (0,3-4,7), Ga (0,5-6,0), Rb (0,5-14,0), Sr (0,1-5,0), Y (0,03-1,5), Zr (0,2-9,0), Nb (0,02-0,9), Cs (0,03-0,1) in *polymetallic* ores; Sc (0,01-0,3), Ti (5,0-41,0), V (4,5-8,0), Cr (0,1-1,8), Co (9,7-36,0), Ni (0,3-10,0), Ga (2,0-10,0), Rb (3,0 -8,0), Sr (2,0-192,0), Y (0,2-1,0), Zr (0,5-3,0), Nb (0,03-0,25), Cs (0,03-0,1) in *barite-polymetallic* ores; Sc (3,5-9,0), Ti (400,0-2100,0), V (8,0-60,0), Cr (10,0-43,0), Co (3,0-25,0), Ni (11,0-32,0), Ga (2,5-12,0), Rb (0,1-0,7), Sr (5,0-11,0), Y (1,5-8,5), Zr (15,0-70,0), Nb (1,5-5,0), Cs (0,01-,015) in *early Paleozoic copper* ores.

Precious metals contents vary from 3,2 to 60 ppm for gold and from 91 to 500 ppm for silver depending on the ore type. There is precious metals (silver and gold) minerals diversity. These elements are present both in bearer minerals and concentrator minerals. Bearer minerals (sulfosalts, intermetallic compounds, native elements) are native silver, electrum, argentite (acanthite), gold and silver amalgams, stromeyerite, hessite, jalpaite. Native silver occurs at all the three deposits (in mostly polymetallic ores); there are both separate inclusions and intergrowth with argentite (acanthite) or chalcocite. There is Te admixture (0,4%) in Zarechenskoe deposit silver. Argentite (acanthite) occurs rather often. Sometimes it forms zonal structures, sometimes it forms separate inclusions and sometimes it forms a «plaque» on fahlore and chalcopyrite.

Hessite (figure 2, e) is present in mostly pyrite ores ore in polymetallic ones. Oftentimes there are also native tellurium inclusions (hessite and native tellurium are found together in galena very often). Silver amalgams are widespread in all types ores, gold amalgams are present in latter bornite-tennantite associations (this association is the most rich with electrum and gold and silver amalgams inclusions). Stromeyerite and jalpaite fill stringers in main ore minerals (chalcopyrite, sphalerite, galena) oftentimes.

Precious metals (silver mostly) concentrator minerals are tetrahedrite (here are the highest Ag contents up to 17%), tennantite, chalcocite and bornite (the lowest contents).

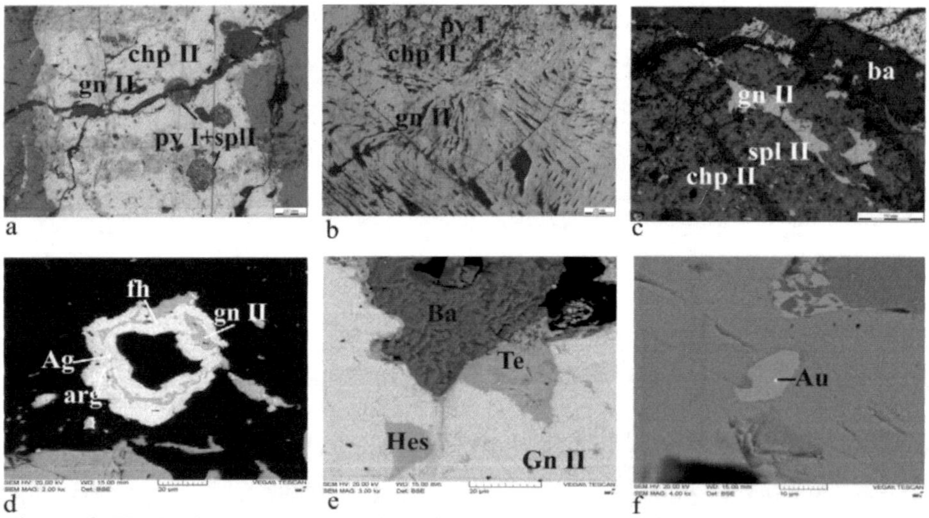

Figure. 2. The basic ore structures. a-brecciated ores, brecciated fragments are represented by aggregates of pyrite I (py) and sphalerite I (spl); b- structure crumpling in galena II (gn); c-emulsion structure of the decomposition of the solid solution of chalcopyrite II (chp) in sphalerite II (spl); d- zonal-ring aggregate, consisting of tetrahedrite, argentite, silver and galena; e- segregation of native tellurium, hessite, in galena; f-isolation of of native gold in fahlore grain.

5. Conclusions

Structural and tectonic position of Au-Ag barite-polymetallic ores of Zmeinogorsky ore district is determined by following circumstances: 1) Caledonian basement and transgressively overlying Devonian fault-lined and slightly plicated volcanogenic-terrigenous-sedimentary cover presence; 2) active disjunctive tectonics in subanticlinorium zone of Zmeinogorsky trough; 3) system of boundary northwest faults and their branch faults, which controlled volcanic structures location during the Devonian magmatites formation, controls ore bodies location in Zmeinogorsky ore district; 4) tectonics determine the barite-polymetallic ores distribution but in case of auspicious conditions lithologic-stratigraphical influence may also become very significant, when rising ore-bearing solutions used stratigraphic niches and became not only vertically distributed but also horizontally distributed. Oregenesis began in early Devonian period (Emsian age in particular).

According to obtained data tectonic zones were activated repeatedly and these activations were followed by simultaneous oregenesis processes, so barite-polymetallic mineralization has multistage and multiphase character (there are two main stages and 5 main phases of the oregenesis). Ores consist of five mineral complexes, which correspond to 5 phases, are same for all studied deposits and reflect pulsation of medium-low-temperature hydrothermal alteration processes. Main stage sulfides deposition took place at 170-190 °C.

According to data on sulfur isotopic composition (from 0 to-3,5‰), one can see that from initial to final stages there was sulfur source change. Thus, there were mantle source in the beginning and crustal source in the end [10]. Significantly lightweighted sulfur in the third complex sulfides (bornite and tennantite) may be considered as a result of partial oxidation in the ore-bearing solution during the oxidation potential increase when isotopes are shared between oxidative and reductive forms of sulfur, at that the latter ones become enriched with more light S^{32} isotopes.

Acknowledgements

This work was conducted as a government task of the Ministry of Education and Science of the Russian Federation, Project Number 5.2352.2017/4.6.

References

[1] Chekalin V M Chernykch A F Belyaev A P *Morphology, genesis and patterns of polymetallic deposits distributions in the northwestern part of the Rudny Altai* [In Russian] Novosibirsk Nauka Publ. 1985 p 56-65

[2] Gaskov I V *Specific features of pyrite ore-magmatic systems development in the island arc environments of Rudny Altai and Southern Urals* [In Russian] Litosfera 2015 no **2** p 17-39

[3] Chekalin V M The main regularities of location and the principal model of the formation of pyrite-polymetallic deposits in the northwestern part of Rudny Altai [In Russian] Geology and Geophysics 1991 No **10** p 75-88

[4] Kozlov M S *Formation conditions of the Rudny Altai metallogenic province* Geology of ore deposits 2015 Vol **57** no 4 p 266-291

[5] Scherba G N *Great Altai* (Geology and metallogeny) [In Russian] Book 1 Almaty Gylym Publ. 1998 p 304

[6] Gorgevski D I Chekvaidze V B *Types of polymetallic deposits of Rudny Altai and their origin and research methods* [In Russian] Moscow Nedra Publ. 1977 p 197

[7] Avdonin V V *Structural-morphological types and wallrock-alterated rocks of pyrite-polymetallic deposits in the northwestern part of the Rudny Altai* [In Russian] Geology of ore deposits 1980 no **2** p 49-66

[8] Bestemianova K V Grinev O M *Steps and stages of ore mineralization of barite–polymetalic deposits at Zmeinogorsk ore district (Rudny Altai)* International School of Earth Sciences Moscow 2016 p 16-17

[9] Bestemianova K V Grinev O M *Wallrock metasomatites of barite-polymetallic deposits Zmeinogorsk ore district (Rudny Altai)* [In Russian] Bulletin of the Tomsk Polytechnic

University Geo Assets Engineering 2017 Vol **328** no 9 p 114-126
[10] Seal R R *Sulfur isotope geochemistry of sulfide minerals* Mineralogy and Geochemistry 2006 Vol **61** p 633 - 677

IOP Conf. Series: Earth and Environmental Science **110** (2017) 012004 doi:10.1088/1755-1315/110/1/012004

Mineralogical-geochemical features of contact zone rocks of the Shivey alkaline granite-quartz-syenite and Chadal gabbroic massifs (Eastern Tuva)

O V Bukharova[1], A M Sugorakova[2], Ch O Kadyr-ool[2], A K Khertek[2]

[1]Department of Geology and Geography, Tomsk State University, Lenin Avenue 36, Tomsk, 634050, Russia
[2]Tuvinian Institute for Exploration of Natural Resources of SB RAS, International Street 117a, Kyzyl, 667007, Russia

E-mail: getina@ggf.tsu.ru

Abstract. The Kaakhem (> 30,000 km^2) is one of the major magmatic areal within the Altai-Sayan folded area (Eastern Tuva, Russia). Formations of the Kaakhemsky area were more than 250 million years with change of geotectonic conditions. The last stage is bound to intraplatform magmatism (300 Ma ago). Massifs of bimodal association – gabbroic and alkaline granitoids particularly the Chadal gabbroic and the Shivey alkaline granite-quartz-syenite massifs were formed as a result of its activities within the Kaakhem magmatic areal. The object of the study are rocks breccia-like and drip-like structurs in the contact zone of the alkali-granitoid Shivey and gabbroid Chadal massifs in the Kaakhem magmatic. Predecessors repeatedly noted that such structurs relationships have typical features of mingling: flexuous, scalloped contours of mafic bodies with orientation of protrusions towards granitoids; oval, rounded form of basal fragments; signs of plastic deformation in the basices and others. Geochemical analysis of rare and scattered elements contents in contact zone rocks showed strongly marked character of hybridization. Results of scanning electron microscope analysis and X-ray microanalysis let consider the probable sequence of mineralization in contact zone of massifs of bimodal association of gabbroic and alkaline granitoids of the Kaakhem magmatic areal. Relationship established between microtexture and geochemical inherity indicate the interaction of basic and granitic magmatic meits and it was formation of middle group rocks.

1. Introduction

The Kaakhem (> 30,000 km^2) is one of the major magmatic areal within the Altai-Sayan folded area. Geochronological studies of rocks of the Kaakhem magmatic areal showed that its formation was a result of progression changes in the island-arc (562 Ma), accretionally-subductional (536 Ma), collisional (512 Ma), transform-shearing (480 - 385 Ma) [1] as well as intraplate (297 – 293 Ma) geodynamic situation [2]. Island-arc stage was accompanied by basalt-andesite-rhyolite volcanism. Accretionally-subductional magmatism begins with the formation of gabbroid massifs of high alkalinity with intraplate geochemical characteristics presenting the Altai-Sayan mantle hot spot (mantle plume). There are 3 manifestation stages of gabbroid magmatism and 6 of granitoid magmatism within 512 to 474 Ma. Predominant part of these granitoids is located in the west of the areal [3, 4, 5]. The next and the main stage of granite formation is post-collisional preceded by formation of gabbroids with intraplate marks after a break of 25 Ma in the interval of 455 – 445 Ma. Granitoids of this stage occupy hundreds of kilometers and form the major part of the Kaakhem areal. Granitoids' geochemical characteristics of this stage outline the features of the substrate composition from which granitic magmas were melted [1, 6, 7]. The last stage of granite formation is associated with mantle plume (intraplate magmatism) which has existed ~ 300 Ma ago. Massifs of bimodal association – gabbroic and alkaline granitoids particularly the Chadal gabbroid and the Shivey alkaline granite-quartz-syenite massifs were formed as a result of its activities within the Kaakhem magmatic areal.

The object of the study are rocks breccia-like and drip-like structurs in the contact zone of the Shivey and Chadal massifs (figure 1). Authors of the earlier research studies repeatedly noted that such texture relationships have typical features of mingling: meandering, scalloped, lobed margins of mafic bodies with orientation of protrusions towards granitoids; oval, rounded form of basic fragments; elements of plastic deformation in the basics (figure 2).

Figure 1. Mingling ratios of the Shivey alkaline granite-quartzsenite massif with the same-aged gabbroic of the Chadal Massif (photo by Khertek A K)

Figure 2. Scalloped edges in the contact zone of the Shivey and Chadal massifs (photo by Khertek A K)

Fragments morphology of acidic composition inside the mafic has an isometric form or relatively large irregularly shaped domains with rounded boundaries. Such case is not typical because viscosity of granitic magma is higher than the mafic one. Classic features of mingling include specific granite «veins» within basic bodies indicating impregnation of acid magma through contraction cracks separating the cooling basic body into fragments. All these features are usually treated as evidence of contact of two liquids with different viscosities, mechanical mixing with the simultaneous intrusion of basic and salic melts [8, 9].

Contact zone of the Shivey and the Chadal massifs with mingling features occupy an area of about 12 km^2 according to geological observations. Granitoids generally save a leucocratic outlook in contact with gabbroic. Features of chemical mixing are widely occurred only in injection bodies intruding inside the basic associations directly in the contact of granitoids with basic where narrow zones with a width of 1–3 cm are occasionally noted where granitoids are enriched in femic minerals and in some cases they change their composition to a more basic one.

Gabbroic have clear reactionary effect indicators of granitic magma unlike granitoids. They are visually expressed in the abundance of bleaching spots and zones, often extending into the gabbroic bodies directly to the continuation of disappeared granitoid apophyses [9 - 12].

2. Object and methods of research

Composition of breccia-like and drop-shaped separate rocks was studied in contact zone of the Shivey and the Chadal massifs. Chemistry of minerals was studied by the electron microprobe analysis on a Tescan Vega II LMU scanning electron microscope with an Oxford INCA Energy 350 EDS system operated at an accelerating voltage of 20 kV, a beam current of 5 nA, a beam size of 1–2 μm and 120s count time. The operation conditions were the same for all minerals. The results were checked against natural minerals and synthetic phases used as standards (MAC 55 Standard Universal Block Layout+F/Cup, Micro-Analysis Consultants Ltd, UK) for calibration. The analyses were carried out at the National Research Tomsk State University (Tomsk). The concentrations of trace elements and REE were determined by ICP-MS on an Agilent 7500cx at the Tomsk University.

3. Petrography and mineralogy.

The Shivey massif has a complex structure and it is mainly composed of syenite-porphyry, leucocratic quartz syenite. The Chadal massif mainly consists of olivine gabbro, gabbro-norite, amphibole gabbro [13].

Petrography of contact zone rocks of the Shivey and the Chadal massifs differs from the petrography of the massifs rocks. Leucocratic rocks of contact zone are represented by biotite-hornblende syenite and quartz monzodiorite, mesocratic-quartz biotite-hornblende monzonite and subalkaline biotite-hornblende diorite.

Generalized data of petrographic analysis and scanning electron microscopy are given in table 1. Composition of rock-forming minerals of contact zone rocks corresponds to the rocks of the middle group of normal and subalkaline series: diorite-monzonite-syenite, the main part of studied rocks is both light and dark belonging to the monzonite series.

It is found that plagioclase is often more acidic than oligoclase-andesine typical for monzonites (№ 30-50) according to the studied chemical features of rock-forming minerals (table 2). There are zonation plagioclase individuals with a characteristic direct zonality (№ 56–35–15) with the marginal zone being insignificant and more similar to metasomatic one in rocks of monzonite series. Porphyry-like separations of orthoclase are corroded by acidic plagioclase (№ 19–20) in syenite. Plagioclase similar in composition is also found in the form of poikilitic inclusions in orthoclase.

There is 9–12% of albite molecule in orthoclase content of studied rocks. The analyzed orthoclase individuals often contain 0.5~1 wt.% BaO. Orthoclase of quartz biotite-hornblende monzodiorite is mostly enriched by barium and there is no Ba at any quartz monzodiorite individuals according to energy-dispersive analysis (table 3).

Dark-colored minerals are represented by amphibole and biotite in the rocks. They form isometric aggregates which make rock slightly porphyry-shaped in subalkaline diorites.

Calcium amphiboles areestablished in the rocks of the contact between the Shivey and the Chadal massifs, the common chemical property of which is increased magnesia. They all correspond to magnesian hornblende. An inverse dependence of Si and Mg/(Mg+Fe^{2+}) content is observed in the structure of amphiboles (figure 3, table 4).

Table 1. Petrography the rocks of contact zone of the Shivey and Chadal massifs.

	Leucocratic		Mesocratic		
Rock type	Syenite biotite-hornblende	Quartz moncodiorite	Subalkaline diorite biotite-hornblende	Quartz moncodiorite biotite-hornblende	Quartz monconite biotite-hornblende
Rock texture	Medium and coarse grained, porphyroid	Medium and coarse grained hypidiomorphic	Fine grained, monzonitic, myrmekite (metasomatic structure)	Porphyroid matrix of fine grained	Porphyroid matrix of hypidiomorphic, monzonitic
Rock-forming minerals	Fsp[a] 50-55 %; Pl[a]$_{№17-20}$ 30 %; Bt[a] 10-12 %; Am[a] 10-12 %; Q[a] ≤ 5 %	Fsp 25 %; Pl$_{№56-14,5}$ 50 % (zoning of crystals) Bt 10-15 %; Am 10-15 %; Q -5 %	Pl $_{№35-26}$ 40 % (zoning of crystals) Bt 20 %; Am 5 %; Q ≤ 5 %	Fsp 20 %; Pl$_{№34 до №20-18}$ 30 % (zoning of crystals) Bt 20 %; Am 20 %; Q < 5 %	Fsp 35 %; Pl $_{№26-19}$ 35 % (zoning of crystals) Bt 15 %; Am 5-7 %; Q 5-7 %
Secondary minerals		Epidote Chlorite Group (Clinochlore)		Epidote	Chlorite Group (Clinochlore)
Accessory minerals	Mgt c V and Ti-Mgt- 1%, Titanite < 1%, Apatite < 1%, Zircon (Hf) < 0,n%.	Titanite ≥ 1%, Mgt and Ti-Mgt < 1%, Apatite < 1%, Zircon – 0,n%, sulfides (pyrite, chalcopyrite).	Apatite (Sr, Cl) < 1%, Mgt and Ti-Mgt < 1%, Titanite ≥ 1%, Zircon – s.g.[b], Baddeleyite, Sulfides (pyrite, chalcopyrite).	Mgt and Ti-Mgt – 5%, Apatite > 1%, Titanite ≥ 1%, Baryte, Sulfides (pyrite, sphalerite).	Mgt c Cr and Ti-Mgt < 1%, Apatite (c F) < 1%, Titanite ≥ 1%, Zircon – s.g., Allanite - s.g., Monazite – s.g., Sulfides (pyrite, phalerite, chalcopyrite).

[a] Fsp = orthoclase, Pl = plagioclases, Bt = biotite, Am = amphibole, Q = quartz.
[b] s.g. = simple grains.

More ferruginous varieties are essential for syenite amphiboles. Mg amount prevails over Fe in amphiboles of diorites and monzodiorites. Silica content in amphiboles from monzonites is higher than in individuals from syenite and diorites.

Biotite is present at rocks in varying amounts and it is a typical biotite because annite and phlogopite component contents are ~ 50–50% (figure 4, table 4). Biotite is low-titanium (Ti=TiO_2/(TiO_2+FeO+MgO+MnO) of 0.10 units from all the presented species of contact zone rocks. More ferruginous biotite are characteristic for syenites, magnesium biotite are characteristic for monzodiorites.

Ti and Fe oxides (ilmenite, magnetite), calcium phosphate (apatite), silicates (titanite, zircon) are present at accessory minerals. Ilmenite and magnetite are present at all contact rocks of the Chadal and the Shivey massifs. Magnetite has 2 generations. Vanadium is present (V_2O_5 0.3–0.5 wt.%) in magnetite of the second late generation in the form of an impurity, chromium (Cr_2O_3 from 0.4–2.0 wt.%) is fixed in diorite individuals. Ilmenite is TiO_2/FeO^* from 0.9 to 1.3 according to type-chemical criteria, the presence of manganese impurities (1.4–4.75 and 8.8–15 wt.%) corresponds to individuals from granitoids and alkaline rocks. A significant amount of pyrophanic molecule ($MnTiO_3$) is characteristic of individuals from quartz monzodiorites and quartz amphibole-biotite monzonites. The exsolution texture is occurred in ilmenite from monzodiorite and diorite, there are lamellas of magnetite (I) subparallel extended, less often they are of irregular shape in ilmenite matrix.

Titanite is present at all the described rocks and can form both hypidiomorphic secretions and can be a substitution product of ilmenite or can form fouling border around ilmenite. Titanite of late generations is often enriched with iron, it may be due to the smallest presence of iron oxide (magnetite) proper in titanite matrix.

Table 2. Representative major-element compositions of plagioclases from the rocks of contact zone of the Shivey and Chadal massifs.

Rock[b]	Leucocratic			Mesocratic				
	1 (zoned mineral)[a]			1 (5)[c]	2 (5)[c]	3 (7)[c]	4 (5)[c]	5 (6)[c]
SiO_2,	54.62	60.22	67.29	64.81	63.68	64.06	62.26	62.2
TiO_2	-[d]	-	-	-	-	-	-	-
Al_2O_3	28.66	25.66	21.07	22.82	22.42	22.42	24.32	23.52
MgO	-	-	-	-	-	-	-	-
MnO	-	-	-	-	-	-	-	-
FeOt	-	0.28	-	-	0.07	0.22	0.07	0.41
CaO	11.79	7.49	3.01	4.21	3.98	3.98	5.97	5.28
Na_2O	4.96	7.51	9.31	9.33	9.12	9.51	8.20	8.41
K_2O	0.21	0.31	0.75	0.20	0.36	0.40	0.28	0.21
BaO	-	-	-	-	-	-	-	-
Total	100.23	101.46	101.41	101.35	99.62	100.58	101.08	99.99
Si, apfu[e]	2.46	2.65	2.94	2.84	2.83	2.815	2.74	2.77
Ti	-	-	-	-	-	-	-	-
Al	1.52	1.33	1.08	1.17	1.17	1.16	1.26	1.23
Mg	-	-	-	-	-	-	-	-
Mn	-	-	-	-	-	-	-	-
Fe		0.01				0.01		0.02
Ca	0.57	0.35	0.14	0.20	0.19	0.19	0.29	0.25
Na	0.43	0.64	0.79	0.79	0.79	0.81	0.70	0.72
K	0.01	0.02	0.04	0.01	0.02	0.02	0.02	0.01
Ba	-	-	-	-	-	-	-	-
Ca (An)[f]	56.13	34.95	14.54	19.71	19.03	18.40	28.26	25.45
Na (Ab)[f]	42.68	63.35	81.18	79.23	78.92	79.42	70.18	73.38
K (Ort)[f]	1.19	1.70	4.29	1.07	2.06	2.19	1.56	1.18

[a] Composite grain.
[b] Numerals in panel stand for: 1=quartz moncodiorite; 2 = syenite biotite-hornblende; 4 = quartz monconite biotite-hornblende; 5 = quartz moncodiorite biotite-hornblende.
[c] Numerals in braces show number of analyses.
[d] (-) is below detection limit.
[e] Atoms per formula units (apfu) are based on 8 oxygens
[f] An = anorthite minale (molecule), Ab = albite, Ort = orthoclase.

Table 3. Representative major-element compositions of orthoclase from the rocks of contact zone of the Shivey and Chadal massifs.

Rock	Leucocratic		Mesocratic	
	1[a] (6) [b]	2 (6)	4 (5)	5 (6)
SiO_2, wt.%	65.84	64.79	64.45	64.67
TiO_2	-	-	-	-
Al_2O_3	18.32	18.33	18.245	18.3
MgO	-[c]	-	-	-
MnO	-	-	-	-
FeOt	-	-	-	0.25
CaO	-	-	-	-
Na_2O	1.21	1.21	1.29	0.98
K_2O	15.57	15.31	15.46	15.70
BaO	-	0.39	0.48	0.70
Total	100.93	100.00	99.94	100.58
Si, apfu[d]	3.00	2.99	2.975	2.98
Ti	-	-	-	-
Al	0.98	1	0.99	0.99
Mg	-	-	-	-
Mn	-	-	-	-
Fe	-	-	-	0.01
Ca	-	-	-	-
Na	0.11	0.11	0.12	0.09
K	0.91	0.90	0.91	0.92
Ba	-	0.01	0.01	0.01
Ca (An)[e]	-	-	-	-
Na (Ab)[e]	10.5	10.56	11.11	8.52
K (Ort)[e]	89.5	88.75	88.06	90.25
Ba (Cel)[e]	0.00	0.69	0.83	1.23

[a] Numerals in panel stand for: 1=quartz moncodiorite; 2 = syenite biotite-hornblende; 4 = quartz monconite biotite-hornblende; 5 = quartz moncodiorite biotite-hornblende.
[b] Numerals in braces show number of analyses.
[c] (-) is below detection limit.
[d] Atoms per formula units (apfu) are based on 8 oxygens.
[e] An = anorthite minale (molecule), Ab = albite, Ort = orthoclase, Cel = celsian.

Apatite is presented by fluoroapatite. It has idiomorphic outlines. Apatite can be poikilitic inclusion in zircon, potassium feldspar, magnetite, ilmenite (at syenite). The apatite in the form of the finest individuals is found in all rock-forming minerals (except quartz) and also take places between parallel thin amphiboles crystals in the quartz monzodiorite.

Zircon forms classical two-pointed crystals. Its chemical composition is close to theoretical one. There is Hf (average value is up to 1.5 wt.% HfO_2) in the form of an impurity element.

There are minerals of rare earths (monazite, allanite) and Zr (baddeleyite) in association with listed mineral species.

Table 4. Representative major-element compositions of amphibole and of mica from the rocks of contact zone of the Shivey and Chadal massifs.

Mineral	Amphibole					Mica				
	Leucocratic		Mesocratic			Leucocratic		Mesocratic		
Rock	1[a] (6)[b]	2 (6)	3 (8)	4 (6)	5 (4)	1 (5)	2 (2)	3 (4)	4 (6)	5 (7)
SiO_2, wt.%	49.18	45.95	49.87	49.49	46.29	37.76	36.7	37.25	37.16	37.18
TiO_2	1.25	1.20	1.11	1.13	1.44	3.97	3.96	4.27	3.30	3.64
Al_2O_3	6.02	6.41	5.40	5.68	7.22	13.73	13.82	13.79	14.2	13.87
MgO	14.9	11.00	14.34	12.69	11.65	13.09	11.25	12.56	12.53	12.22
MnO	0.85	0.73	0.33	0.62	0.43	0.77	0.40	-	0.37	0.03
FeOt	13.5	18.27	14.49	15.96	17.53	18.67	21.4	19.39	20.6	20.21
CaO	11.69	12.06	12.26	12.10	12.16	-[c]	-	0.16	0.07	-
Na_2O	1.28	1.22	0.89	1.18	1.23	-	-	-	-	-
K_2O	0.51	0.74	0.49	0.61	0.75	9.33	8.73	9.25	8.61	9.53
Cl	-[c]	0.02	-	0.03	-	0.06	-	-	-	-
P_2O_5	-	0.47	-	-	-	-	-	-	0.11	-
Total	99.16	98.04	99.15	99.47	98.66	97.36	96.23	96.65	96.92	96.67
Si, apfu[d]	6.99	6.86	7.13	7.16	6.81	2.85	2.83	2.85	2.81	2.85
Ti	0.13	0.13	0.12	0.12	0.16	0.23	0.23	0.24	0.19	0.21
Al	1.00	1.13	0.91	0.97	1.26	1.22	1.25	1.24	1.26	1.25
Mg	3.18	2.46	3.08	2.75	2.57	1.48	1.30	1.44	1.42	1.4
Mn	0.10	0.09	0.04	0.07	0.05	0.05	0.03	-	0.02	-
Fe	1.60	2.28	1.73	1.93	2.16	1.18	1.37	1.24	1.30	1.29
Ca	1.78	1.93	1.88	1.87	1.92	-	-	0.01	0.01	-
Na	0.35	0.36	0.25	0.33	0.35	-	-	-	-	-
K	0.09	0.14	0.09	0.11	0.14	0.9	0.86	0.9	0.83	0.93
Cl	-	0.01	-	0.01	-	0.01	-	-	-	-
P	-	0.06	-	-	-	-	-	-	-	-
O	-	-	-	-	-	0.27	0.22	0.35	0.10	0.29
OH	-	-	-	-	-	1.72	1.79	1.65	1.90	1.71
Σ X						0.90	0.86	0.92	0.84	0.93
Al^{IV}	1.01	1.14	0.87	0.85	1.19	1.15	1.18	1.15	1.19	1.15
Al^{VI}						0.07	0.08	0.08	0.07	0.09
C site total	5.01	4.96	4.96	4.87	4.94					
Na_B	0.22	0.07	0.12	0.13	0.08					
Na_A	0.13	0.28	0.12	0.21	0.27					
A site total	0.22	0.42	0.21	0.32	0.41					
Mg#[e]	0.67	0.52	0.64	0.59	0.54					
Fe (Ann)[fe]						39.14	45.74	41.13	43.2	42.97
Mg (Phl)[f]						49.40	43.28	47.95	47.35	46.79
Al (Mus)[f]						2.34	2.53	2.78	2.41	3.20
Mn (?)[f]						1.63	0.86	-	0.80	0.07

[a] Numerals in panel stand for: 1=quartz moncodiorite; 2 = syenite biotite-hornblende; 4 = quartz monconite biotite-hornblende; 5 = quartz moncodiorite biotite-hornblende.
[b] Numerals in braces show number of analyses.
[c] (-) is below detection limit.

IOP Conf. Series: Earth and Environmental Science **110** (2017) 012004 doi:10.1088/1755-1315/110/1/012004

[d] Atoms per formula units (apfu) are based on $A_{0-1}B_2C_5[T_4O_{11}]_2(OH, O, F, Cl)$ (amphibole) and $XY_{2\text{-}3}$ $[AlSi_3O_{10}](OH, F, Cl)_2$ (mica).
[e] Mg#=Mg/(Mg+ΣFe).
[g] Ann = annite minale (molecule), Phl = phlogopite, Mus = muscovite, (?) = abstract molecule.

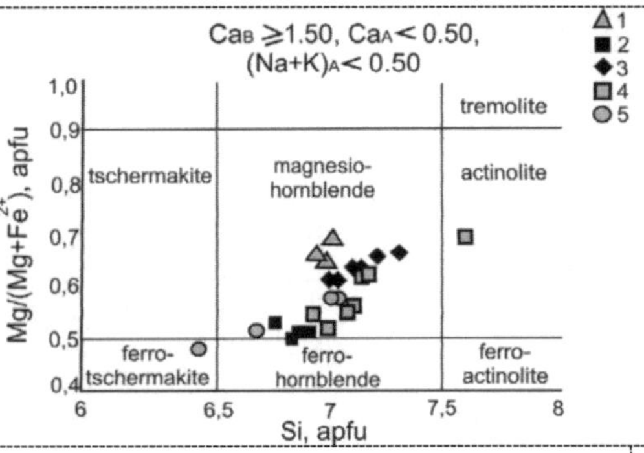

Figure 3. Chemistry of amphiboles from the rocks of contact zone of the Shivey and Chadal massifs. Classification diagram, after [14]

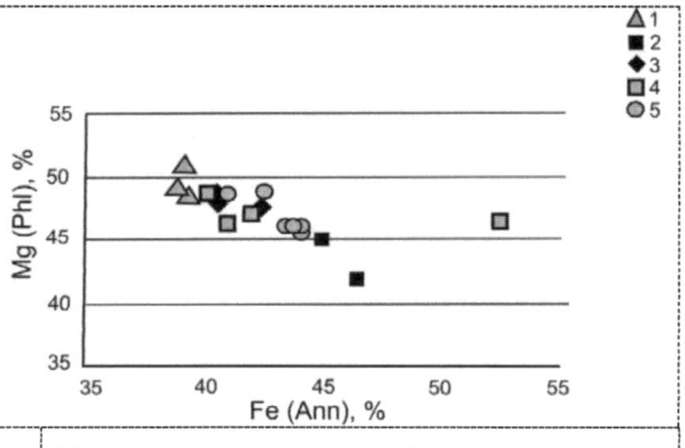

Figure 4. of micas from the rocks of contact zone of the Shivey and Chadal massifs

Numerals in panel figure 3 and 4 stand for: 1=quartz moncodiorite; 2 = syenite biotite-hornblende; 3 = subalkaline diorite biotite-hornblende; 4 = quartz monconite biotite-hornblende; 5 = quartz moncodiorite biotite-hornblende.

4. Geochemical characteristics of contact zone rocks

Geochemistry of contact zone rocks indicates the process of chemical mixing (figure 5). Granitoids of the Shivey massif are enriched with incompatible elements especially Th, K, Zr (up to 1086 ppm) and Hf, less and unevenly with Ta and Nb (0,49–7,07 and 4,94–69,8 ppm) and rare earth elements (ΣРЗЭ = 142–420 ppm) especially with light REE [2, 13, 14]. The presence of positive anomalies Ba, Th, U and K, Sm and negative anomalies Nb-Ta and Zr-Hf are characteristic for contact rocks (both leuco-and mesocratic)

Geochemical feature of contact zone rocks is a weak europium minimum and gentler slope in comparison with the case of REE distribution in rocks of the Shivey massif (figure 5b).

5. Results discussion and conclusion

Analysis of mineral associations of contact rocks of the Shivey and the Chadal massifs and their geochemical features showed that a reactionary effect occurred between the crystallized melt products and melts. Crystallization of minerals is occurred in two ways according to texture relationships. Manifestation of accessory and ore minerals occurred firstly: small crystals apatites, then zircon, ilmenite and magnetite. Aluminosilicates and silicates are abundantly formed later: titanite, orthoclase, medium plagioclase, hornblende, biotite and quartz. Secondly reverse situation is obvious where the rock-forming minerals are crystallized in the beginning and ore minerals are crystallized after.

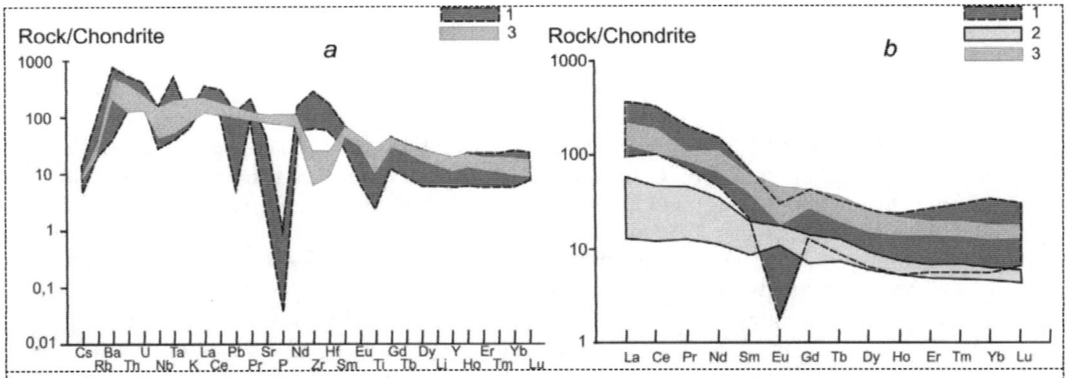

Figure 5. Chondrite-normalized REE patterns and multi-element diagrams (a, b) of the rocks contact zone of the Shivey and Chadal massifs

Numerals in panel figure 3 and 4 stand for: 1= alkaline granite-quartzsenite Shivey massif, 2 = gabbroic Chadal massif, 3 = rocks contact zone of the Shivey and Chadal massifs

Earlier formation of apatite (fluoroapatite) suggests high acidity at initial stages of crystallization because $[PO_4]^{3-}$ is one of the strong acid anions. Another fact indicates on increased acidity of the melt with the presence of orthosilicic acid – feldspars formation is later than apatite and zircon. Aluminum forms an aluminosilicate anion as an amphoteric element in alkaline silicate melts resulting in formation of feldspars and feldspathoids.

Another feature of mineral formation is the presence of exsolution texture in ilmenite which is characteristic of ilmenite from gabbro. It is worth noting that apatite inclusions are not found in these ilmenite individuals. Fact considering this ilmenite as a relic product of crystallization of a basic magma composed of the Chadal massif remains unclear because no other signs have been found. Typochemical characteristic of second-generation magnetite forming isometric crystals and presenting with ilmenite in one paragenesis let suggest that it could crystallize from the melt of basic composition of both normal and increased alkalinity. The increased content of pyrophanic molecule in ilmenite is also a typomorphic feature of crystallization of a mineral species from alkaline melts.

Change of acidic situation to alkaline could occur against the background of an oxygen fugacity increase. This process is occurred in replacement borders of ilmenite with titanite. Ilmenite becomes unstable that could result to its replacement with calcium and orthosilicic acid in the system as well as ferric oxide.

Geochemical analysis of rare and scattered elements contents in contact zone rocks showed strongly marked character of hybridization. These rocks are characterized by a sharp predominance of light over heavy lanthanides. Negative Zr-Hf anomalies and positive anomalies Th and U are especially strongly marked. Contact zone rocks occupy an intermediate position between REE distribution curves in rocks of the Shivey and the Chadal massifs according to REE content.

Results of raster electron microscopy and X-ray microanalysis let consider the probable sequence of mineralization in contact zone of massifs of bimodal association of gabbroic and alkaline granitoids of the Kaakhem magmatic areal. Relationship established between microtexture and geochemical inherity indicate the interaction of basic and acid melts with rock formation of middle group.

References

[1] Rudnev S N, Vladimirov A G and Ponomarchuk V A 2006 Kaakhemskiy polichronous granitoid batholiths (Eastern Tuva): Composition, Are, Sources and geodynamic position Journal of Lithosphere **2** 3–33
[2] Sugorakova A M, Yarmolyuk V V, Lebedev V I and Lykhin D A 2011 Late Paleozoic alkaline-

granitoid magmatism of Tuva and its relationship with intraplate processes within the Siberian paleocontinent Journal of Doklady Earth Siences **439** article 5 641–647
[3] Mongush A A and Sugorakova A M 2013 Age and sources of magmas of post-collisional gabbroids of the Kaakhem magmatic areal, Eastern Tuva: results of the first $^{40}Ar/^{39}Ar$ and Sm-Nd studies *Journal of Geochemistry International* **11** 1042–1047
[4] Salnikova E B, Kozakov I K and Kovach V P 2003 Early Ordovician age of the Mazhalyk complex of Eastern Tuva: U-Pb method on zircon. Isotopic geochronology in solving problems of geodynamics and ore Genesis Isotopic geochronology: Center of information Culture (Saint Petersburg) 450–452 (in Russian)
[5] Sugorakova A M 2011 Summary of the results of the latest geochronological studies of magmatic formations on the territory of Tuva Geodynamic evolution of the lithosphere of the Central Asian mobile belt (from the ocean to the continent) (Irkutsk) 205–207 (in Russian)
[6] Borodina E V, Egorova V V and Izokh A E 2004 Petrology of the Ordovician collision peridotite-gabbro massifs (the Mazhalyk intrusive case study, South-Eastern Tuva) *Journal of Geology and Geophysics* **45** article 9 1075–1091
[7] Rudynev S N and Polyakov G V 2013 Early Paleozoic granitoid magmatism of the Altai-Sayan folded region area and in the Lake Zone in Western Mongolia Russian Academy of Sciences, Siberian Branch (Novosibirsk) (in Russian)
[8] Cook N D J 1988 Diorites and associated rocks in the Anglem Complex at The Neck, northeastern Stewart Island, New Zealand: an example of magma mingling *Journal of Lithos* **21** 247–262
[9] Lowell G R and Young G J 1999 Interaction between coeval mafic and felsic melts in the St. Francois Terrane of Missouri, USA *Precam. Res.* **95** 69–88
[10] Titov A V, Litvinovsky B A, Zanvilevich A N and Shadaev M G 2000 Phenomena of hybridization in combined basite-leucogranite dikes of the Ust-Khilok massif (Transbaikalia) *Journal of Geology and Geophysics* **41** article 12 1714–1728
[11] Titov A V, Khromykh S V, Vladimirov A G and Pospelova L N 2001 Molten inclusions in garnet and quartz from dacite porphyry of the Aktobe structure (Kazakhstan): estimation of generation conditions and composition of primary melts *Journal of Doklady Earth Siences* **377** article 1 86–90
[12] Ludwig K R 1999 Berkeley Geochronology Center Special Publication **1a** 49
[13] Sugorakova A M 2008 Peculiarities of contact between the Chadal gabbroid and the Shivey alkaline granite-quartz-syenite massifs (Eastern Tuva) Geodynamic evolution of the lithosphere of the Central Asian mobile belt (from the ocean to the continent) (Irkutsk) 115–166 (in Russian)
[14] Sugorakova A M and Yarmolyuk V V 2008 The Shivey alkaline granite-quartz-syenite massif (Eastern Tuva) Granites and Earth evolution (Ulan-Ude) 92–97 (in Russian)
[15] Leake B N, Woolley A R, Arps C E S, Birch W D, Gilbert M C, Grice J D, Hawthorne F C, Kato A, Kisch H J, Krivovichev V G 1997 Nomenclature of amphiboles: Report of the Subcommittee on Amphiboles of the International Mineralogical Association, Commission on New Minerals and Mineral Names *American Mineralogist* **82** 1019-1037

IOP Conf. Series: Earth and Environmental Science **110** (2017) 012005 doi:10.1088/1755-1315/110/1/012005

Large Igneous Provinces and Their Mafic-Ultramafic Intrusions

R E Ernst [1,2], S M Jowitt [3], J A Blanchard [1], D A Liikane [1,4]

[1] Department of Earth Sciences, Carleton University, Ottawa, Canada

[2] Faculty of Geology and Geography, Tomsk State University, Tomsk, Russia

[3] Department of Geoscience, University of Nevada, Las Vegas, USA

[4] Department of Earth Sciences, University of Toronto, Toronto, Canada

E-mail: Richard.Ernst@ErnstGeosciences.com

Abstract. Here we provide an overview of the range of settings for mafic-ultramafic layered intrusions as part of the plumbing system of Large Igneous Provinces, and address the metallogenic implications.

1. Definition and Characteristics

LIPs have become an important focus for research in recent years due to their use in paleocontinental reconstructions [1], in exploration targeting [2-3], as planetary analogues [4-5], and as a result of their links to dramatic climate change [5]. LIPs represent large volume (>0.1 Mkm^3; frequently above 1 Mkm^3), mainly mafic (-ultramafic) magmatic events of intraplate affinity (based on tectonic setting and/or geochemistry) that occur in both continental and oceanic settings, and are typically either of short duration (<5 Ma) or consist of multiple short pulses that occur over a maximum of a few 10s of Ma. LIPs comprise volcanic packages (flood basalts) and a plumbing system of dyke swarms, sill complexes, layered intrusions, and a crustal magmatic underplate. LIPs can also be associated with major silicic magmatic events termed Silicic LIPs (SLIPs), as well as carbonatites and kimberlites. LIP events occur at a variable rate through time, averaging every 20–30 Ma, but with possible peaks associated with supercontinent breakup, back to at least 2500 Ma. The rate of LIP occurrences in the Archean is less certain due to its poorer preservation [5, 6].

1.1 Plumbing System of LIPs

Mafic-ultramafic (M-UM) intrusions are an important component of the plumbing system of LIPs, as outlined below. LIP-related regional mafic dyke swarms that can have linear (>2000 km long), radiating (up to >2500 km in radius), and giant circumferential (sub-circular or elliptical) geometries that form swarms up to nearly 2000 km in diameter [7-8]. While linear, radiating and circumferential swarms are all associated with mantle plumes, the linear swarms are uniquely linked to mantle plume-related triple junction rifting [6]. Dolerite sill provinces are common in sedimentary basins, such as the Karoo sills in the Karoo basin of southern Africa and the 2215 Ma Nipissing sills of the Huronian basin of the southern Superior craton of North America, but are also hosted in basement rocks, with the largest sill system being the 3000 km long Ferrar event of Antarctica/Tasmania/Australia [6, 9, 10].

2. Setting of Mafic-Ultramafic Intrusions in lips

Mafic-ultramafic (M-UM) intrusions are key exploration targets for a range of different mineral deposits (e.g. magmatic sulphide Ni-Cu-platinum group element (PGE) deposits, PGE "reef" style deposits) and many are linked to LIPs [11, 2, 5]. Key examples are large stratiform intrusions, such as the Bushveld and Stillwater, which can be the source of Ni-Cu-PGEs and Cr, amongst other commodities. Smaller mafic-ultramafic intrusions such as those with the geometry of chonoliths (e.g. Noril'sk) and giant dyke-like layered intrusions (e.g. the Great Dyke of Zimbabwe) have been proven to host significant amounts of Ni-Cu sulphide mineralization. There is also a link between LIPs and hydrothermal ore deposits, in which LIPs provide the necessary heat and fluids to facilitate hydrothermal circulation that leads to the concentration of desirable metals (e.g. native copper). The 5-part classification of Ernst and Jowitt [3] also identifies links between LIPs and additional types of ore deposits.

Here we provide an overview of the settings of mafic-ultramafic (M-UM) intrusions as part of the plumbing system of LIPs. We are not considering those layered intrusions associated with arc settings, nor those associated with spreading ridges. Instead, we are focussed on those that are linked to intraplate settings, and in particular, those inferred to belong to LIPs.

2.1 Magmatic underplate

The largest M-UM intrusions may correspond to magmatic underplates, located at the base of the crust. In LIP events, these are typically generated by partial melting of the underlying mantle plume head, yielding underplates that may be up to several-hundred-km across and 10-20 km thick [12; associated with the 130-90 Ma High Arctic LIP]. These underplates may be linked with the edge of circumferential swarms, as illustrated in Figure 13e from Makitie et al. [13], for example.

Geophysical studies support the presence of these deep crustal magma bodies, termed underplates, and magnetic maps processed as "pseudo gravity" [12] indicate their overall volume and importance within LIPs. Oakey and Saltus [12] identify a 17 million km^3 underplate associated with the High Arctic LIP (Canada).

The cooling history of these magmatic underplates is poorly understood, but an intriguing idea has been proposed by Xu et al. [14] whereby the final stage of crystallization of the magmatic underplate is associated with the release of metals and fluids into the overlying crust. Speculatively, cooling and crystallization of the magmatic underplate may release fluids and metals 30-100 myr after the LIP event occurs, leading to the formation of hydrothermal Cu and Pb-Zn deposits at shallow crustal levels [14, e.g. associated with the 260 Ma Emeishan LIP].

2.2 Mid-crustal intrusions circumscribing the plume centre region

Mafic intrusive units frequently exhibit geochemical evidence for crustal contamination. For instance, a geochemical/isotopic study in a portion of the Karoo province revealed multiple magma batches that could be related through the progressive AFC contamination of a staging chamber, from which there was periodic expulsion of magma batches (e.g. Fig. 18 in [15] It is generally interpreted that such contamination occurred in a staging magma chamber(s) located at greater depths within the crust, underlying the area where the contaminated units were emplaced [15]. However, the notion of large-scale

IOP Conf. Series: Earth and Environmental Science **110** (2017) 012005 doi:10.1088/1755-1315/110/1/012005

lateral transport of magma through dykes [16, 6] and sills [9] suggests that such staging chambers are not necessarily located directly underneath the crustally contaminated units, but could be located hundreds of km or more away to the side.

One possible location for such chambers is surrounding the mantle plume centre at a distance of hundreds of km. For instance, the plume centre for the 1270 Ma Mackenzie LIP is surrounded by a ring of gravity/aeromagnetic anomalies [16] that represent mid-crustal M-UM intrusions 10s of km across and a few km thick [4]. These intrusions are located at the edge of a magmatic underplate and were likely generated from magmas ascending from the edge of that underplate [17]. The broader application of this model is being tested with geophysical studies around other plume centres [17].

One of the mid-crustal intrusions surrounding the Mackenzie plume centre is geophysically linked to the exposed Muskox layered intrusion that hosts Ni-Cu-PGE mineralization. The other mid-crustal intrusions that circumscribe the Mackenzie plume centre could therefore also be, hypothetically, linked to shallow bodies (as yet undiscovered) with economic magmatic sulfide potential [17].

2.3 Mid-shallow crustal intrusions within rift zones associated with LIPs

Separate intrusions aligned along rift zones: These are likely part of triple junction rift systems [17] and can spawn rift-parallel dyke swarms [18]. In the case of Greenland, there are lines of intrusions spaced along the rift margin. A similar case is associated with the Deccan, where the intrusions are alkaline in composition [17].

2.4. Sill-like (Stratiform) M-UM intrusions emplaced in sedimentary basins

This represents a class of dyke-like intrusions often emplaced at the base of supracrustal sequences. This class includes the exemplary 2060 Ma Bushveld and 2710 Ma Stillwater complexes, both of which host world-class PGE mineralization. The largest such stratiform M-UM intrusions are probably proximal to mantle plume centres, although smaller stratiform M-UM intrusions could also potentially be located more distally from the plume centre and fed via radiating dyke swarms.

2.5. Dyke-like M-UM

Dyke-like layered intrusions (e.g. the 2580 Ma Great Dyke of Zimbabwe, 2420 Ma Jimberlana dyke of western Australia, and the 2730 Ma Ahmeyim Great Dyke of Mauritania) and associated Ni-Cu-PGE mineralization can also be linked to LIP events [6]. For instance, the Jimberlana dyke is part of the widespread 2420-2400 Ma Widgiemooltha swarm of the Yilgarn craton, which can potentially be linked to a plume centre off the eastern margin of the craton [19]. The 2580 Ma Great Dyke of Zimbabwe has several companion dykes, but they are located close by and an overall fanning pattern has not been identified, so a plume centre cannot be currently linked to the Great Dyke of Zimbabwe event.

2.6. M-UM intrusions at a distance from the plume centre

Many LIPs have associated radiating dyke swarms that diverge from the plume centre and can channel magma laterally for distances of up to several 1000 km away from the plume centre [20, 6]. If the radiating dyke swarm intersects a sedimentary basin, the dykes may reorient and be emplaced as sills instead, some of which may be large and prospective hosts for mineralization [21, 6].

2.7. M-UM intrusions associated with giant circumferential dyke swarms

A new class of regional dyke swarms termed "giant circumferential" has now been recognized [22, 13, 23]. These dyke swarms circumscribe the plume centres of LIP events at distances up to more than 1000 km in radius, and are potentially fed vertically from the outer boundary of the plume head. We speculate that local M-UM intrusions may be associated with these giant circumferential swarms.

IOP Conf. Series: Earth and Environmental Science **110** (2017) 012005 doi:10.1088/1755-1315/110/1/012005

DISCUSSION

There are insights that arise from studying the plumbing system of LIPs including the role of mafic-ultramafic intrusions.

Range in geochemical composition

Geochemical studies of some LIPs reveal that the magmatism is very similar throughout the entire LIP, and in other cases, multiple compositions seem to be present [6]. Any variation in chemistry can result from different mantle sources, and mixing of these sources (plume, upper mantle asthenosphere, lithospheric mantle), variation in the depth of partial melting, and potential contamination as magma transects the lithospheric mantle and the crust, with especially significant modification if the magma sits in any chamber for a protracted period of time [24, 15].

Location of source mantle

The most important element related to the magma sources is the plume centre region. It has been established that giant radiating dyke swarms can transport magma from the plume centre region (about 500 km across) out to distances of more than 2000 km away from the plume centre, through laterally emplaced radiating dyke swarms. In such cases, the magma for the entire LIP is generated in the plume centre region rather than from the mantle underlying the entire swarm. This has particular significance when a radiating swarm feeds a sill province. An important example is the feeding of the Nipissing sill province in the Huronian basin by the Senneterre radiating swarm; it has been proposed that the Nipissing sill magmas were emplaced from laterally propagating dykes of the Senneterre radiating swarm and that the magma originated 1300 km to the NE of the Huronian basin, in the vicinity of the Ungava plume centre (at the focus of the Senneterre radiating swarm; [25, 6]). This means that mantle source for the Nipissing sills was not underlying the Huronian basin on the south side of the Superior craton, but was actually in the Ungava plume centre region on the NE side of the Superior craton, 1300 km away.

Integration with geophysics

In regional gravity and aeromagnetic maps, and on seismic cross surveys, there are typically many anomalous features observed. Knowledge of the LIP event units in the vicinity can provide a framework to interpreting the geophysical data, and in particular, to identifying those features which can be linked with LIP-related processes. For instance, plume centre regions can be located through giant radiating dyke swarms and such regions can then be interrogated using geophysics to identify the location of associated layered intrusions and the lower crustal magmatic underplate.

Distribution of associated ore deposits

Mafic-ultramafic intrusions are the host of most Ni-Cu-PGE ore deposits, and the question must be asked: Where are the most favourable locations for such deposits to form within LIP systems. As frequently noted, the footprint of an ore deposit is a very small fraction of the overall size of a LIP. For instance, the Norilsk deposits of the Siberian trap LIP represent a target that is less than 0.1% of the areal extent of the overall LIP itself [2]. It has been argued that proximity to the plume centre [2], along with association with translithospheric faults [26], are important criteria for mineralization potential. Other kinds of hydrothermal ore deposits can be spatially linked with regional extensive sill provinces, which may responsible for providing the heat to drive hydrothermal circulation that upgrade metal concentrations to produce these ore deposits.

References

[1] Ernst, R.E., Bleeker, W., Söderlund, U., & Kerr, A.C. (2013). Large Igneous Provinces and supercontinents: Toward completing the plate tectonic revolution. *Lithos*, **174**: 1–14.

[2] Ernst R.E., Jowitt S.M. (2013). Large igneous provinces (LIPs) and metallogeny. In: Colpron M. et al., eds. Tectonics, metallogeny, and discovery: The North American Cordillera and similar accretionary settings. *Society of Economic Geologists Special Publication*, v. **17**, p. 17-51.

[3] Ernst, R.E. & Jowitt, S.M. (2017) Multi-commodity, multi-scale exploration targeting using the Large Igneous Province record. In: Wyche, S. and Witt, W.K. (eds.) TARGET 2017, Perth, Australia: Abstracts. *Geological Survey of Western Australia*, Record 2017/**6**, p. 41-44

[4] Head, J.W. & Coffin, M.F. (1997). Large Igneous Provinces: a planetary perspective. *AGU Geophysical Monograph*, **100**: 411–438.

[5] Ernst, R.E., Buchan, K.L., Jowitt, S.M., Youbi, N. 2017. Applying the terrestrial Large Igneous Province (LIPs) context to planetary intraplate magmatism. *Lunar and Planetary Science Conference* **XLVIII**, abstract 1373.

[6] Ernst R.E. (2014) *Large Igneous Provinces*. Cambridge University Press. 653 p.

[7] Buchan, K.L., and Ernst, R.E. 2018a. A giant circumferential dyke swarm associated with the High Arctic Large Igneous Province (HALIP). Submitted to Gondwana Research.

[8] Buchan, K.L. and Ernst, R.E. (2018b) Giant circumferential dyke swarms: Catalogu and Characteristics. In: Srivastava, R., Ernst, R.E., Peng, P. (eds.) Dyke Swarms of the World: A Modern Perspective. Springer (submitted).

[9] Magee, C., Muirhead, J.D., Karvelas, A., Jackson, C.A.L., Bastow, I., Holford, S.P., Schofield, N., Stevenson, C.T.E., McLean, C., McCarthy, W., Shtukert, O. (2016). Lateral magma flow in sill-complexes, *Geosphere*, **12**, 1-33, doi:10.1130/GES01256.1.

[10] Elliot, D. H. & Fleming, T. H. 2017. The Ferrar large Igneous Province: field and geochemical constraints on supra-crustal (high-level) emplacement of the magmatic system. In: Sensarma, S. & Storey, B.C. (eds) Large Igneous Provinces from Gondwana and Adjacent Regions. Geological Society, London, Special Publications, 463, https://doi.org/10.1144/SP463.1

[11] Naldrett A.J. (2010). Secular variation of magmatic sulfide deposits and their source magmas. *Economic Geology*, v. **105**, p. 669–688.

[12] Oakey, G.N. & Saltus, R.W. (2016). Geophysical analysis of the Alpha-Mendeleev ridge complex: Characterization of the High Arctic Large Igneous Province. *Tectonophysics*, v. **691**, p. 65-84.

[13] Mäkitie, H., Data, G., Isabirye, E., Mänttäri, I., Huhma, H., Klausen, M.B., Pakkanen, L., Virransalo, P. (2014). Petrology, geochronology and emplacement model of the giant 1.37 Ga Lake Victoria Dyke Swarm on the margin of a large igneous province in eastern Africa. *Journal of Asian Earth Sciences*, v. **97**, p. 273-297.

[14] Xu Y-k., Huang, Z-l., Zhu, D., Luo, T-y. (2014). Origin of hydrothermal deposits related to the Emeishan magmatism. *Ore Geology Reviews*, v. **63**, p. 1-8.

[15] Neumann, E.-R., Svensen, H., Galerne, C.Y., & Planke, S. (2011). Multistage evolution of dolerites in the Karoo Large Igneous Province, Central South Africa. *Journal of Petrology*, **52**: 959–984.

[16] Baragar W.R.A., Ernst R.E., Hulbert L., Peterson T. (1996). Longitudinal petrochemical variation in the Mackenzie dyke swarm, northwestern Canadian Shield. *Journal of Petrology*, v. **37**, p. 317–359.

[17] Blanchard J.A., Ernst R.E., Samson C. (2017). Gravity and magnetic modelling of layered mafic-ultramafic intrusions in large igneous province plume centre regions; Case studies from the: 1.27 Ga Mackenzie, 1.38 Ga Kunene-Kibaran, 0.06 Ga Deccan and 0.13-0.08 Ga High Arctic events. *Canadian Journal of Earth Sciences*, v. **54**, p. 290-310.

[18] Klausen M.B. & Larsen H.C. (2002). East Greenland coast – parallel dike swarm and its role in continental breakup. In: Menzies M.A., Klemperer S.L., Ebinger C.J., Baker J., eds. Volcanic Rifted Margins. *Geological Society of America Special Paper*, v. **362**, p. 133–158.

[19] Pisarevsky, S.A., De Waele, B., Jones, S., Söderlund, U., Ernst, R.E. (2015). Paleomagnetism and U–Pb age of the 2.4 Ga Erayinia mafic dykes in the south-western Yilgarn, Western Australia: Paleogeographic and geodynamic implications. *Precambrian Research*, v. **259**, p. 222-231.

[20] Ernst R.E., Buchan K.L. (1997). Giant radiating dyke swarms: their use in identifying pre-Mesozoic large igneous provinces and mantle plumes. *AGU Geophysical Monograph*, v. **100**, p. 297–333.

[21] Palmer, H.C., Ernst R.E., & Buchan, K.L. (2007) Magnetic fabric studies of the Nipissing sill province and Senneterre dykes, Canadian Shield, and implications for emplacement. *Canadian Journal of Earth Sciences*, v. **44**, p. 507-522.

[22] Ernst R.E., Buchan K.L. (1998). Arcuate dyke swarms associated with mantle plumes on Earth: implications for Venusian coronae. *29th Lunar and Planetary Science Conference*. March, 1998. Houston, Texas. 1998. Extended Abstract #1021.

[23] Buchan K.L., Ernst R.E (2016). Giant circumferential dyke swarms on Earth as possible analogues of coronae on Venus. Lunar and Planetary *Science Conference* **47**, Extended Abstract #1183.

[24] Bryan, S., Ukstins Peate, I., , Peate, D.W., Self, S., Jerram, D.A., Mawby, M.R., Marsh, J.S., Miller, J.A. (2010). The largest volcanic eruptions on Earth. *Earth-Science Reviews*, **102**: 207–229.

[25] Buchan, K.L., Mortensen, J.K., Card, K.D., & Percival, J.A. (1998). Paleomagnetism and U–Pb geochronology of diabase dyke swarms of Minto Block, Superior Province, Quebec, Canada. *Canadian Journal of Earth Sciences*, **35**: 1054–69.

[26] Begg, G.C., Hronsky, J.A.M., Arndt, N.T., Griffin, W.L., O'Reilly, S.Y., Hayward, N. (2010). Lithospheric, cratonic, and geodynamic setting of Ni–Cu–PGE sulfide deposits. *Economic Geology*, **105**: 1057–70.

Specifics of geological composition, geochemistry and geochronology of rocks from the Kresty alkaline-ultrabasic massif (Maimecha-Kotui province, Polar Siberia)

A M Sazonov[1,2], E A Zvyagina[1], I F Gertner[2], T S Krasnova[2], G P Lipenkov[3]

[1] Siberian Federal University, Krasnoyarsk, Russia
[2] Tomsk State University, Tomsk, Russia
[3] All-Russian Geological Institute, Saint-Petersburg, Russia

E-mail: labspm@ggf.tsu.ru

Abstract. In this work, we demonstrate new data that allows us to accurate geochronological ranges of formation of the Kresty alkaline-ultrabasic massif, which is considered to be a satellite of the Gulinian giant pluton. We also interpreted geological, geochemical and isotope-geochemical data obtained earlier for major varieties of this volcanic-plutonic association taking into account new geochronological results, as well as considered new aspects/information on matter source of alkaline-ultrabasic massifs from this province. One of the main aspects is interaction of Siberian super plume matter with hosting substrate of Siberian craton continental crust.

1. Introduction

The Maimecha-Kotui province (MKP) is one of the unique geological objects in Polar Siberia. Its geotectonic position is caused by formation of continental Khatanga rift at the boundary between Siberian Craton and Taimyr orogen. One of the important structural elements of its spatial localization is transform fault complicated by large gravity anomaly with its center in Gulinian giant pluton. Many authors think that opening of Ob-Zaisansk Ocean was initiated by local intrusion of plume material in this particular point/location [1]. However, we think that this episode/event is only a local “finger plume”, which promoted formation of a unique mineral province. Such transform fault of Khatanga rift system controls one more large ore node, which corresponds to Norilsk-Talnakh copper-nickel deposits too. What makes the Kresty province unique is that it is spatially adjacent to Gulinian pluton, whose sizes exceed all known intrusive bodies in the Maimecha-Kotui province. Main intrusive phases/stages of the Kresty Massif are represented by ultramafic rocks with normal alkalinity (olivinites, veprlites, pyroxenites, and its ore-bearing kinds), as well as melilitic rocks like kugdites and melilitolites and monticellitolites. Among dykes we see nepheline and alkaline syenites, as well as alkaline granosyenites and granites. We have obtained new geochronoligcal data on perovskites and zircons from these rock varieties. These data confirmed its formation at Permian-Triassic boundary. An important aspect of this massif formation is industrial concentrations of noble metals (gold and platinoids up to 20 g/t), which are found in some rock varieties undergone secondary metasomatic changes. This aspect is extremely important in further studying of the Kresty Massif.

2. Geological setting and internal structure of studied object

Composition of the Kresty volcanic-plutonic structure (KVPS) is determined by the spatial superposition of volcanic fluxes of melanephelinite, intrusive bodies of olivinite-pyroxenite and melilitolite-monticellitolite series, and the dykes of thachydolerites, subalkaline syenites and granosyenites, alkaline picrites and lamprophyres. By geophysical data, the Kresty Massif has a round shape in plane about 6 km in diameter with a steep dip of contacts (*Fig. 1*) [2]. Like the Guliniam pluton, the massif is covered over 2/3 of its area by Cenozoic deposits of the Khatanga depression.

IOP Conf. Series: Earth and Environmental Science **110** (2017) 012006 doi:10.1088/1755-1315/110/1/012006

The bulk of its volume is formed by ultra- mafic rocks with normal alkalinity. Melilite and monticellite rocks are located over the periphery of the olivinite core and form a series of small autonomous bodies. Their later introduction is confirmed by active contact affect onto ultramafic rocks, which led to production of hybrid variations like melilite and perovskite-enriched olivinites and uncompahgrites. The dyke derivatives break through the massif rocks of zonal extension along the boundary of the plutonic and volcanic associations.

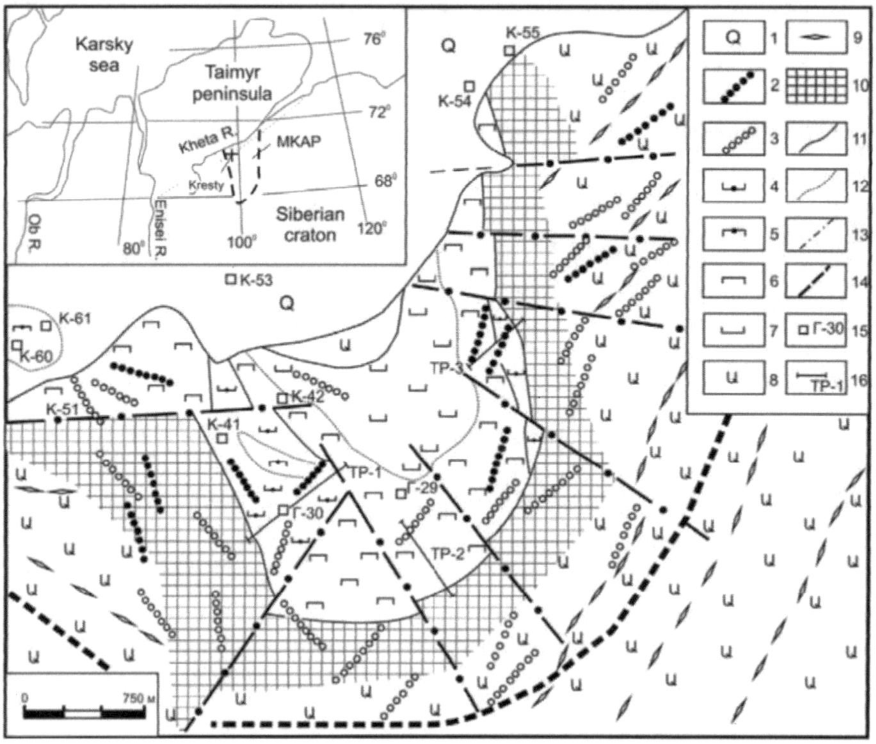

Figure 1. Scheme of geological structure of the Kresty Massif.The inset shows the geographical position of the Maimecha-Kotui province (MKP) and the Kresty volcanoplutonic structure (KVPS) [2].

1- loose deposits, KZ; 2 -9. - KVPS rocks, MZ_1^ dykes of alkalaine syenites and granosyenites (2), of trachydolerites and trachyandesites (3), kugdites and melitholites (4), montichellitolites (5), banded olivinite-wehlite-clinopyroxenite complex (6), olivinites (7), melanephelinites (8), and local fluxes of melanephelinitee clastolavas (9); 10 – chertaureol within the host effusive mass; 11 – geological boundaries; 12 –bound areas between petrophical variations within intrusive bodies; 13 – faults; 15 – KVPS boundaries by gravity data; 15 – boreholes and their Nos; 16 – ditches and their Nos.

By their petrochemical parameters, the KVPS rocks conform to those of tholeiite (dunites, wehrlite, and clinopyroxenites), subalkaline (trachydolerites, trachyandesites, and syenitoids), and alkaline series (melilitolites, monticellitolites, nephelinites, and alkaline picrites), mainly of potassium–sodium specialization. These rocks, as all MKP magmatites, are characterized by high concentrations of most of trace elements comparable to numbers in oceanic island basalts (OIB), which confirms plume nature of magmatism in the province. Heterogeneity of the products of tholeiite and alkaline series is well pronounced in the behavior of the most magmatophilic elements (Nb, Ta, Th, Sr, LREE) concentrated mainly in alkaline melts. The content of REE and La/Yb ratio increase subsequently in series from olivinites and wehrlites to pyroxenites, nephelinites, and melilitolites, reaching maximum in monticellitolites. The dykes of syenites and trachydolerites show moderate concentrations of lanthanides

IOP Conf. Series: Earth and Environmental Science **110** (2017) 012006 doi:10.1088/1755-1315/110/1/012006

comparable to ultrabasic rocks of normal alkalinity, which negates thereby the model of differentiation from a single parent melt. The fractionation degree of rare elements increases in following series: "volcanites→ dyke formation → main intrusive phases", at the same time obvious negative anomalies of Rb, Ba, Sr, Nb, Zr, Hf, and Ti appear in olivine-rich variations.

Geochemical heterogeneity of the KVPS rocks is also traced in the parameters of the neodymium and strontium isotope composition, whose variations indicate autonomous matter sources for associations of normal and high alkalinity [3]. However, a valid Sm–Nd isochrone line was obtained only for the whole rock compositions of olivine, wherlites, and two pyroxenites, yielding the age of 251 ± 20 Ma ($\varepsilon Nd(T)$ = + 2,0). Alkaline ultramafic rocks show higher primary $^{143}Nd/^{144}Nd$ ratios ($\varepsilon Nd(T)$ = + 2,4 ÷ 3,1), whereas decreased values of this parameter (from –0.6 to –15.9) are typical for trachyandesites and syenitoids. The probable heterogeneity of the sources of rock associations is also shown in the values of the T_{DM} model age. Compared to the small intrusions of alkaline syenites (~ 2.36 Ga), younger dates were found for melanephelenites and monticellitolites (~ 590–620 Ma). The isotope composition of strontium shows a wider variability with I_0 values ranging from 0.7035–0.7037 (for nephelinites, melilitolites, and monticellitolites) and 0.7038–0.7041 (for olivinites, wehrlites, and pyroxeniteto 0.705–0.708 (for trachydolerites and alkaline syenites). The ratios of oxygen isotopes in whole-rock compositions and in rock-forming pyroxene ($\delta^{18}O$ = + 1.1 ÷ 4.9 %) suggest considerable interaction of the rocks with meteor waters permeating probably as deep as 500 m.

3. Analytical techniques

New U-Pb isotopic data on zircon grains was studied. It represents two intrusive phases of the Kresty massif, including one of the main rock types like kugdites and vein types of rocks like alkaline syenites. Zircons from this object were analyzed with SHRIMP-II microprobe at the Russian Geological Research Institute named after A.P. Karpinsky (St. Peterburg) using a standard technique with a spot diameter of 18 μm. The internal structure of mineral grains was determined with cathodoluminescent images taken on ABT55 scanning electron microscope in a usual operating mode. Data was processed with SQUID software. The U/Pb ratios were normalized to the value of 0.0668 of the TEMORA standard zircon. The errors of isotope ratio and age measurements were within ± 1σ, and the errors in calculation of concordant ages and intersections with concordia were within ± 2 σ. The plots with concordia were built using the ISOPLOT/Ex software.

4. Results and Discussion

Our study has given new proof of these very short-term intrusive events being connected to Siberian plume. Our new geochronological data is demonstrated in *Table 1 -2 and Fig. 2 -3*.

IOP Conf. Series: Earth and Environmental Science **110** (2017) 012006 doi:10.1088/1755-1315/110/1/012006

Table 1. U-Pb data for kugdites of the Kresty massif

	Spot	% $^{206}Pb_c$	ppm U	ppm Th	$^{232}Th/^{238}U$	ppm $^{206}Pb^*$	(2) $^{206}Pb/^{238}U$ Age		Total $^{238}U/^{206}Pb$	±%	Total $^{207}Pb/^{206}Pb$	±%	(1) $^{238}U/^{206}Pb^*$	±%	(1) $^{207}Pb^*/^{206}Pb^*$	±%	(1) $^{207}Pb^*/^{235}U$	±%	(1) $^{206}Pb^*/^{238}U$	±%	err corr
1	T-3.1.1	20,20	19	213	11,86	0,800	254	5	19.89	1.8	0.2131	1.9	25.73	3,0	0.025	83	0,14	83	0,0391	3,0	,037
2	T-3.2.1	20,84	18	250	14,73	0.751	250	5	20.05	2,0	0.2181	1.9	27.04	3.4					0,0372	3.4	
3	T-3.3.1	62,03	21	140	6,93	1,800	241	10	9,95	1.7	0.5479	1.2	30,70	6.5					0,0338	6.5	
4	T-3.4.1	24,92	13	278	21,58	0.581	242	5	19.67	2,0	0.2505	2,0	26.95	3.6	0.028	92	0,15	92	0,0374	3.6	,039
5	T-3.5.1	21,55	17	243	15,14	0.727	253	5	19,60	1.9	0.2239	2.2	26.88	3.4					0,0375	3.4	
6	T-3.6.1	22,48	48	265	5,73	2,020	242	4	20.31	1.5	0,2310	1.3	26.09	2.1	0.054	22	0.289	23	0,0385	2.1	,091
7	T-3.7.1	23,86	15	266	18,16	0.679	252	6	19.13	2,0	0.2424	2.2	24.91	3.7	0.058	45	0,32	45	0,0404	3.7	,082
8	T-3.8.1	19,81	16	259	16,45	0.693	251	5	20.17	1.9	0.2099	2.3	25.54	3,0	0.039	52	0,21	52	0,0393	3,0	,058
9	T-3.9.1	20,28	16	242	15,24	0.702	251	5	20,10	1.9	0.2136	2.3	26.49	3.3					0,0380	3.3	
10	T-3.10.1	22,35	15	266	18,93	0.642	253	6	19.42	2,0	0.2303	2.2	26.03	3.3					0,0387	3.3	
11	T-3.11.1	16,57	23	223	9,90	0.961	253	5	20.82	1.9	0,1840	2,0	25.55	2.6	0.032	48	0.175	48	0,0393	2.6	,054
12	T-3.12.1	22,82	15	257	18,14	0,650	253	6	19.31	2.1	0.2341	2.2	27,80	3.8					0,0363	3.8	
13	T-3.13.1	23,88	14	311	22,94	0.628	251	6	19.15	2.2	0.2425	2.2	26.48	3.7					0,0380	3.7	
14	T-3.14.1	22,29	16	265	17,65	0.654	241	6	20.39	2.1	0.2295	2.3	26.83	3.3	0.033	68	0,17	68	0,0375	3.3	,049
15	T-3.15.1	22,22	16	259	16,72	0.681	244	6	20.17	2.3	0,2290	2.2	27,90	3.9					0,0362	3.9	

(2) Common Pb corrected by assuming $^{206}Pb/^{238}U$-$^{207}Pb/^{235}U$ age-concordance.

IOP Conf. Series: Earth and Environmental Science **110** (2017) 012006 doi:10.1088/1755-1315/110/1/012006

Kugdite ages were determined on perovskite grains, which is one of the main accessory minerals of this rock. Despite small error, the age (249±3.4 Ma) corresponds fully to the Paleozoic-Mesozoic boundary. More accurate dating was obtained from zircons for another rock variety (alkaline syenite type), which represents one of the latest phases of dyke complex formation in this intrusive massif (*Table 2, Figure. 3*).

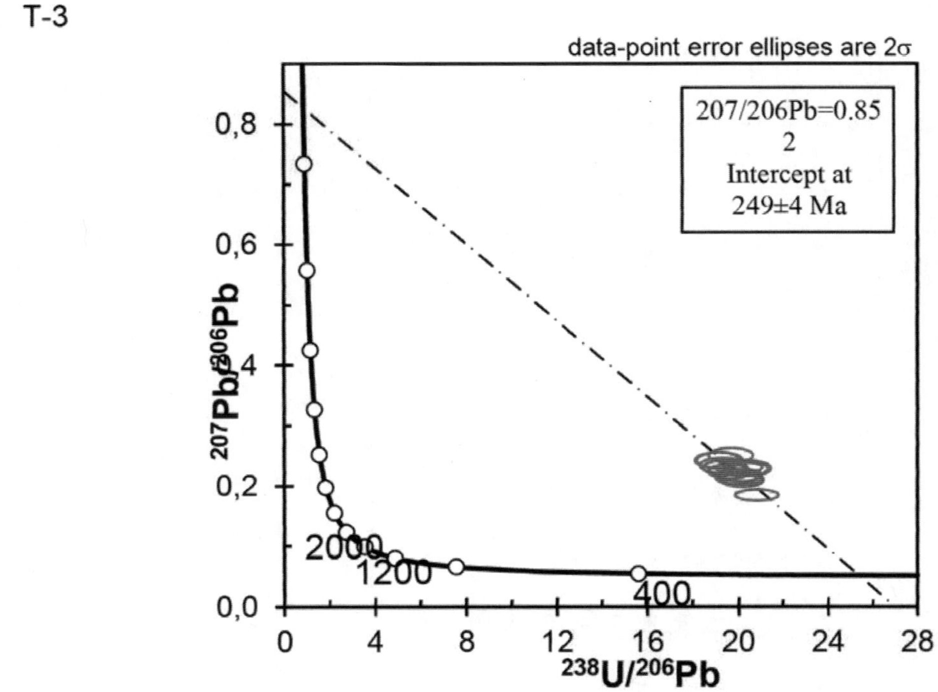

Figure 2. U-Pb plot for kugdites of the Kresty massif.

IOP Conf. Series: Earth and Environmental Science **110** (2017) 012006 doi:10.1088/1755-1315/110/1/012006

Table 2. U-Pb data for alkaline syenites of the Kresty massif

	Spot	% $^{206}Pb_c$	ppm U	ppm Th	$^{232}Th/^{238}U$	ppm $^{206}Pb^*$	(1) $^{206}Pb/^{238}U$ Age		(1) $^{207}Pb/^{206}Pb$ Age		% Dis-cor-dant	(1) $^{238}U/^{206}Pb^*$	± %	(1) $^{207}Pb^*/^{206}Pb^*$	± %	(1) $^{207}Pb^*/^{235}U$	±%	(1) $^{206}Pb^*/^{238}U$	± %	err corr
1	G-29-390.4.1	0,21	735	457	0,64	24.8	248	1,7	198	81	-20	25.52	0,7	0,0501	3.5	0.2705	3.6	0,0392	0,7	,200
2	G-29-390.10.1	0,17	881	636	0,75	29.7	248	1,8	294	66	18	25,5	0,7	0,0522	2.9	0.2822	3,0	0,0392	0,7	,249
3	G-29-390.2.1	0,26	985	641	0,67	33.6	250	1,5	179	73	-29	25.26	0,6	0,0497	3.1	0,2710	3.2	0,0396	0,6	,194
4	G-29-390.6.1	0,26	1198	898	0,77	41.2	252	1,4	321	70	27	25.05	0,6	0,0528	3.1	0.2908	3.1	0,0399	0,6	,184
5	G-29-390.7.1	0,20	703	315	0,46	24.2	253	1,7	231	77	-9	25.01	0,7	0,0508	3.3	0.2799	3.4	0,0400	0,7	,206
6	G-29-390.1.1	0,11	1182	937	0,82	40.7	253	1,2	290	44	15	24.99	0,5	0,0521	1.9	0.2874	2,0	0,0400	0,5	,247
7	G-29-390.9.1	0,20	738	504	0,71	25.4	253	1,7	323	73	27	24.96	0,7	0,0529	3.2	0.2921	3.3	0,0401	0,7	,212
8	G-29-390.5.1	0,11	1299	1063	0,85	44.8	254	1,3	259	51	2	24.93	0,5	0,0514	2.2	0.2843	2.3	0,0401	0,5	,234
9	G-29-390.11.1	0,13	978	694	0,73	34	255	1,3	171	52	-33	24.76	0,5	0,0495	2.2	0.2755	2.3	0,0404	0,5	,237
10	G-29-390.8.1	0,00	1358	1019	0,78	47.4	257	1,5	219	43	-15	24.62	0,6	0,0505	1.9	0.2829	2,0	0,0406	0,6	,302
11	G-29-390.3.1	0,21	111	42	0,39	16.2	1008	10	959	65	-5	5,91	1,1	0,0710	3.2	1,6570	3.3	0,1692	1,1	,328

Errors are 1-sigma; Pb_c and Pb^* indicate the common and radiogenic portions, respectively.

Error in Standard calibration was 0.29%(not included in above errors but required when comparing data from different mounts).

(1) Common Pb corrected using measured ^{204}Pb.

Ages of alkaline syenites determined on accessory zircon grains turned out to be more accurate, considering range of possible errors. Real age values correspond to the same values of 252 ± 1.0 Ma (*Fig. 3*).

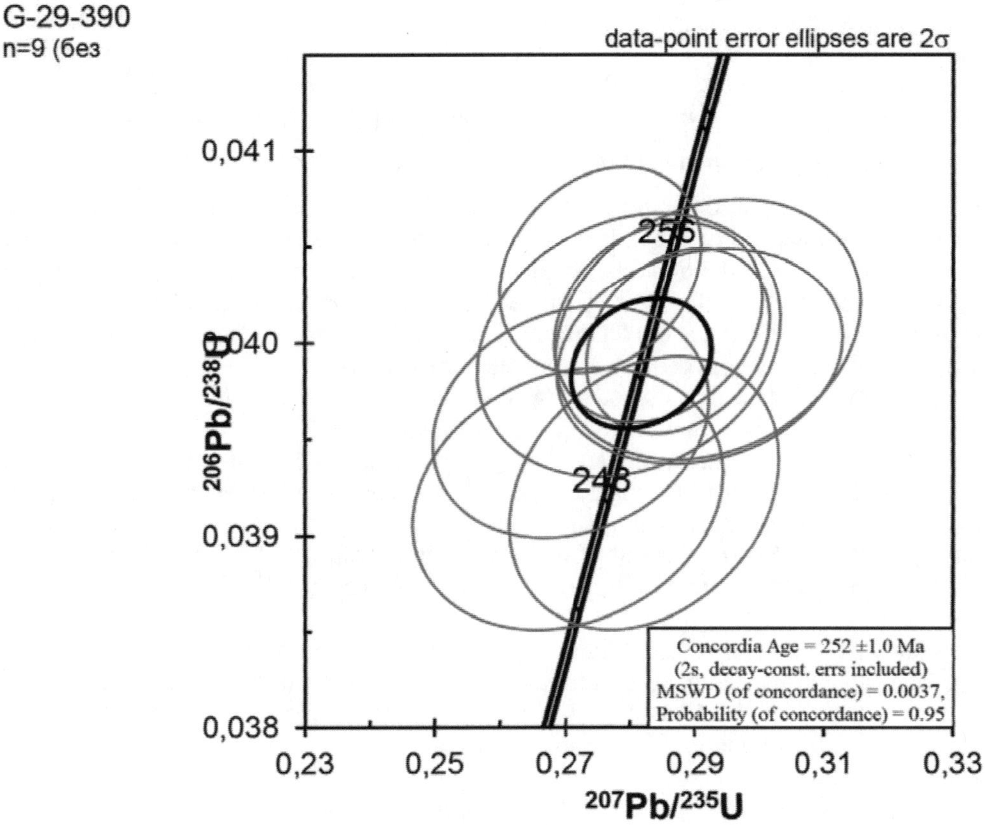

Figure 3. The plot for alkaline syenites of the Kresty Massif.

Patterns in isotopic evolution of KVPS rocks are obviously shown on $\varepsilon_{Nd}(T)$-$\varepsilon_{Sr}(T)$ diagram, where composition points of petrographic varieties of Maimecha-Kotui province form composition field expanding in a fan-like pattern from moderately depleted mantle of HIMU and PREMA type into the area of enriched sources EMI and EMII. Boundaries between them coincide with lines of mixing of lower crust (LC) and upper crust (UC) matters (mixing in ratios from LC:UC = 10:1 to LC:UC = 1:10) [3]. At the same time, isotope compositions of tholeiitic and alkaline magma products show noticeable discreteness only at the base of the Mesozoic volcanogenic mass/thickness (Pravoboyarsk, Onkuchansk and Tyvankitsk series) or in separate intrusive bodies (Kugda massif) distant from assumed "finger plume" center. As they get closer to the "finger plume" center, composition fields of magmatic rocks move closer to each other and subsequently move towards area of depleted mantle substrates. Similar variations of isotope composition with certain level of Sr enrichment are also seen in flood basalts from Norilsk region [9]. We can assume that it might indicate lower permeability of mantle matter and its interaction with brines of sedimentary cover of the platform.

By their isotope composition, volcanic and plutonic KVPS varieties take intermediate position between their analogues in the Gulinian and Kugda massifs in a relatively narrow contrast range between the alkaline and tholeiite derivatives. This is quite consistent with the scale of their spatial distance from the probable center of "finger plume". The most complete coincidence of isotope parameters was observed in ultrabasic and alkaline rocks, as well as in carbonatites of the Gulinian pluton. The approximating point of convergence of variation trends conforms to the values of $\varepsilon_{Nd}(T)$ = +5 and $\varepsilon_{Sr}(T)$ = −15, which are close to the characteristics of the PREMA source [5]. Taking into account extreme depths (>200 km) of the generation of primary alkaline–ultrabasic MKP melts [6], the leading role of the homogenous PREMA-type mantle substrate [10] in the processes of their generation seems to be the most valid. This conclusion may be confirmed by the parameters of lead isotope composition in basalts of the Putorana Plateau representing the geochemical peculiarity of the prevailing component of the Siberian (North Asian) superplume [1]. Despite the fact that the $\varepsilon_{Nd}(T)$ and $\varepsilon_{Sr}(T)$ values in these rocks indicate a chondritic matter source, the $^{206}Pb/^{204}Pb$ (18.3), $^{207}Pb/^{204}Pb$ (15.5), and $^{208}Pb/^{204}Pb$ (38.0) ratios conform precisely to the PREMA matter. A deviation from the defined trend was registered exclusively in the late small intrusions (dolerites, syenites, and granites) formed under active interaction with the crust matter, or with its direct melting. A similar model of the formation of hypabyssal rocks of similar Sr and Nd isotope parameters is accepted for the Early-Mesozoic A-type granitoids, as well as for the lamproites from Taimyr Peninsula and Altai Mountains formed simultaneously with final stages North Asian superplume evolution [6].

Thus, the example of the Kresty volcanoplutonic structure raises a question whether or not two or more plume sources participated in formation of the studied rock associations from the Maimecha–Kotui province. Evidently, a single mantle matter existed conforming to PREMA matter in isotope composition. The multiplicity observed for magmatic complexes may have been caused by different scaleplume–lithosphere interactions during exhumation/raise of melts along the Yenisei– Khatanga rift zone in the northern periphery of Siberian craton. Due to thermal activation of crust under the plume impact, the role of the mantle component in magmatic evolution decreased subsequently/successively [7].

The new data that we obtained proves "short-term" formation of intrusive complexes within Maimecha-Kotui province and their temporal correlation with volcanic eruptions in this region. Close ages for alkaline phase of the Kresty Massif and dykes completing its formation suggest short-term magmatic events, which defined one of the largest changes in organic world of our planet at Paleozoic-Mesozoic boundary.

This work was conducted as a government task of the Ministry of Education and Science of the Russian Federation, project No. 5.2352.2017/4.6

References

[1] R.E. Ernst Large Igneous Provinces / *Cambridge University Press*. **653** p.
[2] A.M. Sazonov, E.A. Zvyagina, S I. Leont'ev, et al., *Platinum-Bearing Alkali-Ultrabasic Intrusions of Polar Siberia* (TsNTI, Tomsk, 2001) [in Russian].
[3] I.F. Gertner et al. Isotope composition and magma sources of the Rresty vulcanopluton in Polar Siberia // *Doklady Earth Sciences*, 2009, Vol., 427A, No 6., pp. **961 – 967**.
[4] S.L. Kamo, G. K. Czamanske, Yu. Amelin, et al., *Earth Planet*. Sci. Lett. 214, **75–91** (2003).
[5] A. Zindler and S.R. Hart, Ann. Rev. *Earth Planet*. Sci. **14**, 493–571 (1986).
[6] L.N. Kogarko, V.A. Turkov, and E.Y. Gulyanitskaya, in *Proceedings of the XXV All-Russian Workshop "Alka- line Magmatism of the Earth"* (Moscow, 2008), pp. **72– 74** [in Russian].
[7] C.I. Hawkesworth, P.C. Lightfoot, V.A. Fedorenko, et al., *Lithos* 34, **61–88** (1995).

IOP Conf. Series: Earth and Environmental Science **110** (2017) 012007 doi:10.1088/1755-1315/110/1/012007

Material composition of the basalt-trachyte series of the early Devonian of the Saralin graben-rift

O M Grinev, R O Grinev, A A Bogorodov, R R Adylbaev

National Research Tomsk State University, Tomsk, Russia

E-mail: tomskgrom@yandex.ru, judestone@yandex.ru, ruslan_ggf@mail.ru, tomaleks@list.ru

Abstract. The article clarifies the structural-tectonic position of the Saralin graben. It is determined that, along with the Balyksinsky graben located to the south and the Goryachegorsky volcanic plateau to the north, they are the connecting structural links between the Kuznetsk-Alatau alkaline province and the adjacent Minusinsk trough. In the early Devonian, the alkaline province was formed as a vaulted-block structure (the "shoulder" of the rift), and the Minusinsk deflection as a depression with the dominant volcanism in it. The boundary between these positive and negative structures was the deep Balyksinsko-Saralinsky fault.

In the stratigraphic section of the graben, the lower molassoid part (Ustkundustylskaya stratum), the middle trachybasaltoid stratum (Bazarskay) and the upper problematic Ashpanian stratum are distinguished. In addition to analcime basalts, a large extrusive-subvolcanic dome-shaped construction of trachytes is considered.

By petrographic and material composition, volcanites of graben are represented by the dominant basalt-trachyte series in the composition of basanites, trachybasalts, trachyandesites, trachytes. The rocks of the basalt-andesibasalt-andesite series are limited. Foidites are rare.

Geochemical data indicate the genetic relationship of the volcanics of the studied series. Geochemically, they are clearly specialized in Sc, Ti, Zn, Zr, Th, U, Mn and P. According to the content of some of these elements, the rocks may have an industrial potential.

In geodynamic and genetic plans, graben and its volcanics were formed during plume-tectonic processes involving mantle plume, enriched mantle, and recycling processes of these formations with the substance of the consolidated PR-PZ_1 cortex.

1. Introduction

The study of the Saralin graben seems topical for a number of reasons: 1) This graben, along with other grabens of the north-eastern part of the Kuznetsk Alatau (Rastayskiy and Talanovsky), is the largest and lies with them in a single structural paragenesis (figure 1); 2) Together with the volcanic plateau located to the north of the Goryachegorsky volcano plateau, the Saralinsky graben is a connecting structural link between the constituent parts of the widely known Kuznetsk-Alatau volcanic-plutonic alkaline province, represented by the volcanic-terrigenous sequences of early Devonian molassoids of the grabens and plateaus listed, a set of alkali-gabbroic plutons, and so the same associated with them and autonomous extended swarms and belts of dikes of alkaline rocks [1] 3) The Saralinsky graben and the Goryachegorsk volcanic plateau are composite structures of the Severo-Minusinsk depression and thus directly testify to the spatio-temporal and structural connection of magmatites and structures of the Kuznetsk-Alatau province with the Minusinsk trough, which in turn is the largest structure of the axial depression zone of Tuvinsko - Minusinsko-Zapadno-Siberian rifto-continental system [2, 3]; 4) Together with Balyksinsky, the Saralinsky graben traces the Balyksin-Saralin deep fault, which is a submeridional branch from the Kuznetsk-Alataus transregional lineament stretching in the axial zone of the Kuznetsk Alatau. This branch of the fracture dissects the

IOP Conf. Series: Earth and Environmental Science **110** (2017) 012007 doi:10.1088/1755-1315/110/1/012007

ridges of Baikalides (R_3, V), salairid, and also contains lenticular fragments of ophiolites. 5) Graben volcanics are represented by a basalt-trachyte series of rocks, considered one of the indicators of the structures of continental reefs. In the northern part of the graben there is a large extrusive-subvolcanic body of trachytes, while in other grabens of the region trachytes are present mainly as dikes; 6) Saralinsky graben is part of the same ore region, in which there are large deposits of indigenous (gold-sulphide-quartz type and Karlin type) and alluvial gold, copper-molybdenum ores, promising ore occurrences of other minerals. A number of deposits and ore occurrences are located in the immediate vicinity or even in the instrumental parts of the graben.

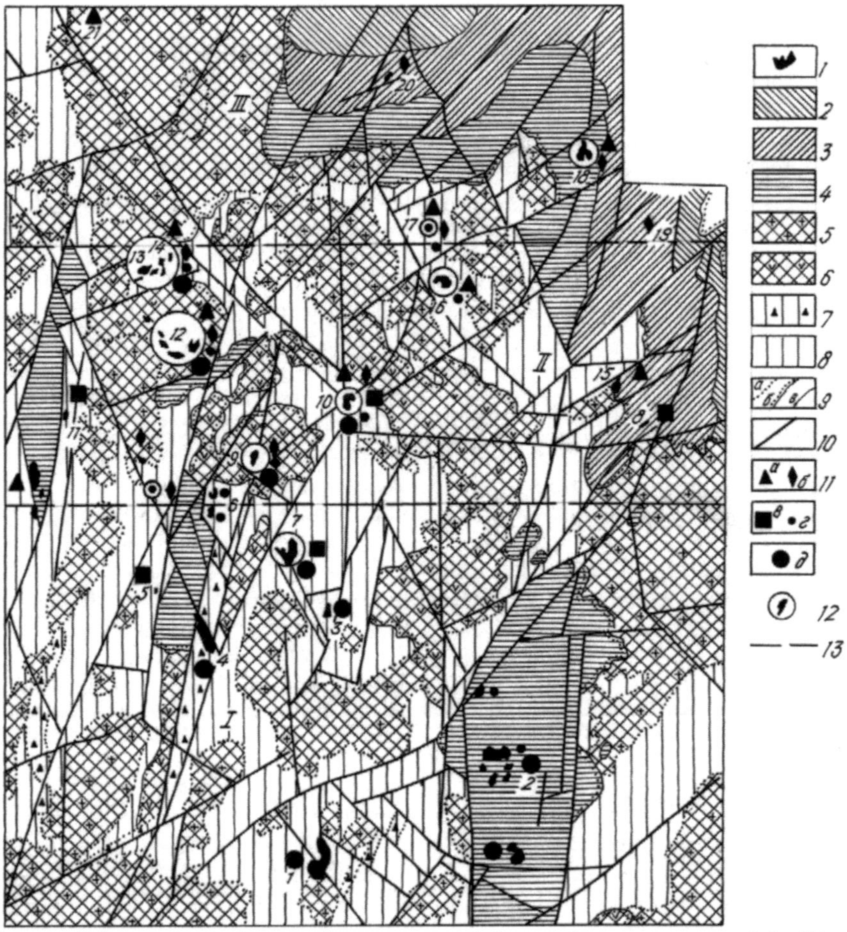

Figure 1. Scheme of the geological-tectonic structure of the north-eastern part of the Kuznetsk Alatau [1]

Upper structural floor (D_{1-2}): 1 – plutons of alkali-gabbroic rocks; 2 – 4 - strata of terrigenous-volcanogenic rocks of the lower Devonian: 2 – Ashken, 3 – Bereshskaya, 4 – Bazar. The average structural floor ($Є_{1-2}$ – O): 5 – gabbro, diorite, granite, syenite intrusions; 6 – stratified formations: thicknesses of the Poltava suite ($Є_{1-2}$) and mainly volcanics of the Berikul suite ($Є_2$). Lower structural floor (R_3 – $Є_1$): 7 – Lower Paleozoic hyperbasite intrusives (ophiolites); 8 – stratified formations of salairid with horstoobrazovannymi protuberances baikalid (R_3 - $Є_1$). 9 – geological boundaries: a – contours of intrusive massifs, b – interformational angular disagreements, c – intraformational angular disagreements, 10 – tectonic disturbances, 11 – main varieties of rocks of alkaline-gabbroic plutons: a – nepheline syenites, b – basic foidolites, ultrabasic foidolites, g – teralites, d – alkaline gabbros. 12 – complex-differentiated alkali-gabbroid plutons. 13 – boundaries of conditionally allocated zones of an alkaline province: I – southern zone, II – central zone, III – northern zone.

Sequence numbers of plutons of alkaline rocks: 1 – p. Dmitrievsky, 2 – Lysoy, 3 – Dedovoy, 4 – Velvet-Kiya, 5 – Kiy outlets, 6 – Malo-Kiya-Shaltyrsky, 7 – Kiya-Shaltyrsky, 8 – r. Podtega, 9 – University, 10 – Belogorsky, 11 –Svetninsky, 12 – Upper Petropavlovsk, 13 – Tulujsky, 14 – Medvedkino, 15 – Uchkuruk, 16 – Kurgusul, 17 – Cheremushinsky, 18 – Goryachegorsky, 19 – Andryushkinoy small rivers, 20 – r. Semenovsky, 21 – r. Wet Berikul.

Specialized petro-geochemical studies of volcanic graben, as well as other volcanic-tectonic structures of the Kuznetsk-Alatau province, were not subjected, unlike the rocks of alkali-gabbroic plutons, which are systematically studied – O M Grinev, I F Gertner, V V Vrublevsky and other employees of the Geological and Geographical Faculty of the TSU [4, 5, 6, 7, 8].

In this communication, the authors give a brief geological description of the Saralin graben and the first results of a geochemical study of its magmatites. More detailed information on the geology of grabens can be found in the works [2, 9, 10, 11].

2. Brief description of the geology of the Saralin graben

Structural-tectonic position. The tectonic position of the Saralinsky graben and its direct analogue of the Balyksinsky graben is determined by their confinement to the zone of the deep Balyksin-Saralinsky fault, which has a complex structure in the form of "stratification" into divergent and converging branches, and complicates their disturbances. In the structural plan, the zone of this fault is the western boundary of the Minusinsk inter-mountain trough. On this fault from the axial zone of the Kuznetsk Alatau, the Batenyovsky Ridge is cut off, which, together with the Belyksky protrusion, divides the Minusinsk trough-the Northern and Southern ones-into two halves. At the same time, the Saralinsky graben with its southern part is structurally connected with the southwestern extremity of the North Minusinsk depression, and Balyksinsky with the same tip of the South Minusinsk depression, thereby emphasizing their unity with the Minusinsk trough.

In terms of Saralinsky graben has a linearly elongated lenticular shape. Its southern, most extended, part is gradually narrowed to complete wedging. In the upper reaches of the river. Black Iyus graben undergoes a knee-like inflection. The northern and middle parts of the graben are most extensive, but in the northern part it also quickly wedges out and is cut off by a diagonal north-eastern fault. Further to the north, along the strike of the graben, there are small remains of basaltic covers, transgressively overlapping the folded strata of the pre-Devonian basin. Their presence suggests that the northern extremity of the graben in its original form had a longer continuation, but in a significant part was destroyed by erosion.

Almost along the entire length of the eastern and western sides, the graben is bounded by the branches of the Balyksinsko-Saralinsky fault and only in the northeastern part of the graben's thickness it partially extends beyond the zone of the instrumental fault, transgressively lying on the rocks to the Devonian base. Throughout, the contacts of the graben with the rocks of the pre-Devonian basin are torn off by tectonics and worked by valleys of rivers. Especially it is typical for its eastern side. It is noteworthy that the river valleys in the southern part of the graben are of a trough character, indicating the manifestation of glacial activity. In the northern part, the graben is divided by transverse and diagonal faults into several slightly moved blocks. Judging by the totality of the characteristics, these transverse disturbances detach from the main body of the graben tectonic block, in which there is a large extrusive-subvolcanic body of trachytes.

The dimensions of the graben along the strike are about 123 km, while the width in the northern part is up to 13 km. In the relief, the graben is expressed in the form of a medium-high mountain ridge, about 600-900 m high, with the presence of adherent ridges 900-1,200 m in height to 1,302.2 m (Lysoy Mountain Range, etc.). Absolute elevations of the ridges from the river rills in the middle and northern parts of the graben vary from 300 to 400 m in the south to 630 m in the north.

3. The material composition of volcanic graben

Petrochemical features of rocks. According to limited data, published in the work [12], the volcanoes of the main composition are mainly basalts and are limited to the pycrobasalates and andesibasaltes.

IOP Conf. Series: Earth and Environmental Science **110** (2017) 012007 doi:10.1088/1755-1315/110/1/012007

The content of SiO_2 in basalts varies within the range of 45.28 - 53.37%, in the picrobasalts it drops to 40.06%, and in the andesibasalt increases to 55.98%. The rocks are characterized by high titanium content (1.72-2.42%), medium and high alumina (15.71 - 18.51%), elevated iron deficiency (Fe_2O_3 - 10.2 - 13.72%, FeO - 3.95 - 7.06%), average calcium content (4.63 - 10.67%), surprisingly low magnesia (0.16-0.24%) and obviously abnormal Mn (4.75 - 8.07%), (possibly this is a technical error, the content of MgO is indicated instead of MnO and vice versa). Unprecedented for volcanoes are also the contents of P_2O_5 (3.0 - 7.29%). The content of alkali in the rocks corresponds mainly to the subcellular series, one analysis is characterized by phoid alkalinity, and the second is Tolleitian. In general, the data on the petrochemical composition of basalts of the Saralya require a re-examination and addition.

In TAS diagrams basalts form separate fields of figurative points. Most of them are located in adjacent fields of tefrites and trachibasalts. By the ratio of alkalis, the basalts are sodium (Figure *3, a, b*).

Extrusive trachetes (14 analyses) Are characterized by the following content of petrogenic oxides: SiO_2 varies from 57.96 to 63.5%, TiO_2 from 0.63 to 1.36%, Al_2O_3 from 15.00 to 17.58%, Fe_2O from 2 , 59 to 11.34%, FeO from 0.96 to 4.91%, MnO from 0.022 to 0.26%, MgO from 0.29 to 2.63%, CaO from 0.48 to 3.52%. It is also characteristic that the content of Fe_2O_3, FeO, MnO, CaO, Na_2O, K_2O and P_2O_5 - the rocks are clearly divided into two groups. Particularly clearly, this separation is manifested in the content of Fe_2O_3, Na_2O, K_2O and P_2O_5. Thus, by their content, the groups with Na_2O content from 5.1 to 6.44% and 0.13-0.27%, K_2O 0.65 - 4.58% and 3.85 - 9.33%, according to P_2O_5 0.05-0.206% and 0.41-4.5%.

In the TAS diagrams (Figure 2 a, b), most of the rocks of the tracheostomy line fall into the field of trachyandesites and trachytes; three analyzes are placed in the field of andesites. The alkali content of the rock refers to the K-Na series.

On the whole, in the diagram (Na_2O + K_2O) -SiO_2, the main bimodal sub-alkaline trend of tephrists, trachibasalts - trachyandesites, trachytes and the Tolleitis differentiated trend of basalt-andesibasalt-andesite are confidently predicted. Noting these trends, it should be emphasized that in the petrochemical plan, the volcanoes of the Saralin Graben need to replenish the data.

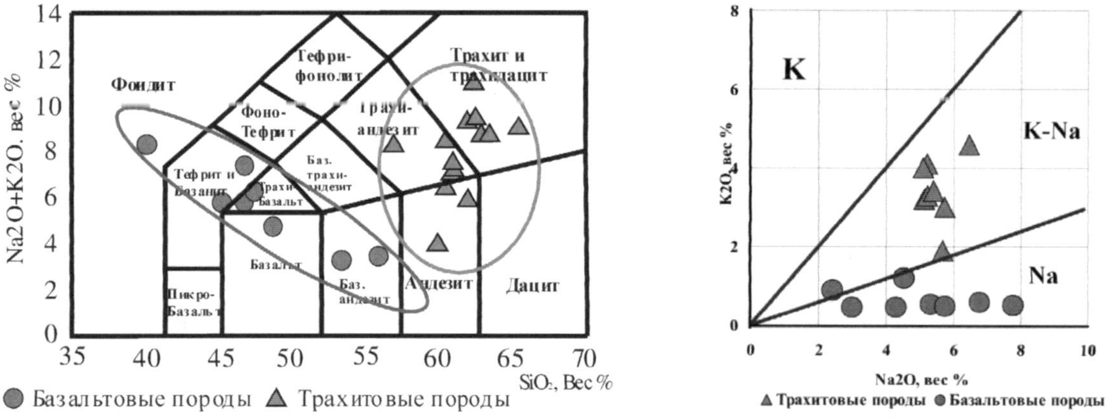

Figure 2, a, b. TAS diagrams (Na_2O + K_2O) -Si_2O (a) and K_2O-Na_2O (b) for the Saralin Graben volcanoes (given by Le Maitre et al., 1989; Peccerillo and Naylor, 1976)

Geochemical features were identified on the basis of data from 36 analyzes performed at the Center for Collective Use "Geochemistry of Natural Systems" of TSU GGF by the ICP MS method.

The most striking geochemical features of the considered bazanit, trachybasalt-trachyanthite, trachyte series of graben rocks are as follows. Basalts are enriched (g/t): Sc (3,2 – 31.8), Ti (1288,4 – 12565,4), V (28,5 – 240,8), Cr (4,0 – 128,4), Ni (3,1 – 50,4), Zn (2,3 – 1283,8), Sn (50,0 – 922.6), Y

(6,5 – 34,4), Zr (38,5 – 193,05); relatively low content are typical for Be, Co, Cu, Ga, Rb, Nb, Cs, Ba (3,2 – 238,05), РЗЭ, Hf, Ta, slightly increased content Th (0,28 – 3,9) and U (0,17 – 2,9).

For trachyandesites, trachytes in high concentrations are contained (g/t): Be (0,25 – 7,8), Sc (0,7 – 38,4), Ti (761,8 – 13331,7), V (0,56 – 401,2), Cr (1,05 – 64,0), Co (0,78 – 45,5), Zn (14,6 – 147,7), Ga (4,6 – 23,3), Rb (5,0 – 148,3), Sn (13,3 – 895,8), Y (7,3 – 97,4), Zr (33,7 – 1019,0), Nb (4,1 – 60,8), Ba (57,5 – 973,3), rare earth elements, Hf (0,03 – 22,4), Th (0,9 – 17,4) и U (0,43 – 7,7).

The enrichment of the rocks of the basalt-trachyte series by some elements is end-to-end, for example, Ti, while others are abrupt, low for basalts and high for trachytes, for example, rare earth elements, and vice versa. As can be seen, elevated concentrations are characteristic mainly for lithophils, some siderophiles (Fe, Ni) and chalcophiles (Zn, Sn). The multi-element spiderdiagrams built based on analytical data, with the N-MORB, OIB, and IAB frames plotted on them, allow us to draw the following conclusions.

In the spider diagram of basalts (Figure 3, a), the distribution spectra of trace elements form their own trend, only partially coinciding with the OIB and IAB benchmarks. According to the configuration of broken curves, the geochemical spectra of basalts on the diagrams coincide most closely with OIB, but with the contents of trace elements from La to Lu reduced relative to this reference. The left part of the trace distribution spectra of the diagram indicates the enrichment of basalts with mobile (LILE) and moderately mobile (HFSE) coarse and high-charge elements, whose content is much closer to OIB basalts than the rest of the components.

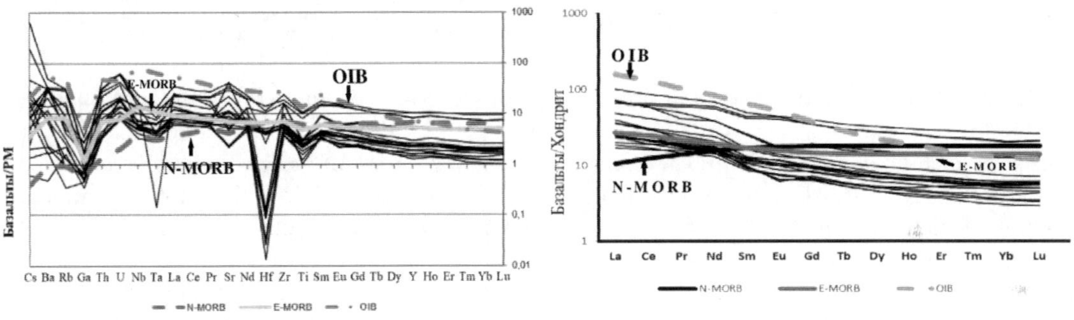

a b

Figure 3, a, b. Multi-element (a) and rare-earth (b) spidergrams of basalts of the Saralin graben

For the rocks studied, with respect to the compositions of the primitive mantle, negative minima of the Ga, Ta, Hf and, to a lesser extent, Ti, contents are clearly manifested. Brightly developed peaks of positive concentric trace elements, relative to the reference frames used, are not observed in basalts.

On the diagram for trachytes, the distribution spectra of microelements are significantly more diverse than basaltic ones with the presence of a number of positive and negative peaks in the contents of most of them relative to basalts (Figure 4, a). However, in general, the conformance of the distribution spectra of microelements within the group of basalts and trachytes is obvious, which indicates their genetic relationship.

Concerning the reference lines of the diagram (Figure 4, a), the distribution spectra of trachytes micronutrients are still closer to OIB, while retaining individual geochemical features and are characterized by higher trace element contents for some trachyte species than their main group. The negative contents of microelements in trachytes characteristic for Ga, Ta, Sr, Hf, Ti, As well as positive peaks for Ba, Rb, Th, U, Nb, La, Zr, Sm и Er.

The individual trend of distribution of lanthanides in basalts relative to the N-MORB, OIB and IAB frames is more clearly visible on the distribution charts of rare-earth elements normalized by chondrite. The configuration of the distribution spectra of the red earth elements on the diagram (Figure 3, b) is conformal to the OIB frame. In this case, according to the content of rare-earth elements, the spectra of basalts are divided into two subgroups. One of them (the smaller one) coincides in their content with the OIB, and the other (the larger one) contains about 10 times less rare-earth elements in smaller quantities. This geochemical feature of rock compositions is correlated with the presence of two main varieties among basalt-tephrites and trachybasalts. In trachybasalts, the Eu minimum is also weak.

The graph of the distribution of rare-earth elements in trachytes (Figure 4, b) is similar to basaltic, with the difference that the content of rare-earth elements in trachytes is about an order of magnitude higher than in basalts. Here, as in the rock basalts, the content of rare earth elements is divided into two subgroups, one of which is characterized by a low content of rare earth elements (subvolcanic trachytes), and the other is clearly elevated - extrusive eruptive trachytes.

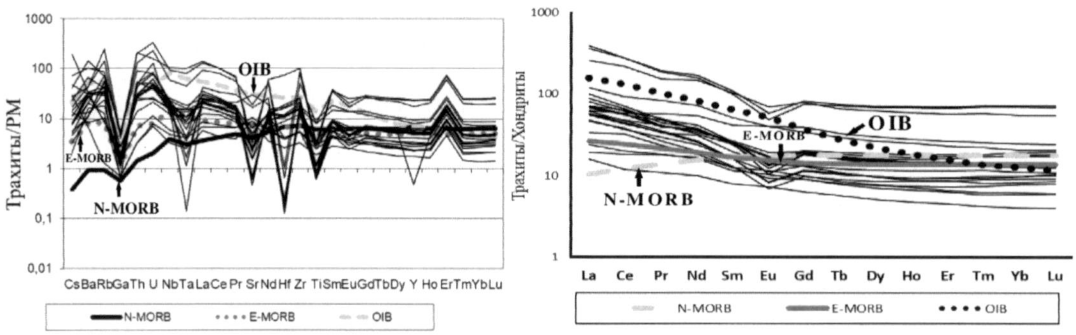

a b

Figure 4, a, b. Multi-element (a) and rare-earth (b) spidergrams of trachytes of the Saralin graben

The peculiarities of the geodynamic conditions for the formation of volcanic make it possible in the first approximation to estimate the discriminant triple diagrams, with fields of figurative points of rock formations depicted on them, characterizing some kind of geodynamic etalons of the basalt volcanic manifestation (Figure 5 a, b, c).

In the La-Y-Nb diagram (Figure 5, a), the figurative points of volcanics fall into the field of continental basalts. In the Zr-Nb-Y diagram (Figure 5, b), the figurative points of volcanites are concentrated in the field of intraplate and island-arc basalts. And on the Th-Hf-Ta diagram (Figure 5, c), the points of the rock formations form an extended trail that captures the lower part of the field A characterizing N-MORB basalts, some of the points are located in field B (E-MORB basalts) and extend to the lower half of the field D corresponding to calc-alkaline basalts of island arcs. A significant part of the points is located on the side of the Th-Ta diagram and lies outside the indicator fields, but gravitating to the C field - intraplate alkaline basalts. This feature is primarily due to the enrichment of the rocks under consideration by radioactive elements, with parallel depletion of their Hf and Ta.

Thus, according to the totality of the data, it can be concluded that the basalts of the Saralin graben are formed in a continental setting close to the continental rifts, involving the formation of N-MORB, E-MORB, and the recycling of bark matter.

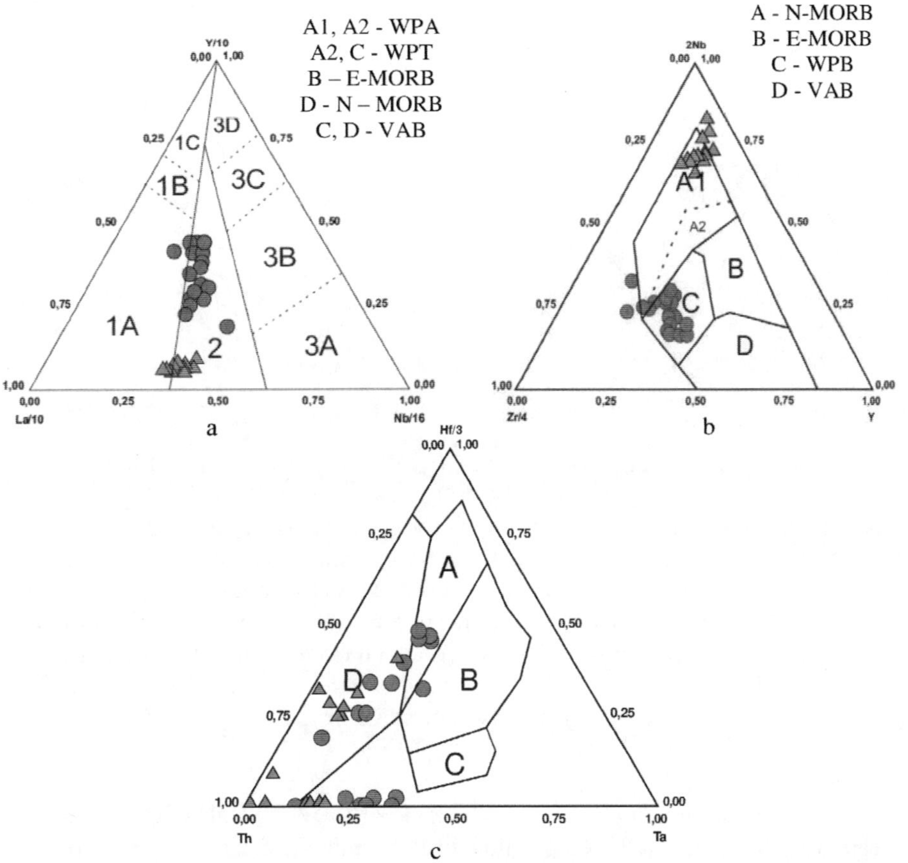

Figure 5 a, b, c. Diagrams of Th-Hf-Nb, Zr-Nb-Y, La-Y-Nb basalts and trachytes of the Saralin graben. The fields in the diagram, a - according to (Wood, 1980): 1 - island arc basalts (1A - calc-alkaline, 1B - overlapping area, 1C - tholeiites, 2 - continental basalts, 3 - ocean basalts, 3A - alkaline basalts of intercontinental rifts, 3B; 3C - enriched and slightly enriched E-MORB; 3D-N-MORB; b - according to (Meschede, 1986); c - according to (Cabanis, Lecolle, 1989)

The source materials of the initial magmas are estimated using the binary diagrams La/Y - Zr/Nb and Nb/Y - Zr/Y (Figure 6 a, b). According to the first diagram (Figure 6, a), the figurative points of the volcanic compositions of the series under investigation form a compact field in the immediate vicinity of the OIB and E-MORB frames, indicating that the source of the primary volcanic material was an enriched mantle converted under hot spot conditions or plume.

The diagram Nb/Y - Zr/Y (Figure 6, b) confirms and supplements the information of the first diagram. Here the figurative points of the rock formations form an elongated, but compact, trail along the dividing line of plume and extra-plume sources, confidently located in the field of plume formations. The trail begins near the relic of the primitive mantle and extends to the EM-2 and EM-1, reaching them.

Thus, a heterogeneous mantle, varying in composition from PM to EM-2-EM-1, with participation of recycling processes of the material of the consolidated crust formed during the Baikal and Salair tectogenesis, should be considered as a starting material of the basalt-trachyte series of volcanics of the Saralin graben.

IOP Conf. Series: Earth and Environmental Science **110** (2017) 012007 doi:10.1088/1755-1315/110/1/012007

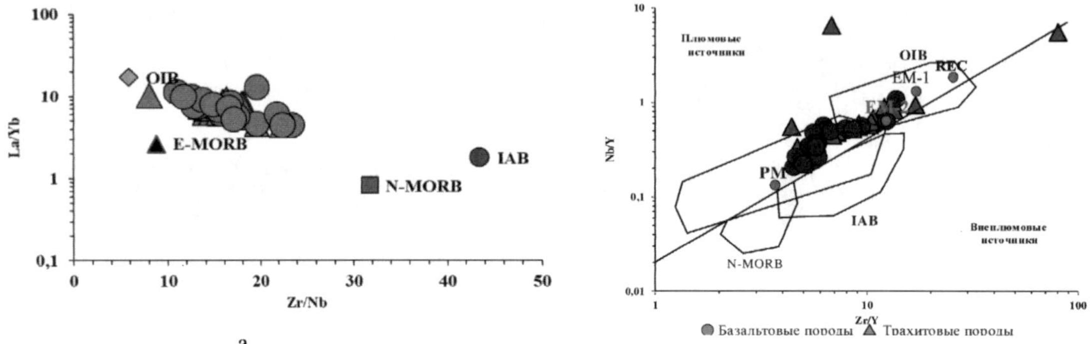

a b

Figure 6, a, b. Geochemical indicators of sources of volcanic substances: a) La/Y-Zn/Nb, b) Nb/Y-Zr/Y

Mugs - basalt of various composition; triangles - rocks of the trachyte series; a - La/Yb - Zr/Nb: b - Nb/Y - Zr/Y; The points and fields of compositions of different types of basalts and igneous sources, according to [Sun, McDonough, 1989; Condi, 2005] N-MORB - mid-oceanic ridges (depleted), E-MORB - mid-ocean ridges (enriched with rare lithophilic elements), basalts of active continental margins and island arcs (IAB), intraplate continental and oceanic basalts (OIB) ; PM - primitive mantle, REC - recycled component, EM-1 - enriched mantle with high Nd/Sm, EM-2 - enriched mantle with high Rb/Sr

4. Conclusions

(1) Formation of the Saralin graben occurred in the zone of the deep Balyksin-Saralinsky fault, which traces the western zone of the riftogenic Minusinsk trough. The situation in the fault zone was determined by two factors. Specific features of the early Devonian tectonic magmatic activation within the north-eastern part of the Kuznetsk Alatau, where the composite structures of the Kuznetsk-Alatau province were formed: volcanic-plutonic associations of grabens, alkaline-gabbroic plutons and dike belt-conditions of the formation within the "shoulder" of the paleorift. While in the complex structure of the Minusinsk trough there was a setting for the formation of the depression zone of the paleorift - deflection and dominant volcanism.

(2) In the situation of arbor formation and extension in Saralinsky prirazlomnom graben, as well as other grabens of the north of the Kuznetsk Alatau, sequences of motley-red-colored volcanogenic-terrigenous molasses were formed. The main features of the structure of the graben sections are: 1) the presence of a basaltic mass of molassoids (Ustkundustuylskaya, Krasnogorskaya strata); 2) formation of thick strata of volcanics corresponding to the composition of the Bazar thickness of the Goryachegor volcanic plateau; 3) and ending the incision of grabens of the ashpan volcanogenic-terrigenous stratum, for which the formation of extrusive-subvolcanic domes of trachyte composition is characteristic.

(3) According to petrographic and petrochemical data, the dominant basalt-trachyte series with the subordinate and fragmented development of nepheline rocks at the top of the sections is characteristic for the Saralinsky and other grabens of the region. According to the specified data, the volcanic series of the Saralin graben is represented by bimodal tephritis, trachybasalt-trachyandesite by the trachyte series, with the subordinate development of the thaleiitic trend of basalt-andesibasalt-andesite differentiates.

(4) The geochemical data cited in the work testify to the genetic unity of the studied series, its conspicuous similarity to the basalts of the oceanic islands (OIB), but with its inherent individual distinctive features. The series volcanics are specialized in a wide range of elements, especially Sc, Ti, Zn, Sn, Zr, Th, U, and possibly Mn and P.

(5) According to geological data and geochemical reference points, graben volcanites were formed under conditions of a consolidated continental crust that underwent destruction in the early Devonian under the influence of a mantle plume that caused continental rifting. Plume-tectonics led to the enrichment of primitive mantle and participation in the process of magma formation of the region's crust. Similar plume-mantle marks have volcanics of the southwestern part of the South Minusinsk depression [13], which closes with Balyksinsky graben.

(6) The explicit and rich mineragenic potential of the rocks of the studied basalt-trachyte series, especially in the part of Sc, Ti, Th, U and possibly P., is of particular interest. The rocks of the Saralin graben require further study.

Acknowledgements
This work was conducted as a government task of the Ministry of Education and Science of the Russian Federation, Project Number 5.2352.2017/4.6.

References
[1] Grinev O M 1990 *Evolution of alkali-gabbroic magmatism of Kuznetsk Alatau. The Author is abstract of the dissertation of the candidate a geol.-mineral. sciences* (Tomsk: Tomsk State University) p 18

[2] Grinev O M 2007 *Rift systems of Siberia: the methodology of study, morphotectonics, mineralogy* (Tomsk: Tomsk State University) p 437

[3] Grinev O M 2007 *Tuvinian-Minusinsko-Western-Siberian rift system: geology, morphotectonics, minerageny* (Tomsk: Tomsk State University) pp 185 – 193

[4] Pokrovsky B G, Vrublevsky V V and Grinev O M 1991 *The role of host rocks in the formation of alkaline-gabbroic intrusions of the north of the Kuznetsk Alatau by isotopic data* (Izvestiya Rossiiskoi Akad. Series geological) pp 81 – 94

[5] Pokrovsky B G, Andreeva E D, Vrublevsky V V and Grinev O M *Nature of the contamination of alkaline-gabbroic intrusions of the southern frame of the Siberian Platform according to the isotope of strontium and oxygen isotope* (Tomsk: Tomsk State University) pp 259 – 273

[6] Vrublevsky V V, Gertner I F, Gutierrez-Alonso G, Hofman M, Grinev O M, Tishyn P A 2013 *U-Pb-isotope age of zircons in alkaline rocks of Kuznetsk Alatau* (Tomsk: Tomsk Polytechnic University) pp 147 – 151

[7] Vrublevsky V V, Gertner I F, Gutierrez-Alonso G, Hofman M, Grinev O M, Tishyn P A 2014 *Isotope (U-Pb, Sm-Nd, Rb-Sr) geochronology of alkaline-basite plutons of Kuznetsk Alatau* (Tomsk: Tomsk State University) pp 1598 – 1614

[8] Vrublevsky V V, Grinev O M, Isoh A E, Travin A V 2016 *Geochemistry, isotope (Nd-Sr-O) triad and 40Ar-39Ar age of Paleozoic alkaline-mafic intrusions of the Kuznetsk Alatau (on the example of the Belogorsky pluton)* (Tomsk: Tomsk State University) pp 592 – 602

[9] Mustafin V Z, Kryukov V G, Lykhina N S 1966 *The main features of the geological structure of the Talanovsk graben (North-West spurs of the Kuznetsk Alatau)* (Tomsk: Tomsk Polytechnic University) p 151

[10] Kryukov V G, Mustafin V Z, Lykina V S *1969 The history of the formation of the Talanovsk graben (northern spurs of the Kuznetsk Alatau)* (Tomsk: Tomsk Polytechnic University) pp 80 – 85

[11] Uvarov A N and Uvarova N M 2010 *Petrotypes of the base-Ashpan trachytephrite-trachybasalt complex* (Novosibirsk: SSRIGGaMC) p 180

[12] Grinev O M 1994 *On the mechanism of formation of graben structures in the northern part of the Kuznetsk Alatau* (Tomsk: Tomsk State University) pp 237 – 259

[13] Grinev O M, Kotelnikov A D, Kaplun M V and Grinev R O 2016 *Plum-riftogenic Early Devonian Volcanic Association of Tashtypsky District Minusinsk*

IOP Conf. Series: Earth and Environmental Science **110** (2017) 012008 doi:10.1088/1755-1315/110/1/012008

Episodic melting and magmatic recycling along 50 Ma in the Variscan belt linked to the orogenic evolution in NW Iberia

G Gutiérrez-Alonso[1,2], A López-Carmona[1,2], G García Acera[1], J Martín Garro[1], J Fernández-Suárez[3], A Gärtner[4], M Hofmann[4]

[1]Department of Geology, University of Salamanca, Plaza de los Caídos s/n, 37008, Salamanca, Spain
[2]Faculty of Geology and Geography, Tomsk State University, 36 Lenina Ave, Tomsk 634050, Russia
[3]Department of Petrology and Geochemistry, Complutense University of Madrid and IGEO-CSIC, 28040 Madrid, Spain
[4]Senckenberg Naturhistorische Sammlungen Dresden, Museum für Mineralogie und Geologie, Koenigsbruecker Landstr. 159, 01109 Dresden, Germany

E-mail: gabi@usal.es

Abstract. The advent of a large amount of more precise U-Pb age data on Variscan granitoids from NW Iberia in recent years has provided a more focused picture of the magmatic history of the Western European Variscan belt (WEVB). Based on these data, three main pulses of magmatic activity seem to be well established.

Variscan granitoid magmatism in NW Iberia is an example of the intimate linkage between granitoid magma production and plate convergence/collisional/post-collisional geodynamic scenarios. In the Western European Variscan belt (WEVB) realm, the major stages of convergent tectonics and its aftermath, are recorded by granitoid suites generated in a time span of about 50 million years (*ca.* 340-290 Ma). There is however some ongoing controversy regarding: i) whether there was a significant magmatic event in the early Carboniferous, for which there is hitherto scarce evidence, and ii) whether the Variscan magmatic activity was continuous or it rather occurred through discrete and relatively short-lived pulses.Unravelling these issues is one of the key elements to better constrain and interpret the different processes involved in the collisional scenarios that lead to the Variscan orogeny in Iberia during Late Paleozoic times [1-5] and the subsequent development of the Ibero-Armorican Arc (IAA) [6].

The advent of a large amount of more precise U-Pb age data on Variscangranitoids (and some volcanic rocks) from NW Iberia in recent years has provided a more focused picture of the magmatic history of the WEVB, providing solid ground for interpretations that link periods of more intense magmatic activity with large-scale crust-mantle processes involved in the collisional orogeny and the subsequent development of the IberoArmorican Arc.

Based on the observation of those more precise U-Pb ages and new data, three main pulses of magmatic activity seem to be well established. These pulses recycle (re-melt) the previously formed granitoid rocks (Figure 1 and 2):

1) Post-orogenicgranitoid suite (POS henceforth) (*ca.* 305-290 Ma), that intrudes all the structural domains of the orogen, including the foreland fold and thrust belt, which makes the WEVB rather unique. The POS includes a large number of volumetrically minor intrusions of mafic and ultramafic rocks. This magmatic event has been extensively studied and dated. The POS has been interpreted as generated by lithospheric delamination triggered by the oroclinal bending of the mountain belt.

IOP Conf. Series: Earth and Environmental Science **110** (2017) 012008 doi:10.1088/1755-1315/110/1/012008

2) Syn-extensional collapse granitoids (*ca.* 325-315 Ma): mostly crustal (S-type) peraluminousleucogranites generated by decompression melting following the extensional collapse of the mountain belt.

3) A third suite of Variscangranitoids, not considered in most of the published models, has been found with ages clustering around 340 Ma. In addition, a significant amount of *ca.* 340 Ma zircon xenocrysts has been found in the *ca.* 320 Ma syntectonicleucogranitoids of the Tormes Dome and surrounding areas, in the *ca.* 305 Ma Toledo Anatectic Complex and the western part of the Gredos Massif, and also as detrital zircons in Variscansyn-orogenic sediments.

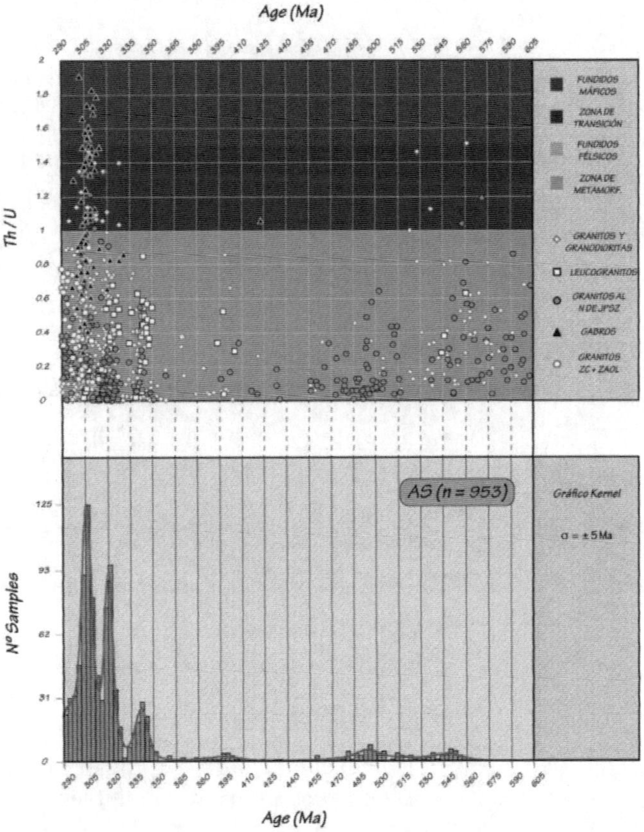

Figure 1.A) Graph [Th/U] vs. Age (Ma).and B) Histogram and Kernel distribution (KDE) of single zircon U-Pb ages of Variscangranitoids in NW Iberia [7]. Data from[8-17].

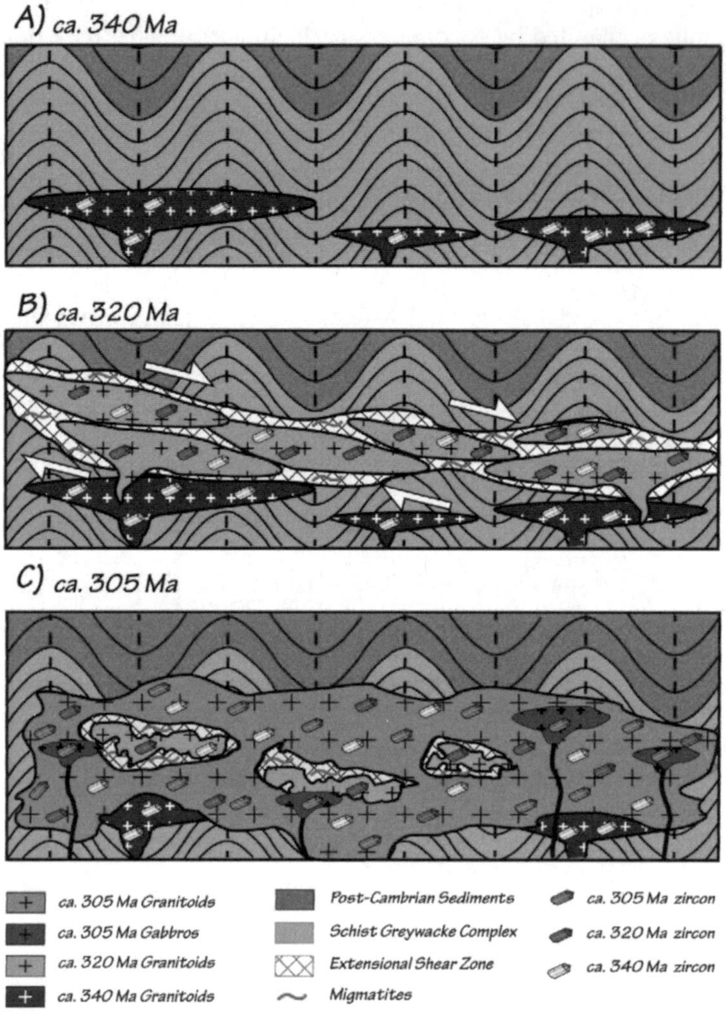

Figure 2. Proposed conceptual model for the recycling and re-melting of the three magmatic events found in NW Iberia according to the ages of the zircon grains found in them. A: Felsic magmatism of uncertain origin at *ca.* 340 Ma. B: Felsic magmatism at *ca.* 320 Ma associated with the orogenic collapse, extensional shear zones and extensive migmatization recycling previous 340 Ma granite bodies. C: Felsic and mafic magmatism of *ca.* 310-295 Ma that produced new zircons and recycled zircons from the two previous magmatic Variscan events. The latter event is interpreted to have occurred due to orocline triggered lithospheric delamination.

This work has been funded by the Spanish Ministry of Economy and Competitiveness under the project ODRE III-Oroclines& Delamination: Relations & Effects (CGL2013-46061-P) and the Russian Ministry of Education and Science under the project «Origin, metallogeny, climitic effects and cyclicity of Large Igneous Provinces» (14.Y26.31.0012).

References

[1] Matte P 2001TheVariscan collage and orogeny (480-290 Ma) and the tectonic definition of the Armorica microplate: a review *Terra Nova* **13** p 122–128

[2] Ballèvre M Bosse VDucassou CPitra P 2009Palaeozoic history of the Armorican Massif: models for the tectonic evolution of the suture zones *Comptes Rendus Geoscience*. **341** p. 174–20

[3] Martínez Catalán J R Arenas R Abati J Martínez S S García F. D Suárez J FClavijo E G 2009 A rootless suture and the loss of the roots of a mountain chain: the Variscan belt of NW Iberia *Comptes Rendus Geoscience* **341(2)**p114-126

[4] Arenas R Sánchez Martínez S Díez Fernández R Gerdes A Abati J Fernández-Suárez J Andonaegui P González Cuadra P López-Carmona A Albert R Fuenlabrada J.M Rubio Pascual F J 2016Allochthonous terranes involved in the Variscan suture of NW Iberia: A review of their origin and tectonothermal evolution *Earth-Science Reviews* **161** p 140–178

[5] DíezFernández R Arenas R Pereira M F Sánchez-Martínez S Albert R Martín Parra L M Rubio Pascual F.J and Matas J 2016Tectonic evolution of Variscan Iberia: Gondwana-Laurussia collision revisited *Earth-Science Reviews* **162** p 269–292

[6] Weil A.B Gutiérrez-Alonso G Johnston S.T Pastor-Galán D 2013Kinematic constraints on buckling a lithospheric-scale orocline along the northern margin of Gondwana: A geologic synthesis *Tectonophysics* **582** p 25–49

[7] García Acera G 2017 Geología y geocronología U-Pb (LA-ICP-MS) de las facies graníticas del Valle del Jerte (Cáceres) BSc Thesis University of Salamanca 32 pp

[8] Gutiérrez-Alonso G Fernández-Suárez J Jeffries T E Johnston S T Pastor-Galán D Murphy J B Franco M P Gonzalo J C 2011Diachronous post-orogenicmagmatism within a developing orocline in Iberia, European Variscides *Tectonics* **30**

[9] Zeck H P Wingate M TPooley G 2006Ion microprobe U-Pbzircón geochronology of a late tectonic granitic-gabbroic rock complex within the Hercynian Iberian belt *Geological Magazine* **144** p 157-177

[10] Díaz-Alvarado J Castro A Fernández C Moreno I 2011Assessing Bulk Assimilation in Cordierite-bearing Granitoids from the Central System Batholith, Spain; Experimental, Geochemical and Geochronological Constraints *Journal of Petrology* **52**p 223-256

[11] Díaz-Alvarado J Fernández C Castro AMoreno I 2013SHRIMP U-Pb zircon geochronology and termal modeling of multilayer granitoid intrusions. Implications for the building and termal evolution of the Central System batholith, Iberian Massif, Spain *Lithos* **175-176** (2013) 104-123

[12] Martín Garro J 2015 Geocronología de U/Pb mediante LA-ICP-MS en circones del Complejo anatéctico de Toledo y el batolito de los Montes de ToledoMSc Thesis Salamanca University 50 pp

[13] Pereira M F Díez-Fernández R Gama C Hofmann M Gärtner A Linnemann U 2017S-type granite generation and emplacement during a regional switch from extensional to contractional deformation (Central Iberian Zone, Iberian autochthonous domain, Variscan Orogeny) International *Journal of Earth Scieces*

[14] Díez-Fernández R Pereira M F 2017Extensional orogenic collapse captured by strike-slip tecotonics: Constrains frm structural geology and U-Pbgechronology the Pinhel shear zone (Variscanorogen, Iberian Massif) *Tectonophysics* **691**p 290-310

[15] Villaseca C Orejana D Belousova E AmstrongR A Perez-Soba C E Jeffries T 2011U-Pb isotopic ages and Hf isotope composition of zircons in Variscangabbros from central Spain: evidence of variable cristal contamination *Mineralogy and Petrolology* **101** p 151-167

[16] Bea F Montero P Molina J F 1999Mafic precursors, peraluminousgranitoids, and late lamprophyres in the Avila batholith: a model for the generation of Variscan batholiths in Iberia *The Journal of Geology* **107(4)** p399-419

[17] Montero P Bea F Zinger T 2004 Edad ^{207}Pb/^{206}Pb en cristal único de circón de las rocas máficas y ultramáficas del sector de Gredos, Batolito de Ávila (Sistema Central Español) *Revista de la Sociedad Geológica de España* **17(3-4)** p157-167

Water in melt inclusions from phenocrysts of dacite pumice of the Vetrovoy Isthmus (Iturup Island, Southern Kuriles)

A A Kotov[1,2], S Z Smirnov[1,3], I A Maksimovich[1,2], P Yu Plechov[4], N V Chertkova[5] and A I Befus[6]

[1] V S Sobolev Institute of Geology and Mineralogy SB RAS, pr. ak. Koptyga, 3, 630090, Novosibirsk, Russia
[2] Novosibirsk State University, Pirogova, 1/2, 630090, Novosibirsk, Russia
[3] Tomsk State University, pr. Lenina, 36, 634050, Tomsk, Russia
[4] A E Fersman Mineralogical Museum RAS, Leninsky pr, 18 building 2, Moscow, Russia
[5] Institute of Experimental Mineralogy RAS, ak. Osipyana str., 4, 142432, Chernogolovka, Moscow area, Russia
[6] University of South Florida, School of Geosciences, Tampa, Florida, USA

Abstract. This work is devoted to the study of one of the largest caldera eruptions of the Kurile-Kamchatka island-arc system that occurred on the island of Iturup. The object of investigation of this work are phenocrysts of quartz and plagioclase from dacite pumice of the Isthmus of the Isthmus, which is located on the island of Iturup. The purpose of this work is to determine the water content in the melts that participated in the caldera eruption of the Vetrovoy Isthmus and the patterns of their changes during the crystallization of magma. In the course of the work, the following were carried out: 1) adaptation and calibration of the Raman spectroscopy method for determining water in rhyolite melt's inclusions glasses in quartz and plagioclase from pumice stone; 2) determination of composition and estimation of water content in melt inclusions in quartz and plagioclase according to x-ray spectral analysis; 3) establishment of the regularities of the change in the water content during the evolution of the magmatic melt; 4) evaluation of fluid pressure by comparison with experimental data

1. Introduction

Catastrophic eruptions, such as Santorin [1], Toba [2-3], Pinatubo [4], etc., are usually associated with reservoirs of acid magmas with a high content of volatile components. Studies of volcanic emanations and melt inclusions in minerals showed that the main volatiles are water and carbon dioxide [5].

The Vetrovoy Isthmus (VI) is 12 km wide graben depression in the northern part of the Iturup Island of the Kurile island arc. Graben is filled with late Pleistocene pumice-pyroclastic deposits with a thickness of more than 260 meters and a volume of about 100 km^3 [6-7]. It is presumed that the eruptive center is located approximately in the middle of the isthmus. The pyroclastic deposits are the product of one of the largest eruptions or series of near-simultaneous eruptions in the Kurile-Kamchatka arc in the late Pleistocene and the Early Holocene.

Previous studies demonstrated that the magmatic reservoir of the Vetrovoy Isthmus in the early stages of its development was degassed, with the release of a substantially aqueous fluid. Studies of melt and fluid inclusions showed that water is the main volatile, while CO_2, F and sulfur compounds play a subordinate role [8]. This conclusion raises the question of how high pressure, which eventually led to a large-scale eruption, could develop in the early degassed reservoir. This question cannot be answered without reconstructing the behavior of water in the magmatic reservoir and determining the change in H_2O content in the magmatic melt. This article presents the results of a detailed study of the water content in melt inclusions in quartz and plagioclase from the pumice of the Vetrovoy Isthmus at Iturup Island.

2. Mineralogy and petrography of the Vetrovoy Isthmus pumices

The rocks of the Vetrovoy Isthmus are represented by tuffs with fragments of light pumice. Pumices contain a large number of porphyritic phenocrysts (25-30%), which are found in the matrix of felsic glass (SiO_2 73-76 wt. %). The phenocrysts are represented by plagioclase, quartz, augite, hypersthene, and Fe-Ti oxides. Amphibole does not appear as phenocrysts, nevertheless it is found as crystalline inclusions in pyroxenes and has a high alumina composition (tschermakite-hornblende series).

Plagioclase, femic minerals, and Fe-Ti oxides compose the early paragenesis of the felsic pumice. Formation of femic minerals ceased immediately before the eruption, and only plagioclase and quartz crystallized then from the melt [8]. An unusual feature of plagioclase is that its crystallization begins with intermediate compositions, then Ca content increases to maximum values (An_{95}), and the latest zones again have an intermediate composition.

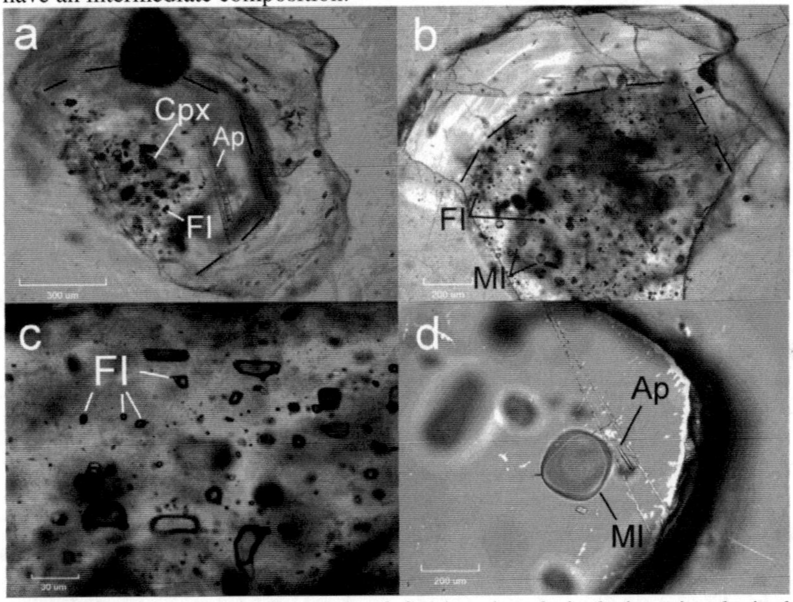

Figure 1. Zoning structure and location of fluid and melt inclusions in plagioclase and quartz: a) plagioclase with crystalline inclusions of apatite (Ap), clinopyroxene (CPx), melt (MI) and fluid inclusions (FI); b) separation of internal core with melt and fluid inclusions and concentrically zoned external parts of plagioclase; c) FI in plagioclase; d) quartz with MI and apatite with a drop of parent melt.

3. Melt and fluid inclusions

Plagioclase and quartz contain abundant melt inclusions (MI), which are located either along growth zones, form irregularly distributed groups or are trapped together with other minerals. Nonheated melt inclusions can be divided into three groups: 1) single-phase, completely vitreous (figure 1 d); 2) two-phase, containing glass and gas bubble; 3) multiphase, containing glass, crystals and / or gas bubble. Crystals in inclusions are apatite (figure 1 a, d), Fe-Ti oxide minerals, augite (figure 1 a), hypersthene and amphibole. Melt inclusions with crystals are the result of simultaneous capture of the melt and other minerals coexisting with plagioclase. Only first-type inclusions without signs of decrepitation were selected for the study as they are naturally quenched and their composition has undergone post-entrapment transformations to the least extent.

Fluid inclusions (FI) are found only in plagioclase, located in zones and areas with the maximum Ca contents. They are represented by two-phase essentially gaseous inclusions of CO_2 with a rim of liquid at room temperature aqueous solution (figure 1 a, b, c). FI form associations with melt inclusions. In this case, some MIs contain a two-phase gas-liquid separation. These indicates the formation of high calcium zones of plagioclase under degassing conditions.

4. Analytical methods and results

4.1. Electron-microprobe analysis

The composition of MI glass was determined by the energy dispersive X-ray spectral analysis (EDS) on the TESCAN MIRA-3 LMU SEM equipped with a microanalysis system with the INCA Energy 450+ software and the X-Max 80 analyzer (IGM SB RAS, Novosibirsk). The accelerating voltage was 20 kV, the probe current was 1.4 - 1.6 nA, the diameter of the focused electron beam was 10 nm. The counting time is 60 seconds. With these parameters, the detection limits of different elements vary within 0.1-0.3 wt. %. The analysis of the MI glass was carried out by scanning the 10x10 μm site to reduce the effect of sodium loss [9-11]. For each inclusion, two glass analyzes and one analysis of the host mineral near the inclusion were carried out. The mineral analysis was carried out to verify the correctness and to account for instrument drift. If necessary, a correction was applied.

4.2. Raman spectroscopy

The study of melt inclusions by Raman spectroscopy was carried out using the LabRAM HR 800 spectrometer at IGM SB RAS. The laser beam was positioned on the analyzed inclusion using an Olympus BX-41 polarization microscope. In the course of work, a 532 nm laser beam with output power of 75 mW was used for excitation. The registration was carried out using a liquid nitrogen cooled CCD detector with an operating temperature of -120° C. The analysis was carried out in the backscattering geometry. To collect the scattered light, a 100x lens was used. The spectra were obtained in the range 100-1250 cm^{-1} and 2800-4000 cm^{-1} (figure 2). The confocal pinhole and the acquisition time were regulated depending on the size and depth of the inclusion. In the course of the work it was established that a change in the size of the confocal pinhole and the acquisition time lead to an error not exceeding ± 0.2 abs. %. The acquisition time varied from 25 sec per spectral window for large inclusions to 400 sec per spectral window for small ones.

5. Determination of the water content using Raman spectroscopy

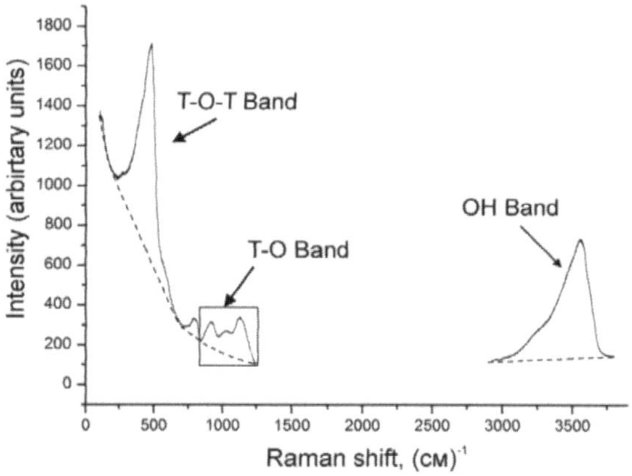

IOP Conf. Series: Earth and Environmental Science **110** (2017) 012009 doi:10.1088/1755-1315/110/1/012009

Figure 2. The Raman spectrum of the reference rhyolite glass with 6 wt. % H_2O. The dashed line is the baseline.

The method of quantitative determination of the water content is based on the calculation of the peak areas ratio in the range 2900-3800 cm^{-1} (Aw), which corresponds to vibrations of OH bonds in water molecules and hydroxyl groups in the silicate glasses [12] and the sum of peak in the range 850-1250 cm^{-1} (As) corresponding to vibrations of Si-O-Si, Al-O-Si and Si-O bonds in the structure of aluminosilicate glasses (figure 2) [13-14]. This ratio (Aw / As) directly depends on water content of the analyzed glass. In the course of this study five reference synthetic glasses of rhyolite composition with water contents from 2 to 6 wt. %, in which the water content was determined by different methods were used as calibration standards.

It is known that the spectra of silicate and aluminosilicate glasses differ significantly depending on their composition [13]. In this regard, the calibration results were checked on a series of reference felsic glasses from the Moscow State University. These glasses were obtained by remelting of natural obsidian in presence of water at specified temperatures and pressures [15]. The water contents in these glasses were measured with secondary ion mass spectrometry (figure 3 b), and Raman spectroscopy (figure 3 a) using the procedure described in [16]. In our study 16 measurements were obtained for four standard glasses by Raman spectrometry. These measurements demonstrated that the results obtained from glasses with less than 8 wt. % H_2O coincide well with the results obtained by ion probe and Raman spectrometry at Moscow State University. Significant deviation was obtained for the glass with 10 wt. % H_2O (figure 3 a, b). We believe that the reason of deviation is related to the compositional difference between the obsidian and synthetic reference glasses. In this regard, we assume that our calibration can be successfully applied to natural glasses of similar composition with a less than 8 wt. % water content.

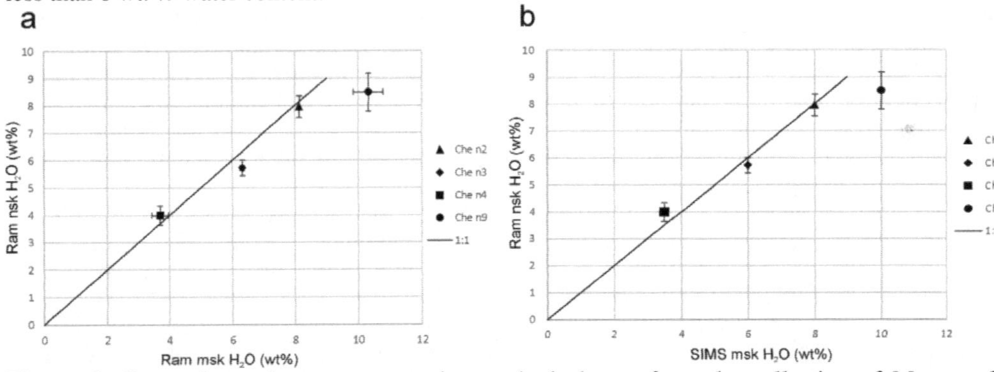

Figure 3. Comparison of water content in standard glasses from the collection of Moscow State University: a) obtained by Raman spectroscopy on the basis of IGM SB RAS (Ram nsk) and Moscow State University (Ram msk); (b) obtained by Raman spectroscopy on the basis of IGM SB RAS (Ram nsk) and secondary-ion mass spectrometry (SIMS msk).

For this study more than 50 inclusions in quartz and plagioclase from the pumices of the Vetrovoy Isthmus were measured by Raman spectroscopy. Each measured inclusion was then analyzed by the EDS method. The water content was calculated from the residual oxygen content. At first, the amount of oxygen coming from the stoichiometry to the oxides of metals and silicon was determined. It was assumed that all the iron in the melt is in a divalent form. Then this amount was subtracted from the measured amount of oxygen. The rest was recalculated to H_2O.

6. Water content in met inclusions

A comparison of the results that were obtained by SEM EDS and Raman spectroscopy is shown in the figure 4.

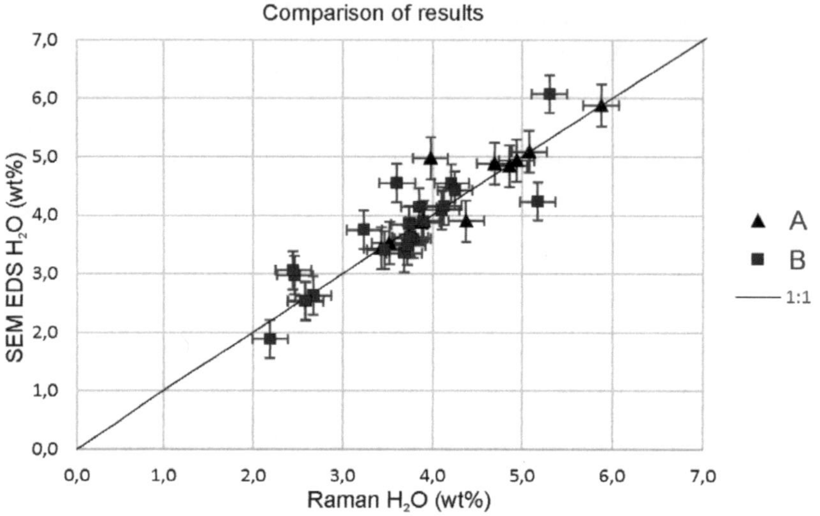

Figure 4. H_2O contents in melt inclusions in phenocrysts: (A) of plagioclase and (B) quartz, determined by Raman spectroscopy and energy dispersive X-ray spectral analysis (SEM EDS).

MI in quartz are the most suitable for analysis as there is a large number of single-phase vitreous melt inclusions, often more than 200 microns in size and the spectrum of the host mineral do not overlap those parts of the spectrum that are measured to assess the water content. Plagioclase lines overlap on the measured parts of the spectrum, and this decrease precision in processing of spectra. On the basis of the data obtained, it can be stated that the water content in quartz varies from 2.2 to 5.3 wt. %. The figure shows that most of the studied inclusions have concentrations of 3.2-2.2 wt. %. A separate group consists of inclusions with a water content of 2.2-2.7 wt. %. And only two inclusions have a concentration of more than 5 wt. % and deviate strongly from the line 1: 1. Concentrations of water in the MI belonging to one association, as a rule, have the same values.

Melt inclusions in plagioclase have water contents from 3.5 to 6.0 wt. %. Inclusions with the highest water content are few and occur in plagioclase with a highest Ca content up to An_{95}. Melt inclusions with water content of 3.5-4.5 wt. % are found only in the outer concentrically zoned plagioclase with andesine or labrador compositions.

7. Discussion

The results showed that the melts of the caldera eruption of the Vetrovoy Isthmus contained water close to saturation for magmas of dacite and rhyolite composition. The presence of amphibole in the form of inclusions in other minerals and the absence in the form of intrinsic phenocrysts suggests that the pressure in the magmatic chamber was below the stability threshold of this water-containing mineral. In these conditions, the saturation of the melt with water should lead to degassing-the separation of the aqueous fluid. Degassing of magma in the focus of the caldera eruption is fixed in the form of associations of fluid and melt inclusions in the highly calcareous zones and plagioclase phenocrysts. Formation of high-calcium plagioclase with An >> 70 mol. % of the rhyolite melt requires specific conditions. Such plagioclase is more typical for magmas of basic composition. However, in addition to the chemistry of the magma itself, the basicity of the plagioclase can also increase with an increase in the partial pressure of water, as demonstrated by experiments [17]. Investigations of fluid and melt inclusions show that the formation of high-calcium zones of plagioclase in pumice is associated with the degassing stage, which should be preceded by an increase in the partial pressure of water in the source. This stage corresponds to a water content in the melt of not less than 6.0 wt. %, as evidenced by the maximum water concentrations in the MI from the high-calcium areas of phenocrysts. A further decrease in the calcium content indicates a re-balancing of the plagioclase with the melt at a lower pressure. The growth of the concentrically-zonal plagioclase came from a melt with a water content of 3.5-4.5 wt. %.

The absence of fluid inclusions in phenocrysts of quartz indicates that it crystallized after degassing. Probably, the lowest water concentrations (<3.5 wt %) correspond to the initial stages of the formation of this mineral. However, most of the inclusions have a water content in the range of 3.2 to 4.2 wt. %. This is close in magnitude to the water concentrations in the inclusions in the late zones of plagioclase that formed after degassing.

The rhyolite melts have a very high viscosity, which is even greater when degassed. It follows that in different parts the magma chamber degassed to different degrees, which could lead to significant differences in H_2O content. High water content can be attributed to the sections of the chamber, which retained their tightness and did not completely remove the fluid phase. In addition, a scenario is possible, suggesting an increase in the water content after degassing due to the crystallization of anhydrous minerals or the entry into its magmatic focus from the enclosing rocks due to the thermal dehydration of water-containing minerals. This enrichment could serve as a reason for re-enriching the melt with water before the final catastrophic eruption. It is suggested that melt inclusions in quartz with water contents of more than 5 wt %, were captured shortly before him.

8. Conclusion

Compositions of melt inclusions permit us to state that the crystallization of phenocrysts of dacite pumice from a catastrophic VI eruption occurred from a rhyolite melt with a water content of 2.0-6.0 wt. %. Analysis of mineral associations of pumice clearly indicates that amphibole was not stable in the magmatic focus and decomposed. In these conditions, saturation of the melt with water should lead to degassing, which occurred in the early stages of its development. This led to a significant decrease in the water content (up to 2 - 3% by weight). However, after this degassing, a new enrichment of the melt with water has probably occurred, up to saturation concentrations just before the catastrophic eruption.

Acknowledgements

The work was carried out within the framework of projects 0330-2016-0005 of the state assignment of the IGM SB RAS and with the support of the RFBR grant 16-05-00894. Analytical studies were carried out in the Center for Multielement and Isotope Studies of the SB RAS.

References

[1] Druitt T H 2014 New insights into the initiation and venting of the Bronze-Age eruption of Santorini (Greece), from component analysis *Bull. Volcanol.* **76** p 794

[2] Chesner C A 1998 Petrogenesis of the Toba Tuffs, Sumatra, Indonesia *Journal of Petrology* **39** Iss.3. pp 397-438

[3] Chesner C A and Luhr J F A 2010 melt inclusion study of the Toba Tuffs, Sumatra, Indonesia *Journal of Volcanology and Geothermal Research* **197** pp 259-278

[4] Borisova A Yu 2005 Constrains on dacite magma degassing and regime of the June 15, 1991, climatic eruption of Mount Pinatubo (Philippines): new data on melt and crystal inclusions in quartz *Journal of Volcanology and Geothermal Research* **145** pp 35-67

[5] Wallace P J 2005 Volatiles in subduction zone magmas: concentrations and fluxes based on melt inclusion and volcanic gas data *Journal of Volcanology and Geothermal Research* **140** pp 217-240

[6] Melekestsev I V 1980 Volcanism and Relief Formation (Moscow: Nauka) p 212 (in Russian)

[7] Melekestsev I V Braitseva O A and Sulerzhickiy L D 1988 Catastrophic explosive eruptions of volcanoes of the Kurile-Kamchatka region at the end of the Pleistocene-Early Holocene *Transaction (Doklady) of the Russian Academy of Science/Earth Science Section* **300 (3)** pp 175-181 (in Russian)

[8] Smirnov S Z Rybin A V Sokolova E N Kuzmin D V Degterev A V and Timina T Y 2017 Felsic magmas of the caldera-forming eruptions on the Iturup Island: the first results of studies of melt inclusions in phenocrysts from pumices of the Lvinaya Past and Vetrovoy Isthmus calderas *Russian Journal of Pacific Geology* **11(1)** pp 46-63

[9] Lineweaver J L 1963 Oxygen outgassing caused by the electron bombardment of glass *J. Appl. Phys* **34** pp 1786-1791

[10] Morgan G B and London D 1996 Optimizing of electron microprobe analysis of hydrous alkali glasses *Am. Mineral* **81** pp 1176-1185

[11] Morgan G B and London D 2005 Effect of current density on the electron microprobe analysis of alkali aluminosilicate glasses *Am. Mineral* **90** pp 1131-1138

[12] Anfilogov V N and Bykov V N 1998 Silica melts. Structure, thermodynamics and physical properties (Miass: Ural Branch of RUS Press) p 160 (in Russian)

[13] Zajacz Z Halter W Malfait W.J Bachmann O Bodnar R.J Hirschmann M.M Mandeville C.W Morizet Y Muntener O Ulmer P and Webster J.D 2005 A composition-independent quantitative determination of the water content in silicate glasses and silicate melt inclusions by confocal Raman spectroscopy *Contrib. Mineral. Petrol* **150** pp 631-642

[14] Le Losq C Neuville D R Moretti R and Roux J 2012 Determination of water content in silicate glasses using Raman spectrometry: Implications for the study of explosive volcanism *Am. Mineral* **97(5-6)** pp 779-790

[15] Stepanova O V, Chertkova N V, Plechov P Yum Persikov E S and Buhtiyarov P G 2016 Analytical methods for determining high water concentrations in obsidian glasses *Annual Seminar on Experimental Mineralogy, Petrology and Geochemistry* (Moscow: GEOKhi) pp 50-51 (in Russian)

[16] Plechov P.Yu Blundy J Nekrylov N Melekhova E Shcherbakov V and Tikhonova M.S Petrology and volatile content of magmas erupted from Tolbachik Volcano, Kamchatka, 2012-13 2015 *Journal of Volcanology and Geothermal Research*

[17] Sisson T W and Grove T L 1993 Experimental investigations of the role of H_2O in calc-alkaline differentiation and subduction zone magmatism *Contrib. Mineral. Petrol* **113** pp 143–166

IOP Conf. Series: Earth and Environmental Science **110** (2017) 012010 doi:10.1088/1755-1315/110/1/012010

Metamorphism, *P-T-t* Conditions of Formation, and Prospects for the Practical Use of Al_2SiO_5 Polymorphs, Chloritoid, and Staurolite (Yenisei Ridge)

P S Kozlov

Institute of Geology and Geochemistry, Ural Branch of the Russian Academy of Sciences, Yekaterinburg, Russia

E-mail: kozlov@igg.uran.ru

Abstract. The Yenisei Ridge is an accretion–collisional orogen located in the southwestern frame of the Siberian Craton in the interfluve between Podkamennaya Tunguska, Angara, Kan, and Yenisei rivers. The Precambrian mono- and polymetamorphic complexes composed predominantly of the Mesoarchean–Neoproterozoic metapelitic rocks have been studied.
Based on the typification of metamorphic complexes by pressure, temperature, metamorphic gradient, as well as age of metamorphism, the location scheme of the fields of the Precambrian sedimentary–metamorphic rock which are prospective for searching deposits of high-alumina metamorphic minerals (andalusite, kyanite, and sillimanite, chloritoid, and staurolite) in the Trans-Angara segment of the Yenisei Region, was compiled. The Teya sillimanite and Panimbinsk andalusite deposits, which are confined to the fields of regional metamorphic complexes of iron-alumina metapelites of the And–Sill facies series, are recommended as a priority for the organization of prospecting works and the subsequent involvement to the metallurgical industry. These metapelites are classified as monomineral. Owing to widespread occurrence and abundance of andalusite and sillimanite, the above deposits have significant inferred resources. Stratiform deposits of garnet–staurolite and chloritoid high-alumina rocks are still insufficiently studied and should be investigated further.
The prospects for the possible use of high-alumina andalusite and sillimanite together with Middle Tatarka and Kiya nepheline syenite massifs and the bauxites of the Chadobets uplift, already being explored in the region, for production of aluminum oxide, silumin, and aluminum, as well as, the prospects for the expansion of the raw material base of the Boguchansk Electrometallurgical Complex, brought into operation in 2016 in the Lower Angara region, are considered.

1. Introduction

The Yenisei Ridge is an accretion–collisional orogen located in the southwestern frame of the Siberian Craton in the interfluve between Podkamennaya Tunguska, Angara, Kan, and Yenisei rivers (Krasnoyarsk Krai). The Yenisei Ridge is composed of Precambrian mono- and polymetamorphic complexes, consisting predominantly of metapelitic rocks. The latter provides valuable information about the history of geological development of the studied region in the interval from Neoarchean to Late Vendian. The mapping and typification of metamorphic complexes is of practical importance for effective target search for high-alumina schists (HAS) among ferruginous–aluminous metapelites.

These schists contain large amounts of such minerals as Al_2SiO_5 polymorphs (andalusite, kyanite, and sillimanite), chloritoid, and staurolite, which are widely used in aluminum industry worldwide.

Over the last 20-25 years, new data on the metamorphism of the Yenisei Ridge based on detailed geological mapping, petrological–geochemical and geochronological studies have been obtained ([5, 23] and references therein; etc.). The regional metamorphic complexes of the region are identified in the structure of the paleocontinental sector (1) (PCS, the age of polymetamorphism in a range of 1950–600 Ma) belonging to the western marginal part of the Siberian Craton, the paleoceanic sector (2) (POS, NP_3) of the northern flank of the Sayan–Yenisei accretion belt at the boundary with the Phanerozoic cover of the West Siberian Plate, and the Cis-Yenisei Regional Shear Zone (3) (CYRSZ, 620–600 Ma) between the above sectors [6]. Mono- and polymetamorphic complexes of the paleocontinental sector of the marginal part of the Siberian Craton, characterized below, are considered to be prospective for paleoreconstruction of the prolonged evolution (~1.5 Ga) of metamorphism (PP–NP_3) and for the practical application for petrological studies of metamorphic complexes. Metamorphic complexes are typified by pressure, metamorphic gradient, and age of metamorphism.

2. Brief description of metamorphism and high-alumina schists of the PCS

The following regional polymetamorphic complexes are mapped here (from ancient to young): *Kan*, *Garevka*, *Teya*, and *Cis-Angara*, *Angara–Pit*, and *Angara–Tissa* complexes of green schist metamorphism.

The ancient *Kan prograde charnokite–granulite polymetamorphic complex* is distinguished in the southern segment of the Yenisei Ridge within the Angara–Kan uplift [11]. *P-T* conditions of the metamorphic evolution of this complex are as follows: *P* up to 9–10 kbar, T = 850–950°C, age is 2.0–1.75 Ga ([25] and references therein); there are local occurrences of the UHT metamorphism (P = 6.8–7.8 kbar, T > 930°C, age is 1.85–1.75 Ga, dT/dH=100–200°C/km) [24]. High-alumina rocks (Al_2O_3, up to 28 wt %) are widespread in the Kan complex [25]. However, there are no commercial sillimanite accumulations among iron-alumina metapelites. The geodynamic setting of the lower crust extension is constrained by the occurrence of orthopyroxene–sillimanite paragenesis (Al_2O_3 up to 9 wt %). However, the orthopyroxene–sillimanite gneisses formed under UHT metamorphism (P = 6.8–7.8 kbar, T = 930°C, 1850–1750 Ma) and the andalusite-bearing rocks developed on granulites in the exocontact aureole of the Tarak massif of granitoids (1950–1750 Ma), do not form industrial accumulations of sillimanite and andalusite. The *Yenisei migmatite–amphibolite–gneiss complex* is formed under the upper epidote-amphibolite and amphibolite facies of metamorphism (P = 7.3–9.2 kbar, T = 633–770°C; age is 750 Ma [16], framing from west and north the Kan complex. There are no industrial accumulations of sillimanite here.

In the Trans-Angara Region are regional metamorphic complexes, which evidence the manifestation of the Grenville Orogeny under the extension–compression conditions (~1.1–0.85 Ga) in the Cis-Yenisei and Central uplifts (Fig.1). The *Garevka deeper metamorphic complex*, which was formed under moderate to higher pressure conditions (moderate–upper amphibolite facies) (P = 7.2–8.64 kbar, T = 582–630°C, age is 1050–850 Ma, dT/dH = 20-25°C/km) and locally under the epidote–amphibolite facies of the exhumation dynamometamorphism (P = 3.9–4.9 kbar, T = 461–547°C, age is 880 Ma, dT/dH=10°C/km) [9]. Within the CYRSZ, kyanite–staurolite–garnet gneisses (up to 5 m thick) are locally developed. The latter are genetically connected with accretion of the POS to the Siberian Craton (600 Ma) and suprasubduction metamorphism.

The Garevka complex is accreted by the *Teya* (P = 3.5–5 kbar, T = 400–650°C, age is 973–953 Ma, dT/dH = 20–30°C/km) and the *Cis-Angara* (P = 4.1–6.6 kbar, T = 500–650°C, age is 1100–850 Ma, dT/dH = 20–30°C/km) [22] zonal metamorphic complexes of And–Sill type formed under *P–T* conditions varying from the green schist to amphibolite facies of metamorphism. Within the zonal Teya metamorphic complex of andalusite–sillimanite (And–Sill) type (P = 3.5–5 kbar, T = 400–650°C; age is 973–953 Ma) and the Central Uplift there are promising areas of HAS rich in andalusite and sillimanite. Owing to mining works the Teya sillimanite and Panimbinsk andalusite deposits

IOP Conf. Series: Earth and Environmental Science **110** (2017) 012010 doi:10.1088/1755-1315/110/1/012010

formed under the *P-T* conditions of the amphibolite and epidote-amphibolite facies were explored. The following areas of garnet–staurolite HAS with a staurolite content of up to 30% and more (Al_2O_3, up to 30 wt%) were distinguished in the field of metapelites of the epidote-amphibolite facies: Cis-Angara complex (Kulakovskoe uplift), northward of the Kan complex, the area of the mining works Slyudorudnik explored, in the lower reaches of the Taseeva River (a left tributary of the Angara River).

Blastomylonites of kyanite–sillimanite (Ky–Sill) type (P = 3.9–8 kbar, T = 450–650°C, the metamorphic gradient – 2.5 to 12°C/km, age is 864–851 to 600 Ma) are locally developed after andalusite–sillimanite schists of the Teya complex in fault (thrust) zones of the Ishimbino–Tatarka regional strike-slip zone (3–5 to 7 km thick). As a result, the HAS occurrences containing andalusite (relicts) + kyanite (Chirimbinskoe, Mayakon) or andalusite (relicts) + kyanite + sillimanite (fibrolite) (Kolorominskoe, Nerazgadannoe, etc.) associations were formed.

The *P-T* conditions and *P-T-t* evolution trends of metamorphic complexes of the Trans-Angara are plotted in Fig. 1; the position of the Teya complex relative to the Garevka one is shown.

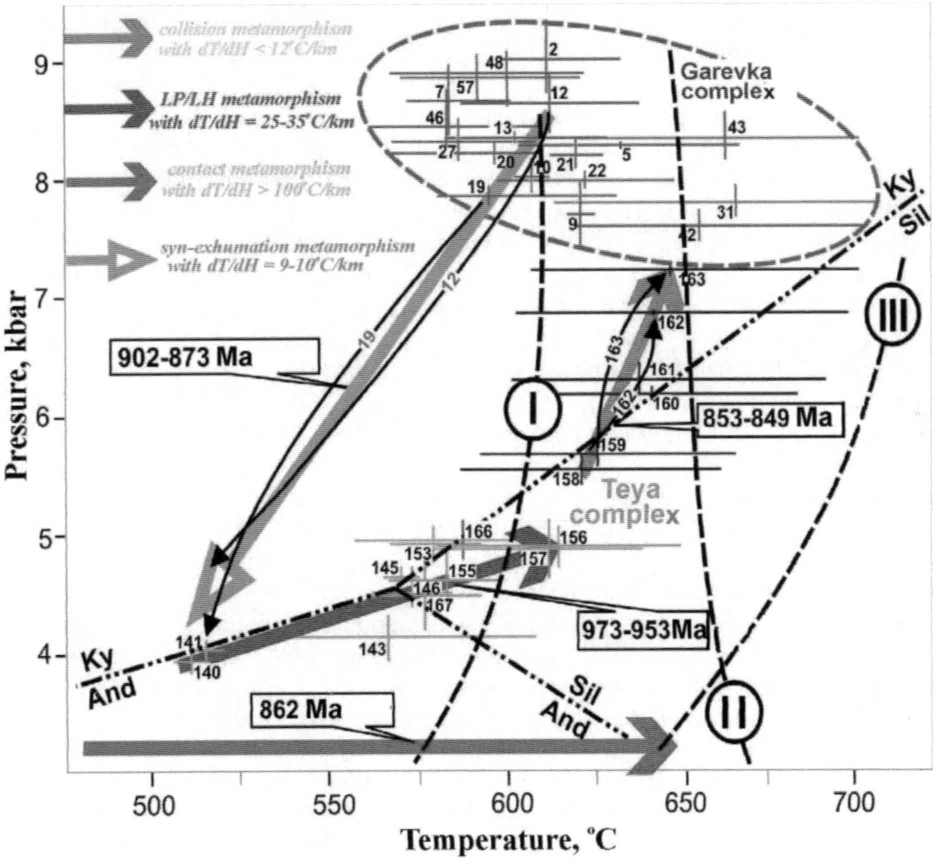

Figure 1. The *P-T* conditions and *P-T-t* metamorphism evolution trends of rocks of the Teya and Garevka metamorphic complexes (Yenisei Ridge) of the Trans-Angara segment of the Yenisei Ridge ([23] and references therein):

Average *P-T* values and their spread, obtained using different geothermobarometers are shown for every sample by crosses (color corresponds the type of metamorphism). Numbers by crosses are numbers of samples. The evolutionary *P-T* trends of metamorphism, calculated based on zonal metamorphic minerals in metapelites are shown by black curves with arrows. Dashed lines with Roman numerals show known mineral equilibriums for the metapelite system: I, Chl + Ms + St + Qz =

Als + Bt + H_2O (Pattison, 2001); II, solidus line of metapelites in water-saturated system (Le Breton and Thompson, 1988); III, (Chatterjee and Johannes, 1974). Coordinates of the triple point and lines of monovariant equilibriums of Al_2SiO_5 polymorphs are shown by dash-and-point lines after (Pattison, 1992).

The polymetamorphic complexes are surrounded by Late–Neoproterozoic *Angara–Pit* and *Angara–Tissa* complexes of regional greenschist metamorphism (NP_3). Chloritoid-bearing schists of the stratiform type with the content of chloritoid of up to 30–40% are mapped in these complexes, but they are insufficiently studied. In the zones of low-*P* (contact) plutonometamorphism (P = 2.5–3.5 kbar, T = 450–650°C, age is 864–861 Ma), superimposed on low-temperature metapelites of the Angara–Pit complex (the middle reaches of the Pit River [34], a weak high-alumina mineralization was established near granitoids. The following mineral zoning is distinguished: chloritoid (outer zone), andalusite (middle zone), and sillimanite (inner zone) (Fig.1).
The moderate-*P* blastomylonites of kyanite–sillimanite (Ky–Sill) type of the collision metamorphism (P = 3.9–8 kbar, T = 450–650°C; the metamorphic gradient is – 2.5 to 12°C/km; age is 864–851 to 600 Ma) are locally developed after andalusite–sillimanite schists in Cis-Yenisei and Tatarka–Ishimbino-Kan regional fault (thrust) zones [20, 35).

3. The evaluation the prospects for practical use of HAS of the Yenisei Ridge

The formation of the most promising deposits and occurrences of the Precambrian HAS in the Trans-Angara segment of the Yenisei Ridge is genetically related to the manifestations of the metamorphism of the And–Sill and Ky–Sill types. The prospects for the possible use of HAS together with nepheline syenites and bauxites, already being explored in the region, to produce aluminum oxide, silumin, and aluminum, as well as, prospects of the expansion of the raw material base of the Boguchansk Electrometallurgical Complex, brought into operation in 2016 in the Lower Angara Region (the joint project of the United Company RUSAL (the world's second largest aluminum company) and the Russian state-controlled hydroelectric power producer RusHydro) are considered. The Boguchansk electrometallurgical complex includes Boguchansk Aluminum Smelter (BAS) with a design capacity of about 600 thousand tons of aluminum per year and Boguchansk Hydroelectric Power Plant (BHPP) with a capacity of 3000 MW [2]. Over the long term, the main domestic raw materials for the production of aluminum are expected to be local ferruginous bauxites–the complex waste-free and environmentally friendly iron-aluminum raw materials of the Chadobets uplift, applicable for production low-silica ferrosilicon (grades F20 and F25), aluminum oxide, and high-quality cement raw materials [3, 33]. According to technological tests, no all bauxites can be applicable for complex processing in direct way. For this purpose, a furnace charge corresponding to the parameters of the iron-aluminum raw material is prepared. Apart from the basic components (ferruginous bauxites and aluminous iron ores), other alumina rocks can be used as additives: allites, kaolines, chloritoids, and high-alumina schists [3], that is, nonbauxite aluminum raw materials [10].

In recent years, such minerals as Al_2SiO_5 polymorphs (andalusite, kyanite, and sillimanite), which are constituents of high-alumina schists (HAS) containing more than 60% of alumina, have become increasingly important in practical terms. The products made of them are used in metallurgical, glass, electrochemical, and chemical industries due to high refractoriness and mechanical strength at high temperatures, chemical inertness with respect to acids and alkalis. An issue of using the concentrates of minerals of the sillimanite group (MSG) for the production of aluminum and modern alloys (silumin, etc.) is urgent [13, 16].

Within the Trans-Angara segment of the Yenisei Ridge, targeted prospecting works for high-alumina andalusite, kyanite, and sillimanite schists were started in the 1956–1960s of the XX century by the Goltsovskaya and Panimbinskaya geological crews of the Angarsk Geological Survey Expedition of the Krasnoyarsk Geological Survey Expedition. As a result, the Golets, Panimbinsk, and Chirimbinsk occurrences of andalusite schists were discovered (1956), and, later, Teya and Noiba occurrences of sillimanite schists (1958–1960). According to the results of prospecting works,

andalusite and sillimanite schists of the Panimbinsk and Teya occurrences, correspondingly are promising for the expansion of aluminum raw material. Of these, two technological samples with a weight of 500 kg each were collected. Samples were examined at the Leningrad Institute of Refractory Materials and these occurrences were recommended for geological appraisal works.
The searching works for HAS were resumed in 1978 in connection with the expansion of the raw material base of Achinsk Alumina Combine. The works were based on the results of studying the processing of nepheline ores at the Krasnoyarsk Institute of Non-Ferrous Metals. It was found that the addition of high-alumina concentrates (andalusite, kyanite, and sillimanite) significantly reduced the specific consumption of nepheline ores, limestones and fuel [14]. By this time, two massifs of nepheline syenites, Kiya and Tatarka, were discovered and explored in the Trans-Angara Region. Prospects for their joint use are shown with confidence in ([12], etc.).

In the 1979–1980s of the XX century, the Nemchan geological crew of the Angarsk GSE PGO "Krasnoyarskgeologiya" carried out prospecting and mining works in 21 sites within the central part of the Yenisei Ridge (North Yenisei and Motygino districts, as well as the Evenk Autonomous Okrug of the Krasnoyarsk Krai). As a result, the prospects of the Panimbinsk and Teya deposits were confirmed with the reserves calculated of 3.1 billion tons and 90.3 million tons of aluminum oxide, respectively. However, this evaluation of resources seems somewhat overestimated, since the ore occurrences with drilling works. Moreover, it was not possible to achieve significant results during this period due to lack of special research works on the metamorphism of the region as an effective methodological basis for searching metamorphogenic high-alumina raw materials. In subsequent years, after the end of the searching works, the legend and map of the metamorphism of the Trans-Angara segment of the Yenisei Ridge were developed, typification of metamorphic complexes was carried out, as well as geological and *P-T* formation conditions and chemistry of high-alumina schists of the most promising Teya polymetamorphic complex were analyzed ([5] and references therein). As a result, the location scheme of distribution areas of Precambrian sedimentary–metamorphic rocks, potential for searching high-alumina minerals, within the Trans-Angara segment of the Yenisei Ridge was compiled (Fig. 2) [7].

From the point of view of the production of alumina, silumin and aluminum, minerals of the sillimanite group or MSG (Al_2SiO_5: andalusite, sillimanite, kyanite) are of particular interest. The HAS of the region are divided into mono-, bi- and polymineral, depending on the occurrence in its composition one or more Al_2SiO_5 polymorphs. The andalusite- (Panimbinsk deposit) and sillimanite-bearing (Teya deposit) HAS formed under the conditions of regional metamorphism of And–Sill type in the Teya and chloritoid schists formed under the greenschist facies of metamorphism in the Angara–Pit metamorphic complexes are classified as monomineral [7].

3.1. Panimbinsk deposit

In this area, first of all, of practical interest are andalusite occurrences of the Panimbinsk deposit, found in the distribution area of Mesoproterozoic carbonaceous metapelites of the Kordin Formation (over 10 km^2 [4]), altered under the *P–T* conditions of the epidote–amphibolite facies of regional metamorphism. Lens-shaped ore bodies extend in submeridional direction in accordance with a general strike of rocks. The northern and southern bodies with parameters (1200×(50–750) and (650×40) m, correspondingly were distinguished. The mineralogical composition of ores: quartz, andalusite (chiastolite) (5–30%), sericite, graphite. Based on results of chemical analysis of four samples, Al_2O_3 content in ore bodies: 19.08%, 19.32%, 21.16, and 19.06 wt % (20.7 wt %, on average), correspondingly. Inferred resources of the deposit on commercial component basis (andalusite, 16.1% on average) to a depth 50 м are estimated to be 70 million tons. The HAS, similar in composition to those of the Golets ore occurrence contain 7.4 % of andalusite, on average, at Al_2O_3 content of 18.1 wt %.

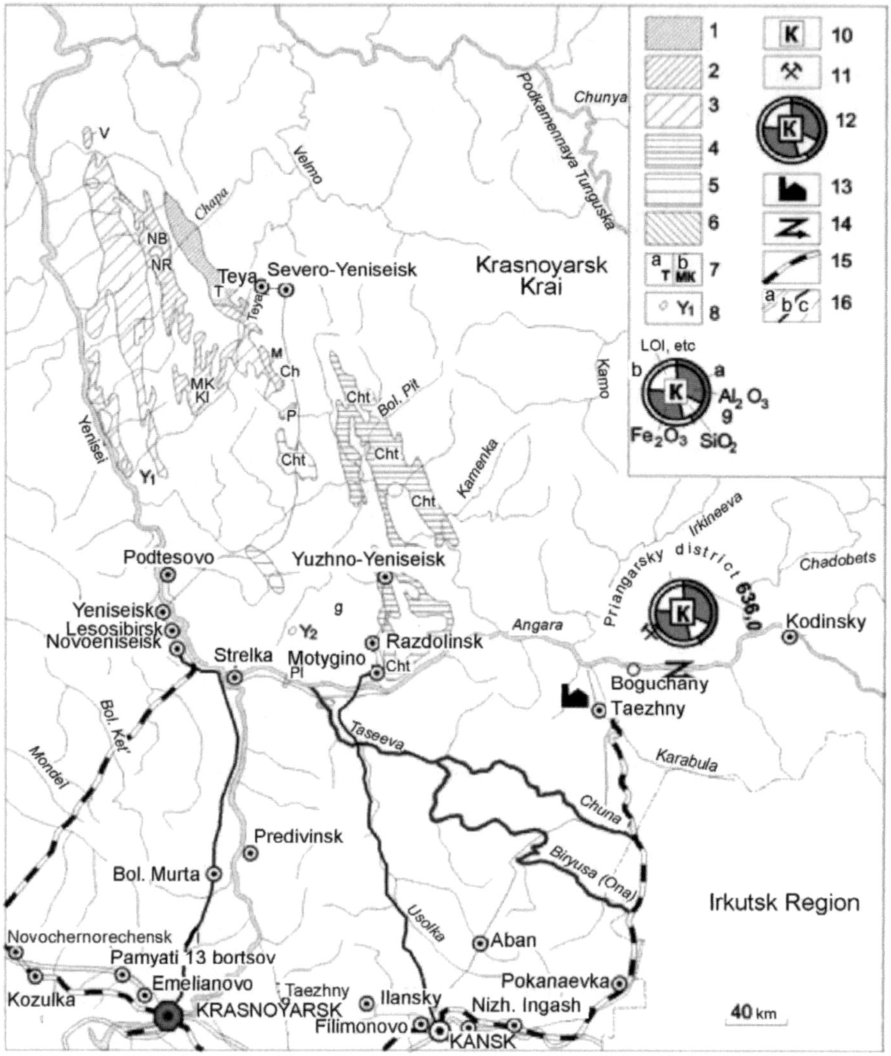

Figure 2. The location scheme of distribution areas of Precambrian sedimentary–metamorphic rocks, potential for searching high-alumina minerals, within the Trans-Angara segment of the Yenisei Ridge: (1) areas, potential for sillimanite, (2) finds of andalusite, sillimanite, and kyanite, (3) ore occurrences, (4) potentially chloritoid-bearing occurrences, (5) finds of chloritoid, (6) mica–garnet–staurolite schists, (7) deposits (a) (T – Teya, P – Panimbinsk); (b) occurrences of high-alumina minerals (V – Vorogovskoe, NZ – Nerazgadannoe, MK – Malokiiskoe, KU – Kiya, C – Chirimbinskoe, N – Noiba, G – Golets, Pl – Polovinkinskoe), (8) massifs of alkaline syenites (1, Kiya, 2, Middle Tatarka); (9–17) Cis-Angara area with reserves 636.0 mln t [33]: (9) reserves (a) and inferred resources (b); (10) Cretaceous age of the ore formation; (11) possible open-cut mining, (12) composition of ores, (13) Boguchansk Smelter, (14) Boguchansk hydroelectrostation, (15) railways, (16) highways of different level.

3.2. The Mayakon ore occurrence

This occurrence is located to the north of the Panimbinsk deposit and is similar to the latter in the age and the composition of host rocks. This occurrence consists of the products of polymetamorphism due to local superimposition of dislocation metamorphism of kyanite–sillimanite facies series on andalusite–sillimanite one. As a result, ore bodies contain andalusite and kyanite and are classified as bimineral. Lens-shaped ore bodies extend for a distance of up to 1000 m at a thickness of 80 m. High-alumina schists are composed of quartz, biotite, graphite, andalusite (6.5–15%), staurolite, muscovite, and kyanite (up to 3.5%). Beyond the local "kyanite" zone of kyanite–sillimanite type of metamorphism andalusite (chiastolite) crystals reach 8 cm in length and up to 1.5 cm wide. The Al_2O_3 content in ores reaches 23.9 wt % ($\overline{X}$ = 20.3%). Inferred resources on the commercial component basis for two ore bodies (andalusite + kyanite 15%) are traced to a depth of 50 m and estimated to be 2.5 million tons.

3.3. Teya deposit

Of the sillimanite occurrences, the Teya deposit, which is confined to the Paleoproterozoic HAS (metapelites, the amphibolite facies of the regional metamorphism) of the Karpinsky Ridge Formation is the most promising. The HAS occupies an area of 14 km^2. Schists of the productive unit consist of quartz (15–50%), biotite (20–60%), sillimanite (fibrolite) (10–40%; 16.5%, on average). An average Al_2O_3 content is 19.16 wt %. The largest ore body is lens-shaped and extends in sublatitudinal direction; a total length is 2.9 km; thickness is 130–190 m; a distribution area is 0.4 km^2. At the concentration of sillimanite of 22% the inferred resources, traced to a conventional depth of 50 m, is estimated to be 100 million tons.

In the 1960s of the XX century, the technological tests on the concentration of two samples of 500 kg each from the Panimbinsk andalusite and Teya sillimanite deposits were performed at the Leningrad Institute of Refractory Materials. The mineralogical composition of the first sample: andalusite, quartz, biotite, muscovite, sericite, kaolinite; the chemical composition is as follows (wt %): SiO_2 = 60.3–61.99, TiO_2 = 0.95, Al_2O_3 = 22.27, Fe_2O_3 = 7.59, CaO = 0.41–0.55, MgO = 1.51–2.21, Na_2O + K_2O = 2.93–3.12, LOI = 1.97–2.8; refractoriness value is 1350°C. As a result the flotation concentrate of the following composition was obtained: SiO_2 = 36.48, Al_2O_3 = 57.14, TiO_2 = 1.6, Fe_2O_3 = 2.22.

The mineralogical composition of the sillimanite-bearing samples: sillimanite, quartz, biotite, graphite, muscovite; chemical composition: SiO_2 = 60.25, TiO_2 = 1.42, Al_2O_3 = 18.2, Fe_2O_3 = 6.69, MgO = 1.58, CaO = 0.42, Na_2O + K_2O = 2.68, SO_3 = 0.35, LOI = 2.48; refractoriness value is 1430°C. The chemical composition of the flotation concentrate: Al_2O_3 = 54.57%, SiO_2 = 41.00%, TiO_2 = 0.32%, Fe_2O_3 = 1.77%, at output of 13.95% and extraction of sillimanite 79%. The concentration was made following the flotation scheme.

In the subsequent years, on the basis of the metamorphism map of the Trans-Angara Region, metamorphic rocks were typified and the promising areas of the MSG distribution in the Teya polymetamorphic complex were identified [5]. In practical terms, first of all, sillimanite-bearing high-alumina metamorphic rocks of the amphibolite facies of prograde regional metamorphism are of most interesting for the mining. A noticeable increase in reserves can be expected in a zone about 6 km wide and 75–80 km long to the northeast and south-west from the Teya deposit (the watersheds of the Chapa–Teya–Enashimo Rivers). A site near the Kalamino granitoid massif (the watershed between the Elizavetinsky and Borzetsovsky streams) in the distribution area of the Paleoproterozoic rocks of the Karpinsky Ridge Formation is recommended as a priority for the organization of prospecting works. Here, the "sillimanite" distribution area of about 40 km^2 in square is outlined; the content of alumina in schists (hand specimen samples) reaches 30 wt %. Quartz–sillimanite quartzites, similar in composition to ores of Kyakhta and Bazybai deposits, are also found among HAS.

3.4. Chloritoid-bearing schists

These schists are widespread among the regional metamorphosed sedimentary rocks of the Neoproterozoic Tungusik Group within the Angara–Pit greenschist complex in the interfluve of the Gorbilka, Pit, and Angara rivers. The chloritoid of HAS (content Al_2O_3 is 23 wt % and more) is closely associated with metasandstones, carbonaceous phyllites, dolomites, and stromatolitic limestones, forming stratified deposits with subhorizontal bedding. The stratiform bedding, a low degree of greenschist facies metamorphism, and the occurrence of marker horizons in the sections make it possible to trace and map them over many-kilometer distance along the strike. The mineralogical composition of chloritoid schists: quartz, sericite, chlorite, chloritoid. The proportion of the latter reaches 30–40%; chloritoid porphyroblasts are 2 mm or less in diameter. The model content of Al_2O_3 in chloritoid is as high as 40 wt %.

3.5. Chloritoid–kyanite schists

These schists occur in the Neoproterozoic rocks of the Sukhoi Ridge Formation in the Lower Angara Region (the lower reaches of the Angara River basin). In the Cis-Angara metamorphic complex they are associated with the manifestation of the local late Ky–Sill metamorphism in the Tatarka lineament zone [9]. A total width of "chloritoid" and "kyanite" zones is about 2 km. The distribution of these zones is traced in Early Mesoproterozoic garnet–staurolite metapelites Sukhoi Pit Group in the Kulakovo uplift (south of the Angara River) by drilling works at the deep geological mapping in the southwestern direction for a distance of over 20 km under the cover of the Phanerozoic deposits up to 50 m thick.

3.6. Garnet–staurolite crystalline schists

These schists were first distinguished in the Taseeva River mouth (a left tributary of the Angara River) in the Cis-Angara polymetamorphic complex [9]. The distribution areas of two-mica garnet–staurolite HAS (SiO_2 – 57.91–50.98, Al_2O_3 – 22.39–27. 95 wt %, n = 2), which contain abundant idiomorphic large (3–5cm) staurolite porphyroblasts (up to 40–50% in the rock matrix), are confined to the crystalline schists of the lower part of the Sukhoi Pit Group of the epidote-amphibolite facies of the regional metamorphism (And–Sill facies series). They form a sublatitudinal "staurolite" zone in metapelites in the area of the settlement Slyudorudnik and in the Kulakovo uplift.

4. Possible variants for processing high-alumina raw materials from the Trans-Angara Region.

Apart from the above-described metapelitic schists containing high-alumina metamorphic minerals (andalusite, kyanite, and sillimanite, chloritoid, and staurolite), other promising high-alumina raw materials: bauxites and nepheline syenites have been found in the Trans-Angara Region.

4.1. Bauxites

The aluminum industry of the Russian Federation is kept with own produced alumina by approximately 30%, the rest is imported from the CIS countries and far abroad. This problem is especially acute for metallurgical smelters of Siberia [1]. In the future, local ferruginous bauxites are proposed to use as the main domestic raw materials for the production of aluminum in the region. Bauxites are the complex waste-free iron-aluminum raw material of the Chadobets structure (uplift), applicable for obtaining low-silica ferrosilicon (grade F20 and F25), aluminum oxide, and high-quality cement raw materials [1, 32].

The Chadobets deposit, located on the right bank of the Angara River near the BAS is regarded to be applicable for industrial operation. The reserves of Chadobets bauxites are 106.4 million tons with average contents of components (%): Al_2O_3 = 29.14, SiO_2 = 11.70, Fe_2O_3 = 32.68, TiO_2 = 8.66, LOI = 15.45. Deposits of bauxite of this type are widely distributed in the Lower Angara Region and the Yenisei Ridge (Turukhansk, Tatarka, Kamenka, Cis-Angara groups, etc.). Chadobets bauxites are characterized by a high content of iron oxide (up to 33 wt %). Due to this, the Bayer-sintering method and the sintering method are not applicable to the processing of these bauxites. The study of the possibilities of the integrated processing of the Chadobets bauxite of the Central deposit using the electrothermal method was carried out in 1989–1990s of the XX century at Siberian Research Institute of Geology, Geophysics and Mineral Resources and State Geological Enterprise "Krasnoyarskgeologiya". According to technological tests no all bauxites can be applicable for complex processing in direct way. For this purpose, a furnace charge corresponding to the parameters of the iron-aluminum raw material is prepared. According to [3], other alumina rocks (for example, chloritoid-bearing and high-alumina schists) can be used as additives for ferruginous bauxite and aluminous iron ores.

In addition, when leaching bauxites at T = 105–125°C, the concentration of rare elements in the red mud is 1.5–2 times more, since all of them, except for gallium, do not dissolve in alkaline solutions. Being accumulated in circulating alkaline solutions to a certain concentration, gallium can be extracted with known methods [3]. The most difficult operation in the processing of bauxites is the extraction of iron. In view of the fact that the iron oxide in the red mud occurs is finely dispersed and forms close intergrowths with other components of the mud, the effective separation of iron is possible only with reducing red smelting [30]. In this case, the second commodity product is produced, viz. cast iron, and the content of rare elements in slags after their additional processing can be increased by 3-4 times in comparison with their contents in the raw material. Preliminary calculations have shown that at the integrated processing of Chadobets bauxite to produce alumina, cast iron, titanium dioxide, and oxides of rare and rare-earth metals, the total value of commodity output is sharply increasing compared to traditionally produced products. Accordingly, an enterprise that performs complex bauxite processing will be highly profitable.

4.2. Nepheline syenites

The searching works for HAS were resumed in 1978 in connection with the expansion of the raw material base of Achinsk Alumina Combine. The works were based on the results of studying the processing of nepheline ores at the Krasnoyarsk Institute of Non-Ferrous Metals. By this time, two massifs of nepheline syenites, Kiya and Tatarka, were discovered and explored in the Trans-Angara Region.

The Middle Tatarka deposit of nepheline syenite of the Vendian age is located on the watershed between the Tatarka and Pogromnaya rivers in 30 km to the NNE from the settlement of Novoangarsk and is represented by the Northern and Southern bodies among the Neoproterozoic carbonate rocks of the Gorevka Formation [33].

Within the Northern body, geological prospecting works (trenching, pit sampling at 100x100 m grid, and three core drilling profile with a drilling depth of 100–300 m) were carried out. According to the catalog of deposits and occurrences of the Krasnoyarsk Krai, it is designated as the object "Massif A no. 100". Within the field there are 3 sites: Central one composed of urtites and ijolites; Western and Eastern composed of nepheline syenites. The Western body is characterized by ores of the highest quality with a uniform distribution of useful components; it is referred to the deposits of the first group.

Mineralogical composition of ores within the deposit varies (numbers in brackets are given for the Western body): nepheline from 10 to 85% (30.5%), aegirine 4–30% (4.5%), feldspar (microperthite, albite, microcline 65%). Al_2O_3 content in ores is 20–24% (average 22.73%), Na_2O = 3.1–11.1%, K_2O = 4–6%. The reserves of the nepheline syenite of the Northern body to the horizon + 132.5 m (the water level of the Tatarka River) in category $C_1 + C_2$ are 3.9 billion tons. The Southern body similar in composition and structure has estimated ore resources at the P1 category of about 1.5 billion tons [33]. In the immediate vicinity of the deposit there is an 110kV transmission line. South of the object, on the left bank of the Angara River are the settlement of Novoangarsk and the pier. The Kiya massif of nepheline syenites (Ordovician–Silurian) is located in the Cis-Yenisei part of the Yenisei Ridge, 12 km above the mouth of the Kiya River, a right tributary of the Yenisei River, 135 km to NNW from the district center and the pier on the Yenisei River. Compared to the Middle Tatar deposit, this object is less studied and is at maximum distance from the BAS. Due to this, it is not given in the description.

5. Prospects for the aluminum oxide production by sintering technology based on the combined processing of nepheline ores and MSG concentrates

Nepheline syenites of the Angara region are characterized by a low alumina content (20–23 wt %). Due to this they are of little use for the production of aluminum oxide following the traditional sintering technology, as is done at the Achinsk Alumina Smelter. The variant of the combined processing of nepheline syenites and MSG was proposed for the Achinsk Alumina Smelter by G.G. Lepezin and V.D. Semin [14]. Addition of 30% MSG (Al_2O_3 = 57 wt %) to the undressed nepheline rock (Al_2O_3 = 22 wt %) will increase the proportion of aluminum oxide in the mixture to 32%. The Kola nepheline ores of the highest quality contain 28–29% Al_2O_3; Kiya–Shaltyr ores – 27%. The best preparation methods yield an output of 27-30%, the theoretical content of Al_2O_3 in nepheline is 35.89%. If the mixture is composed of 60% MSG concentrate and 40% nepheline ore, then the proportion of Al_2O_3 reaches 43% reaching its concentration in bauxite. The practical implementation of this proposal would allow using nepheline ores with relatively low alumina content without enrichment.

6. Possibilities of application of electrothermy of MSG for the silumin production

In recent years, minerals of the sillimanite group have become increasingly important. In addition to the production of refractories, ceramics, proppants, and other high-tech products, minerals of the sillimanite group can be used to produce silumin. In addition, it is promising to introduce arc-plasma technology for the processing of mechanically activated minerals of the sillimanite group in the production of silumin ([15] and references therein).

Silumin is an aluminum–silicon alloy. Silumin is currently obtained by fusion of crystalline silicon and aluminum in electric or combustion furnaces. Electrothermy is an alternative technology. A detailed review of the development of electrothermy in Russia and abroad as a whole is given in ([28] and references therein). According to numerous expert estimates, this method of silumin production, and, then, aluminum, has a number of advantages, leading to reduction in specific and capital costs and a significant economic effect. The advantages of the electrothermal method for the production of silicoaluminum are compared with the currently used technology of obtaining silumin by fusing electrolytic aluminum and silicon. Comparative calculations of the economic efficiency of the production of aluminum–silicon alloys of the AK12M2MN and AK18 grades using the traditional method for fusing the alumina A0, A7 and crystalline silicon and the electrothermal method of aluminum production from kaolin and sillimanite were carried out. If a new technology (plasma arc melting furnace is used, the approximate electric energy consumption will decrease by approximately 25-30% and amount to 10-12 thousand kW/h per one ton of aluminum–silicon alloy.

7. Conclusions

Thus, metamorphogenic deposits and numerous ore occurrences of the minerals of sillimanite group and high-alumina chloritoid and garnet–staurolite schists have been found in the Trans-Angara segment of the Yenisei Ridge. They belong to the aluminous formation of the fold belts of the Siberian Craton frame and to the class of orthometamorphic high-alumina rocks, which was formed from the protolith of iron-alumina metapelites under the regional metamorphism [10]. In the future, after the additional appraisal survey, including drilling works (the top-priority Teya and Panimbinsk deposits), they can be used as high-alumina raw materials in combination with bauxites of the Chadobets uplift and nepheline ores of the Middle Tatarka and Kiya massifs. Over the long term, the urgency of such an assessment for the Krasnoyarsk Krai is important for expanding the raw material base of the, brought into operation in 2016 in the Lower Angara region (Fig.2). In terms of involvement of the HAS in the production of the Boguchansk electrometallurgical complex, the investment project "Integrated Development of the Lower Angara Area" (2005), developed by the Institute for Regional Strategy is actual [1]. According to this project, the infrastructure development of the region (construction of a bridge over the Angara River, construction of roads in the Trans-Angara region, etc.) is under consideration. According to the adopted program of the development of the Lower Angara region, the "Pikhtovaya" railway station is under construction on the Karabula–Reshety railway branch for the transport support of the Boguchansk Electrometallurgical Complex. It is planned that this railway branch connects the general railways with railway lines belonging to the BEC.

According to our ideas, the research results, and the infrastructure development program in the region [1], the high-alumina schists will be in demand in connection with the development of production capacities and the expansion of the BEC product range, which will require the involvement of domestic iron-alumina raw materials and minerals of the sillimanite group together with nepheline syenites and bauxites in the metallurgical production.

Acknowledgments

I am grateful to Doctor in Geology and Mineralogy, Professor G.G. Lepezin and Candidate in Geology and Mineralogy I.I. Likhanov for support, consultations, and attention to the studies of metamorphism and metamorphogenic ore formation in the Yenisei Ridge region.
This work was supported by the Russian Science Foundation (project no. 16-17-10201).

References

[1] Bandman N K 2014 Selected Works and the Continuation of Initiated Works (Novosibirsk: Izd. Inst. Econ. Org. Prom. Proizv. SO RAN)

[2] Boguchansky electrometallurgical complex. https://boaz-zavod.ru (Cited December 5, 2017)

[3] Zhabin V V 2012 Development prospects of the mineral resource base of the aluminum industry of Siberia by using the non-waste and ecologically clean iron-aluminum raw material *State and Development Perspectives of Mineral Resource Base* **4** 78

[4] Kozlov P S 1989 Problems of petrology and petrochemistry of andalusite schists of the Trans-Angara segment of the Yenisei Ridge (on the example of the Panimbinsk deposit) *Problems of Geology and Metallogeny of the Krasnoyarsk Krai* (Novosibirsk: Nauka) pp 89–100

[5] Kozlov P S and Lepezin G G 1995 Petrology, petrochemistry and metamorphism of the Trans-Angara segment of the Yenisei Ridge *Geol. Geofiz.* **36** 3

[6] Kozlov P S 2017a Estimation of the potential of high-alumina metamorphic rocks of the Trans-Angara segment of the Yenisei Ridge for searching minerals of the sillimanite group, chloritoid and staurolite *Vestn Ural. Otd. Ross. Min. Ob-va* **14** 69

[7] Kozlov P S 2017b Metamorphic complexes of the Yenisei Ridge: *P-T* conditions, age, and geodynamics, Proc. 5th Russ. Conf. Geol. Geodynam. Precambrian "Geodynamic environments of Precambrian and Phanerozoic metamorphism", (St. Petersburg: Svoe Izd.) pp 79–80

[8] Kozlov P S, Likhanov I I, Reverdatto V V and Zinov'ev S V 2012 Tectonometamorphic evolution of the Garevka polymetamorphic complex (Yenisei Ridge) *Russ. Geol. Geophys.* **53** 1133

[9] Kozlov P S, Likhanov I I, Zinov'ev S V and Khiller V V 2014 The Cis-Angara metamorphic complex (Yenisei Ridge): geology, *P-T* conditions and age of metamorphism *Litosfera* **6** 141

[10] Koroteev V A, Ogorodnikov V N, Voitekhovsky Yu L, Polenov Yu A, Savichev A N, Shchiptsov V V, Sazonov V N and Koroteev D V 2011 Nonbauxite Aluminum Raw Material (Yekaterinburg: IGG UB RAS)

[11] Kuznetsov Yu A 1941 The Precambrian petrology of the South Yenisei Ridge (Tomsk: Izd. ZSGU)

[12] Lepezin G G, Sherman M L, Semin V D and Kravtsov I S 1979 Prospects for application metamorphic rocks of the Altai–Sayan fold area and the Yenisei Ridge as a source of the high-alumuna raw material *Geol. Geofiz.* **11** 35

[13] Lepezin G G and Goryunov V A 1988 Application area of minerals of the sillimanite group *Geol. Geofiz.* **2** 89

[14] Lepezin G G and Semin V D 1989 Prospects for development of raw mineral base for the alumina industry of Siberia *Geol. Geofiz.* **2** 89

[15] Lepezin G G 1997 Deposits and occurrences of minerals of the sillimanite group in Russia and prospects for development of industrial production of concentrates on its basis *Ogneupory Tekhn. Keram.* **8** 29

[16] Lepezin G G, Kargopolov S A and Zhurakovsky V A 2010 The minerals of the sillimanite group as new promising raw material for the Russian aluminum-producing industry *Russ. Geol. Geophys.* **12** 1247
[17] Lepezin G G, Anshakov A S, Faleev V A, et al. 2014a Preparation of silumin through arc-plasmous processing mechanically activated minerals of the sillimanite group, *Chem. Sustain. Dev.* **22** 131
[18] Lepezin G G, Anshakov A S, Faleev V A, et al. 2014b Plasma chemical method for producing silumin from the minerals of the sillimanite group *Dokl Earth Sci.* **456** 110
[19] Lepezin G G 2016 Minerals of the sillimanite group as a promising import-substituting raw material for the Russian refractory industry *Nov. Ogneupory*, **5** 6
[20] Likhanov I I, Reverdatto V V and Kozlov P S 2011 Collision-related metamorphic complexes of the Yenisei Ridge: their evolution, ages, and exhumation rate *Russ. Geol. Geophys.* **52** 1256
[21] Likhanov I I, Reverdatto V V, Kozlov P S and Khiller V V 2013 Neoproterozoic metamorphic evolution in the Transangarian Yenisei Ridge: evidence from monazite and xenotime geochronology *Dokl. Earth Sci.* **453** 556
[22] Likhanov I I, Nozhkin A D, Reverdatto V V and Kozlov P S 2014a Grenville tectonic events and evolution of the Yenisei Ridge at the western margin of the Siberian Craton *Geotectonics* 2014a **48** 371
[23] Likhanov I I and Reverdatto V V 2014b *P-T-t* constraints on the metamorphic evolution of the Transangarian Yenisei Ridge: Geodynamic and petrological implications *Russ. Geol. Geophys.* **55** 299
[24] Likhanov I I, Nozhkin A D, Reverdatto V V, Krylov A A, Kozlov P S and Khiller V V 2016 Metamorphic evolution of ultrahigh-temperature Fe- and Al-rich granulites in the south Yenisei Ridge and tectonic implications *Petrology* **24** 392
[25] Nozhkin A D and Turkina O M 1993 Geochemistry of granulites of the Kan and Sharyzhalgai Complexes (Novosibirsk: OIGGM SO RAN)
[26] Nozhkin A D, Turkina O M, Likhanov I I and Dmitrieva N V 2016 Late Paleoproterozoic volcanic associations in the southwestern Siberian Craton (Angara-Kan block) *Russ. Geol. Geophys.* **57** 247
[27] Nozhkin A D, Dmitrieva N V, Likhanov I I, Serov P A and Kozlov P S 2016 Geochemical, isotopic, and geochronological evidence for subsynchronous island-arc magmatism and terrigenous sedimentation (Predivinsk terrane of the Yenisei Ridge) *Russ. Geol. Geophys.* **57** 1570
[28] Odokiy E N, Voropaeva N P, Leonenko I N, et al. 1988 The development of the mineral resources base of aluminum industry by using complex -iron-aluminum ores *Bauxites and Other Ores for Aluminum Industry* (Moscow: Nauka) pp 47–51
[29] Saltykov A M and Baimakov A Yu 2003 The establishment and development of thermalelectric production of aluminum-silicon alloys *Tsvetn. Met.* **7** 101
[30] Semin V D, Ponomarenko V V and Kruk Yu M 1968 The complex use of high-silica iron bauxites and alumina-iron concentrates *Izv. Vuzov Tsvetn. Met.* **1** 45
[31] Cherkasov G N 2015 Bauxites: Methods of Genetic Reconstructions and the Bauxite Potential of Russia (Novosibirsk: SNIIGiMS)
[32] Shibistov B V 2002 Regularities of formation and concentration of continental bauxites. Doctoral Diss. In Geology and Mineralogy (Tomsk)
[33] Tselykovskii A F 2004 The 1 : 200000 Geological Additional Appraisal of the Gorevka Ore Mining Area within Sheets O-46-XV, VI (Ust-Angara Area) (Activity Report of Geological Survey Team over 1991–2000) (Motygino)
[34] Likhanov I I, Reverdatto V V, Sheplev V S, Verschinin A E, Kozlov P S 2001 *Lithos* **58** 55
[35] Likhanov I I, Reverdatto V V, Kozlov P S, Khiller V V and Sukhorukov V P 2015 *Asian J. Earth Sci.* **113** 391

Composition and conditions of the late Mesozoic volcanism occurrence of the northern part of the Borschovochny ridge (Eastern Transbaikalia)

Ivan Kremer, Platon Tishin

Tomsk State University, Tomsk 634050, Russia

E-mail: stilcev@yandex.ru

Abstract. On the basis of geochemical study results composition heterogeneity and formation conditions of late Mesozoic subvolcanic rocks from northern part of Borschovochny ridge by an example of shadoronsky and abgatuysky complexes are discussed. Obtained results show that volcanic rocks of both complexes have heterogeneous composition and differ from each other in level and other features of alkalinity. Rare and trace elements geochemistry study results show that there is a significant enrichment with LILE (Cs, Rb, Ba) and some depletion with HFSE (Nb, Ta), meanwhile the correlation between Th and Yb indicates conditions of active continental margin. Gd/Yb and La/Sm ratios allow to conclude that shadoronsky complex rocks originated from garnet-stable mantle source (partial melting of 5% of garnet peridotite), while abagatuysky complex rocks originated from transitional garnet-spinel mantle.

1. Introduction

Eastern Transbaikalia has complicated geological structure. One may consider this region as combination of geotectonic zones of different age considered from the point of view of plates and and terrains geodynamics. According to data on tectonic zoning, there are following structural elements taking part in the structure of the region: active continental margin of Siberian craton (zone of Stanovoy range) at the north, Argunsky microcontinent at the south, in the central part [1, 2].

Mongol-Okhotsk fold belt is the suture zone that appeared at the junction of Siberian craton and Argunsky microcontinent (figure 1) [3]. On the territory of studied area it's presented by Hantay-Daursky and Aginsky terrains [1]. Aginsky terrain is bounded by disjunctive faults system from all sides, which formed as a result of collision processes in Siberian craton – Argunsky microcontinent system. Such activation ongoing throughout all Mesozoic period conditioned complexity of the terrain structure. There is a series of blocks overthrusted on Argunsky microcontinent. In such a way, that collision stage was ended with general rise of the terrain, followed by major granitization and paleoocean closure.

Borschovochny ridge, which is made up of granitic rocks, and some dissimilar rocks complexes together are known as Borschovochnoe upland. This upland takes important role in the whole structure of Aginsky terrain. Northern margin of the upland is a conjunction of metamorphic rocks spread zones (urulginsky complex, PR_1), local dynamometamorphites belts (aginsko-borschovochny complex, PZ_3) and superposed mesozoic depressions (K_1). Herewith structure of the last two is complicated with appearance of subvolcanic rocks of shadoronsky (J_{2-3}) and abagatuysky (K_1) complexes.

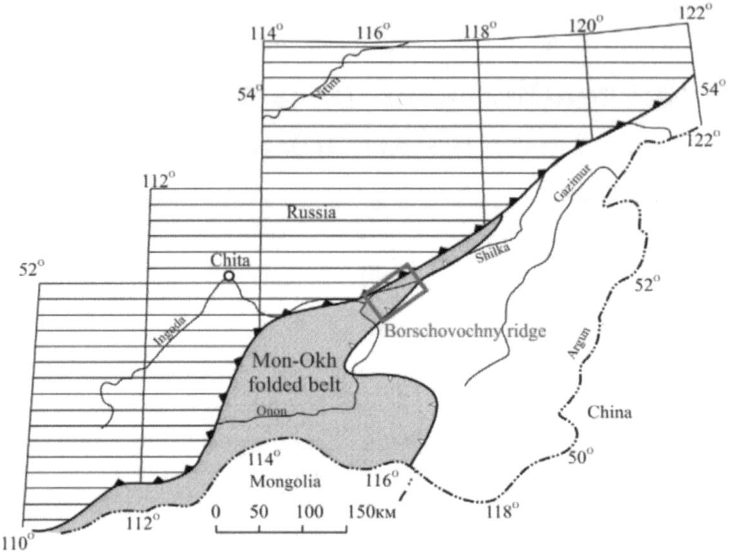

Figure 1. Location of Borschovochny ridge in structures Mongol-Okhotsky folded belt

2. Geological setting and composition of subvolcanic rocks

Subvolcanic rocks of shadoronsky andesite-dacite complex are localized in area of distribution of dynamometamorphic rocks of aginsko-borschovochny complex at the northern part of Borschovochny ridge framing. There are sill-like bodies and high and medium angle dipping dikes of dacites and andesidacites. Vast majority of subvolcanic bodies has north-east strike. Some single dikes have latitudinal strike. Transgressive magmatic bodies have smaller thickness (varying from a dozens of centimeters to 10-15 meters) in comparison with subconcordant ones.

As a result of petrographic study of the shadoronsky complex rocks two groups of them were distinguished. The first group includes porphyritic and aphyric andesites. The second group includes more felsic kinds of rocks – dacites and dacite porphyries.

Andesites are massive or poorly foliated cryptocrystalline rocks with brownish grey colour, often with shades of green. Presence of short-prismatic grains 1-2 mm size of white-coloured plagioclase on a background of cryptocrystalline bulk is the characteristic feature of the rocks. Overall amount of phenocrysts (5-7 mm size) in porphyry kinds of rocks doesn't exceed 10-15%.

Dacites are light brown and homogeneous in the bulk with up to 2 mm long phenocrysts made up of quartz, feldspar, muscovite, less often – biotite. Dacite porphyries differ from dacites by significant increase of plagioclase phenocrysts size up to 5 mm. The bulk of rock is composed of quartz, feldspar, mica, hematite and iron hydroxides and has microlitic/felsitic structure.

Trachybasalts and rhyolites of abagatuysky complex are located within bounds of mesozoic superimposed depressions, which are located at the north-west edge of Borschovochny ridge and filled with turginskaya suite rocks (K_1). Extrusive rocks lie at the bottom of turginskaya suite. There are dikes, thin veins, small stock-like and funnel-like bodies with basalt-andesite and dacite-rhyolite compositions, less often there are their alkaline kinds. Abagatuysky complex rocks act as feeders for turginskaya suite formation and break through its rocks both in the mesozoic depression. Dikes and dike-like bodies are from first meters to 250 meters thick, their strike length may reach 1,5 km. Stock-like and funnel-like bodies sizes are from hundreds of meters to 1-1,5 km across [4].

In terms of petrography there are two kinds of rocks among abagatuysky complex rocks: andesites and microgranites. At that there is possibility of porphyric and aphyric kinds of meta-andesites appearance as among shadoronsky complex rocks.

Microgranites belong to felsic kinds of abagatuysky complex rocks. There are microporphyry xenomorphic quartz phenocrysts not more than 0.8 mm size, which may occupy to 15% of rock volume in the number of samples. The bulk of the rock is composed of plagioclase, quartz and potassium feldspar.

In terms of chemical composition shadoronsky and abagatuysky complexes rocks are well differentiated. There are low TiO_2 (0,3-1,4 wt. %), P_2O_5 (0,09-0,7 wt. %) contents, rather high Al_2O_3 (12-17 wt. %) content. Total silica amount varies within 50-72 wt. % range. Alkali amount varies отwithin 3,9-5,7 wt. % range (3,7-9,4 wt. % in abagatuysky complex rocks). It is shown on the $Na_2O + K_2O - SiO_2$ diagram (figure 2a) that there are two groups of points for abagatuysky complex. First group is located within bounds of from trachyandesibasalts to trachytes field. The second one is within from fields of basalts to fields of rhyodacites. Shadoronsky complex rocks match with medium alkaline rocks of abagatuysky complex. At the same time both complexes differ significantly by kind of alkalinity. Shadoronsky complex rocks mainly are potassium-sodium containing while abagatuysky complex rocks mainly are sodium containing (figure 2b).

Rare and trace elements geochemistry study reveals that accumulation of most of LILE (Cs, Rb, etc.), Th and U and moderate HFSE contents are characteristic for both complexes (figure 3b, d). Elements behavior pattern corresponds mostly to peculiarities of rare elements composition CIAB type source [5] with spectral minima Nb-Ta and Zr-Hf (for a number of samples from shadoronsky and abagatuysky complexes). Concentrations of Cs, Rb and Ba are increased due to their rather high solubility in aqueous fluid in subduction zones. Maxima of Th and U (which are HFSE elements) are there because both the elements behave in subduction environment as «non- conservative» ones [6]. Since Th solubility degree in subduction zone fluid is particularly low, it is considered that this element was taken from sedimentary component of submerging plate. Sr concentrations and rather high total LILE amount are peculiarities of both complexes. In comparison with primitive mantle, in shadoronsky complex samples there is higher enrichment with large ion lithophile elements and positive Sr correlation. At the same time there is an opposite tendency in abagatuysky complex, medium LILE concentrations and Sr minimum. Such differences probably are related to tectonic evolution of the region that expressed in subduction processes replacement by rifting processes and, as a result, decrease of subduction component in abagatuysky complex geochemistry.

Rare earth elements are widespread in both complexes (figure 3a, c). Shadoronsky complex rocks have medium total REE content (ΣREE = 80-130 ppm) with significant predominance of LREE over HREE ($La/Yb_{(N)}$ = 19-29, REE contents are chondrite-normalized according to [7]). For abagatuysky complex rocks total amount of REE varies in 113-240 ppm range; $La/Yb_{(N)}$ ratio is 2,6-10, spectra disposition is complicated by slight negative Eu anomaly (Eu/Eu* = 0,67-0,8).

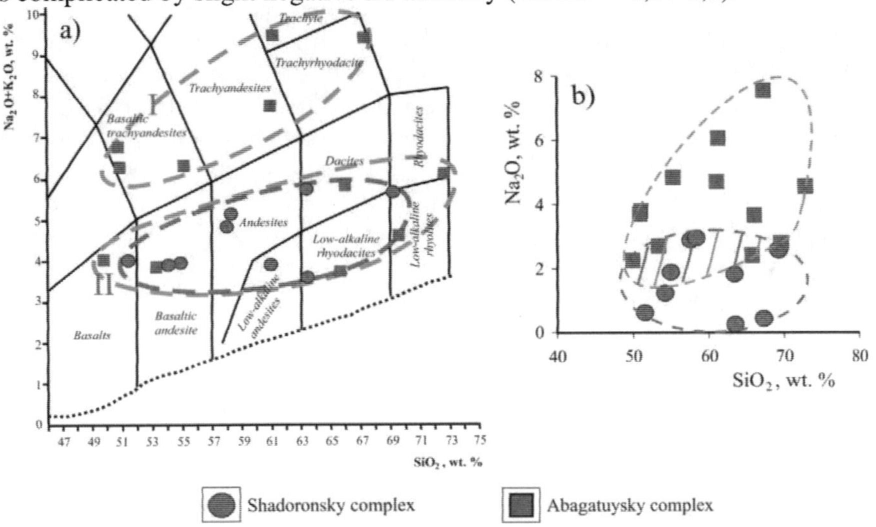

Figure 2. The classification diagrams (Na_2O+K_2O) – SiO_2 (a) [8] and $K_2O - SiO_2$ (b) for the volcanics of the shadoronsky and abagatuysky complexes

IOP Conf. Series: Earth and Environmental Science **110** (2017) 012011 doi:10.1088/1755-1315/110/1/012011

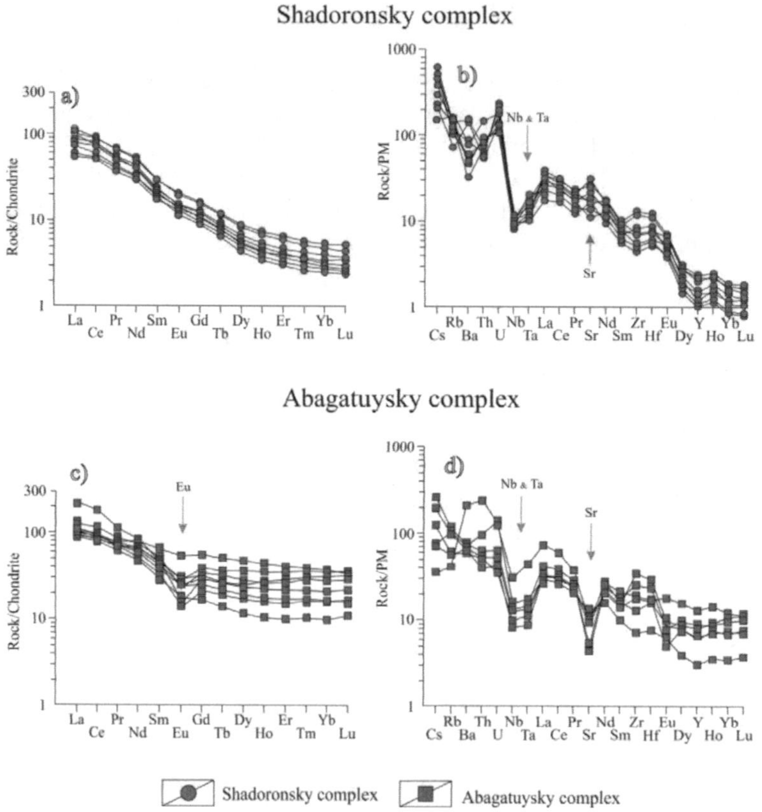

Figure. 3. Chondrite-normalized [12] REE patterns (a - shadoronsky formation, c - abagatuysky formation), and primitive mantle-normalized [7] multi-element spectra (b - shadoronsky formation, d - abagatuysky formation)

3. Volcanism conditions and source composition

Geodynamic nature of shadoronsky and abagatuysky complexes was considered by variation diagrams composing in Th/Yb - Ta/Yb, Th/Ta – Yb coordinates (figure 4a, b). Shadoronsky complex rocks belong to active continental margin field (ACM), while abagatuysky complex rocks have their source shifting from ACM field to within plate volcanic zone (WPVZ) field. Such points shifting over the ACM-WPVZ boundary may be caused by significant influence of crustal component. Besides that, such shifting may reflect complicacy of tectonic nature of the region where evolution from low angle subduction to extension conditions took place. Since that volcanic complexes, which are related to low angle subduction, have their points located within ACM field, while other volcanic complexes, which are related to extensional volcanism phase, have their points lying within WPVZ field. In our case abagatuysky complex rocks are likely to be transitional ones and include andesites (related to subduction) and felsic rocks (related to rifting) [9]

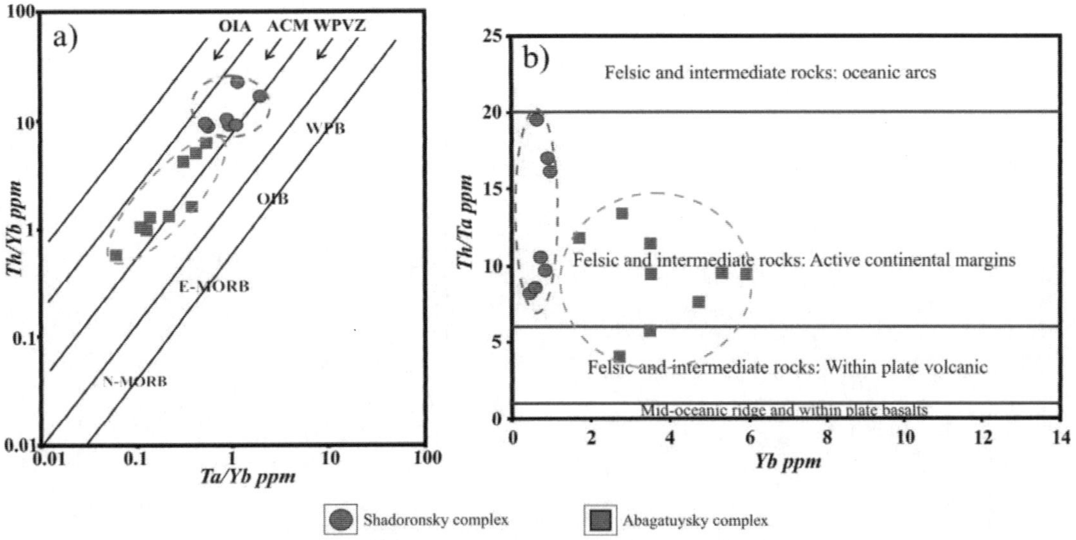

Figure 4. Geochemistry diagrams: a) Binary diagram Th/Yb-Ta/Yb (ACM (active continental margin), OIA (ocean island arcs), OIB (basalts of oceanic islands), E - MORB (basalts of mid-ocean ridges), WPVZ (Intraplate volcanic zone), WPB (within plate basalts)) [9]; b) Binary diagram Th/Ta – Yb [9]

Partial melting degree of studied complexes may be estimated by rare earth elements composition. Heavy rare earth elements (HREE) are compatible in garnet, while middle rare earth elements (MREE) are imcompatible or poorly compatible [10]. On the contrary, all REE are imcompatible in spinel [11].

In such a way, magmas from stable garnet source have higher Gd/Yb ratios than those of magmas from stable spinel source. On the other hand, Gd/Yb and La/Sm ratios may be increased while fractional crystallization of mafic minerals and plagioclase, so low Gd/Yb and La/Sm ratios would become rather close to initial ratios of primary magmas [12].

Calculations show that abagatuysky complex rocks were formed within the overlapping zone between garnet mantle and spinel mantle about 80-110 km deep (figure 5). Yet it is hard to estimate partial melting degree with use of positions of points on the diagram. In accordance with Gd/Yb relations shadoronsky complex rocks could formed of mantle source, which has high portion of garnet in its composition (about 200-300 km deep). Moreover, such modeling shows that these rocks formed by partial (5 %) melting of garnet peridotite.

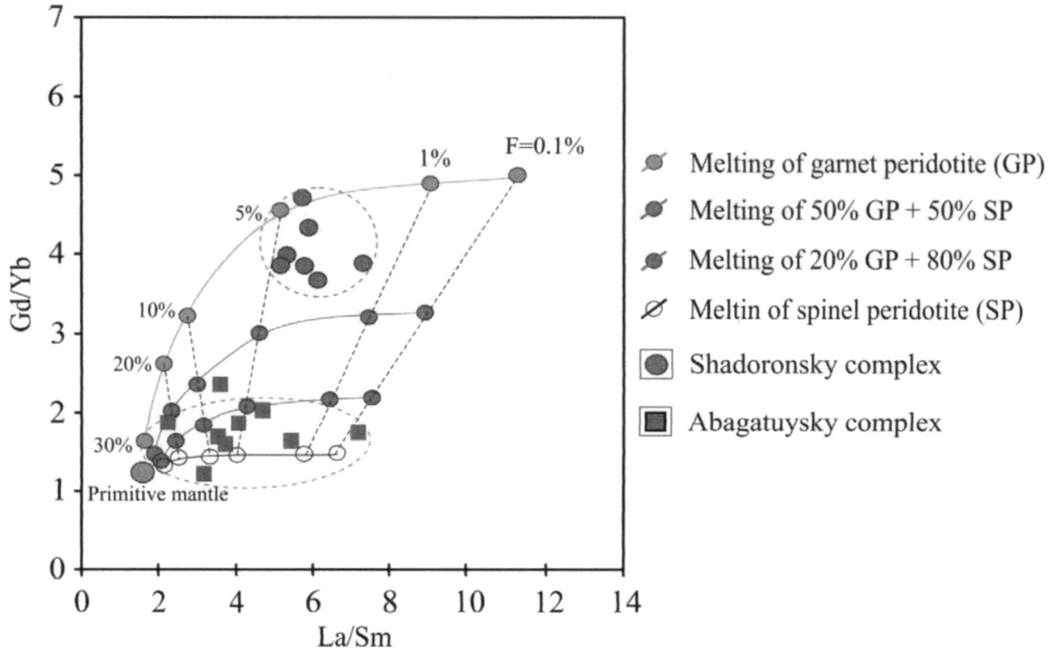

Figure 5. Scattergram of Gd/Yb ratios vs La/Sm ratios of the rocks shadoronsky and abagatuysky complexes. Primitive mantle and the lower crust compositions are from Sun and McDonough (1989) [12]

4. Conclusions

(1) Studied subvolcanic rocks of shadoronsky and abagatuysky complexes are differentiated with SiO2 contents in the 50-72 wt. % range and alkali contents in 3,7-9,4 wt. % range. At that increased alkalinity (6,2-9,4 wt. %) with sodium specificity is observed for abagatuysky complex rocks.

(2) Rare and trace elements geochemistry study reveals significant enrichment with LILE, what expresses in positive anomalies of Cs, Rb и Ba contents, which are caused by influence of aqueous environment of subduction zones. Lower HFSE contents with clearly expressed negative Nb-Ta anomaly, as it's supposed, are attributes of mantle source with composition alike to that of volcanic rocks of island arcs. Direct correlation of Th and U contents is because of involvement of sedimentary component of oceanic crust in melting process. Divergence between Zr-Hf and Sr contents may be related to abyssal changes of melting conditions, in other words – to geodynamic environment changes.

(3) Th/Yb relation indicates that studied rocks formed in conditions of active continental margin. However, emerging trend of abagatuysky complex samples shift to intraplate volcanic zones field means that in terms of geodynamic model the region transformed from subduction zone to riftogenic situation. Similar model had been suggested for west coast of USA, Rocky Mountains region. Many researchers draw the analogy between that territory and territory of Eastern Transbaikalia, which undergone similar transformations.

(4) Findings on studied rocks source composition indicate that shadoronsky complex rocks probably formed as a result of partial melting of garnet peridotite from up to 300 km deep, while abagatuysky complex rocks formed as a result of partial melting of spinel peridotite in depleted lithospheric mantle in conditions of relatively low pressure (80-110 km).

Acknowledgements
This work was conducted as a government task of the Ministry of Education and Science of the Russian Federation, Project Number 5.2352.2017/4.6. The examinations of samples were performed on the equipment of Analytic Center of Natural Systems Geochemistry of Tomsk State University.

References
[1] Mazukabzov A M Sklyarov E V Donskaya T V Gladkochub D P Fedorovsky V S 2011 Metamorphic core complexes of the Transbaikalia: Review *Geodynamics & Tectonophysics* (Russian geodynamics & tectonophysics) Vol **2** p 95-125
[2] Sklyarov E V Mazukabzov A M Melnikov A I 1997 Cordilleran-Type Metamorphic-Core Complexes [in Russian] Izd. SO RAN Novosibirsk p 197
[3] Parfenov L M Popeko L I Tomurtogoo O 1999 Problems of tectonics of the Mongol–Okhotsk orogenic belt. *Geology of the Pacific Ocean* Vol **18** № 5 p 24-43
[4] Rutshtein, I G 2013 Geology of ore-bearing structures dynamometamorphic of the East Transbaikalia [in Russian] Izd. Geokart-Geos Moscow p 154
[5] Kelemen P B Hanghøj K Greene A R 2003 One view of the geochemistry of subduction-related magmatic arcs, with an emphasis on primitive andesite and lower crust, In: Treatise on Geochemistry (Eds. Holland Y.D. and Turekian K.K.) Elsevier Ltd. **3** p 593–659
[6] Pearce J M Peate D W 1995 Tectonic implications of the composition of volcanic arc magmatism *Annual Review of Earth and Planetary Sciences* Vol **23** p 251-285
[7] Sun S S McDonough W F 1989 Chemical and isotopic systematics of oceanic basalts: implications for mantle composition and processes *Geol. Soc. Spec. Publ.* Vol **42** p 313-345
[8] Bogatikov O A Petrov O V 2008, Petrographic code of Russia. Second edition reworked and supplemented [in Russian] St. Petersburg VSEGEI p 204
[9] Gorton M P Schandl E S From continents to island arcs: a geochemical index of tectonic setting for arc-related and within-plate felsic to intermediate volcanic rocks *Canadian Mineralogist* Vol **38** 2000 p 1065-1073
[10] Irving A J Frey F A 1987 Distribution of trace elements between garnet megacrysts and host volcanic liquids of kimberlitic to rhyolitic composition *Geochim. Cosmochim. Act* **42** p 771-787
[11] Kelemen P B 1990 Reaction between ultramafic rock and fractionating basaltic magma I. Phase relations, the origin of calc-alkaline magma series, and the formation of discordant dunite *J. Petrol.* **311** p 51–98
[12] Song X Y Keays R R Xiao L Qi H W Ihlenfeld C 2009, Platinum-group element geochemistry of the continental flood basalts in the central Emeisihan Large Igneous Province, SW China *Chemical Geology* p 246-261

IOP Conf. Series: Earth and Environmental Science **110** (2017) 012012 doi:10.1088/1755-1315/110/1/012012

Sedimentary deposits and petrography of the basalts of the island of Tenerife of the Canary archipelago (Spain)

V. A. Krylova and I. F. Gertner

Department of Geology and Geography, National Research Tomsk State University, Lenin Avenue 36, Tomsk, 634050, Russia

E-mail: krylova.vera1994@yandex.ru, labspm@ggf.tsu.ru

Abstract. The article is devoted to the study of sedimentary deposits on the slopes of the Teide volcano and petrography of the volcanic rocks of the island of Tenerife of the Canary archipelago (Spain). A description of the sediments, which are presented on the island's territory, and an analysis of the features of the material composition of rocks are given.

1. Introduction

The island of Tenerife is one of the largest islands of the Canary archipelago, belonging to Spain and located in the Atlantic Ocean. The area of the island is 2058 km^2. The height of its main volcano Teide is 3,718 m above sea level.

The name of the island comes from the words "tene" (mountain) and "ife" (white). It is the name, which was given to this island by the locals of the Guancho tribe, since in the winter the top of the Teide volcano is covered with snow. When the two words were combined, the letter "r" was added and the modern name Tenerife was obtained [1]. The Teide volcano and the archipelago itself have a number of other features. This is the highest point in Spain. Taking into account the base of the volcano under the ocean, its height is about 6000 meters (Fig. 1).

Figure. 1. Exterior view of the Teide volcano today [photo by Gertner I F].

The geological history of this island and the archipelago, in general, is a unique subject of research for many scientists. For example, only in the history of mankind were recorded two eruptions of the Teide volcano itself. The tectonic position of the Canary Islands corresponds to the classical scheme of manifestation of intraplate magmatism in the conditions of the formation of the present-day ocean and is characterized by a wide range of igneous rocks, including tholeiitic and subalkaline basalts, tephrite, phonolites and xenoliths of ultramafites. In this paper, only some aspects of the geological structure of the Teide volcano and the features of the chemical and petrographic composition of the main varieties of rocks of a given geological object are considered.

2. Sedimentary deposits on the slopes of the Teide volcano.

Sedimentary deposits on the island are represented by weathering products of volcanics. One of the attractions of this resort is black sand, which is the result of the washout of basalt flows in the tidal zones. Directly in the Las Canajades caldera, the products of aeolian erosion are observed with the formation of bizarre forms of outcrops and eluvial-deluvial sediments. Partly they have features of lake sediments, in the springtime due to the melting of snow on the peak of Teide, the caldera is filled with water. During the construction of the trunk road on the western slope of the volcano, was discovered a zone of chemical weathering of volcanic deposits, called a "layered pie" (Fig. 2).

Figure. 2. Exposure along the road on the western slope of the Teide volcano, revealing products of chemical weathering of volcanic flows [Photo by Gertner I F].

Its structure is due to the almost horizontal deposition of layers, painted in dark gray and light colors with a greenish, bluish, yellowish or pinkish hue. The bizarre alternation of multi-colored layers allowed the builders to call this outcrop a "layered pie". The rocks in this outcrop are represented by fine-grained aleuro-clay material [2]. The different coloration of thin layers is due to variations in the

IOP Conf. Series: Earth and Environmental Science **110** (2017) 012012 doi:10.1088/1755-1315/110/1/012012

transformation conditions and the chemical composition of the products of volcanic eruptions, in particular, the enrichment of certain gases and the results of fumarolic activity. It should be noted that the width of this outcrop is only a few hundred meters and in the geomorphological terms is confined to the intermountain tract formed as a result of the "plowing" of the volcanic stream during the eruption of Teide in 1706 [3]. A similar type of exposure, but with a less spectacular variety of colors, is on the coast near the city of Garachico. Given that on this slope periodically forms a lake in the caldera of Las Canajes, it can be assumed that these rocks have a mixed origin, i.e. may be alluvial deposits of a locally occurring water stream. One of the arguments in favor of this point of view is another type of cemented boulder deposits, whose outcrops are observed literally 200 meters to the south in the same bay (Fig. 3).

Figure. 3. Outcrops of cemented boulder-pebble deposits [photo by Gertner I F].

This sedimentary complex is most likely formed due to the erosion of deluvium of the outcrops in the coastal zone, followed by cementation by clay material from the adjacent weathering zone. Probably, this was taking place against the background of a general uplift of the island. The strength of cementation of such deposits is not high, therefore in the most crowded places their outcrops are strengthened by a metal grid.

3. Petrographic description of the basalts of the island of Tenerife.

At the disposal of the author was a small number of samples that corresponded to the most widespread facies of eruptions of the island of Tenerife, namely, basalts. Among them, the following main varieties were identified, reflecting the different degree of crystallization and the likely formation conditions.

For the petrographic description of samples, the manual [4,5,6,7] was used as a definition of the optical constants of the minerals under study.

Olivine basalt - a dark-gray rock, cryptocrystalline, dense with numerous almonds and pronounced pores. In the pores fragments of organic remains are noted. Under the microscope, a porphyry structure and a disseminated homogeneous texture are established. The phenocrysts are represented by olivine and augite (with grain sizes in the average: 1-4 mm long, 0.5-1.5 mm wide). The bulk has a pilotate structure, consisting of microlites of olivine, plagioclase, augite, and magnetite. Mineral composition: olivine - 70%, augite - 5%, plagioclase - 15%, magnetite - 10%. Olivine is represented mainly in the form of phenocrysts. The phenocrysts have an elongated, isometric form of grains. At one nikole it is colorless with a high relief and slightly pronounced cleavage at an angle of 90 °. In crossed nicols has bright colors of interference: the birefringence index is Ng-Np = 0.34, the extinction angle c: Ng = 0°. Olivine is also observed in the form of microliths. In this sample, there are grains in the form of "windows", where the olivine grain and the bulk of it, this indicates that rapid crystallization has occurred. Augite is represented in phenocrysts with angular, diamond-shaped, hexagonal grains with dimensions on average: along the length up to 2.5 mm and in width up to 1 mm. In crossed nicols, a slightly greenish color with a shade of gray is observed, the refractive index Ng-Np = 0.25 with an extinction angle c: Ng = 42. In one nikole, it is colorless. The characteristic extinction is in the form of a figure of an hourglass, which allows us to call it augite. Plagioclase is represented by small, elongated grains with average dimensions: 1-5 mm in length and a width of up to 0.5 mm. In one nicole, it is colorless. In crossed nicols, it has dark gray interference colors with a refractive index of Ng-Np = 0.007. Characterized by polysynthetic twins. To determine the number of plagioclase, more than five measurements of different grains were performed. As a result, andesine with the number 37 was determined. Magnetite is observed as a fine expressed rash, isotropic.

Similar to this breed are samples №2, №4 and №6, but there are small differences. In sample No. 2 the rock is dense with less pronounced porosity, and sample No. 4 in terms of density and porosity corresponds to sample No. 2 and has two differences from sample No. 1. Here olivine has more rounded grains, and augite is green in color, in sample No. 6 large elongated grains of olivine are observed. In all samples, the amount of phenocrysts reaches 10%.

The basalt conglomerate is composed of pebbles of gray, dark gray and brown color, cemented with light gray cement. The structure is small-pebble. The size of the wreckage varies from 0.5mm to 2cm in the diameter. Pebbles are well-rounded (3rd class of roundness according to A.V. Khabakov). The texture is spotty due to the presence of darker fragments on the background of light-colored cement. Microscopically the rock has a light gray, brown color. The bulk is poorly crystallized with a hyalopilite structure. Pebbles are represented by basalt, phenocrysts of olivine and pyroxene are separately observed. Olivine has a rounded shape of grains, in one nikole it is colorless with a pronounced relief and slightly pronounced cleavage. In crossed nicols has bright colors of interference, the refractive index Ng-Np = 0.036. We observe a direct extinction angle c: Ng = 0 °. Augite is presented in the form of angular, hexagonal shape. In one nikol - colorless, brownish. In crossed nicols, the extinction angle c: Ng = 42 °, the refractive index Ng-Np = 0.025. In pebbles, small porphyry grains of plagioclase are observed (with dimensions of up to 0.5 mm in width and up to 3 cm in length), as well as a small number of fine augite grains with a dimension of up to 1.5 cm in length and a width of up to 0.5 cm. Plagioclase in one nikole is colorless, and in crossed nicols has a light gray color with a refractive index of Ng-Np = 0.009. Based on the results of several measurements, plagioclase was determined under number 39 - andesine.

The sintered tuff - a brown rock, which is composed of rounded fragments of volcanic glass with dimensions: up to 4 cm in length and up to a width of 1.5 cm, isotropic. Cement is iron-chlorite. In addition, there are grains of olivine and augite. A shell-shaped form with a brownish color in crossed nicols prisms and black in one, isotropic, represents volcanic glass. Olivine is presented in the form of small diamond-shaped, elongated grains with dimensions: length up to 3mm and width to 0.05mm. In one nicole is colorless, and in crossed nicols has bright colors of interference, has a birefringence index Ng-Np = 0.037. Augite is observed in the form of elongated grains with dimensions on average: 1 mm in length and in width to 0.02 mm. In crossed nicols, the refractive index: Ng-Np = 0.025 and the extinction angle c: Ng = 42, and in one nikole is colorless.

Olivine is a pyroxene porphyrite, which is characterized by a dark gray color. In the breed, there is a large number of phenocrysts, which are represented by olivine and pyroxene. Under the microscope, the structure of the main mass is intergranular; the texture is interspersed. The phenocrysts are represented by olivine and augite. Olivine is observed in the form of rounded and diamond-shaped grains with an average length of up to 8 mm and a width of up to 3 mm. In one Nikole it is colorless with a pronounced relief, in crossed nicols prisms has bright colors of interference with a birefringence index Ng-Np = 0.038 from yellow to blue-green color. The grains are large and heavily "cracked." Augite is represented in the form of elongated grains with dimensions, on average, in length up to 1cm and in width up to 0.5mm. In one nikole it is colorless, while gray-violet color with a zonal structure is observed in the crossed ones, the refractive index Ng-Np = 0.05, c: Ng = 42.

Gialobasalt - presumably the basic composition. The structure is hyalopilitic. The whole rock is composed of a glassy, isotropic mass. In the form of microlites and small phenocrysts, plagioclase and biotite grains are fixed, which makes it possible to compose this rock with basaltoids of increased alkalinity. Plagioclase is represented in the form of small, elongated grains (microlites) with polysynthetic twins. In the crossed nicols it has the light gray interference colors; the refractive index Ng-Np = 1.530. In one nicole it is colorless. The size of grains on average up to 2 cm in length and the width to 0.5 cm. Biotite is observed in the form of elongated leaves. In one nikole, it has brown colors interfere with pleochis, in crossed - yellowish. The birefringence index is Ng-Np = 0.050, the extinction angle is c: Ng = 0. Laminar motion is observed in this rock.

4. Features of the chemical composition of volcanics of the Canary Archipelago.
The igneous rocks of a given geological object have a wide range of compositions, reflecting variations from ultrabasic to medium magma derivatives. The main mass of the rocks corresponds to the series of increased alkalinity (Figure 4).

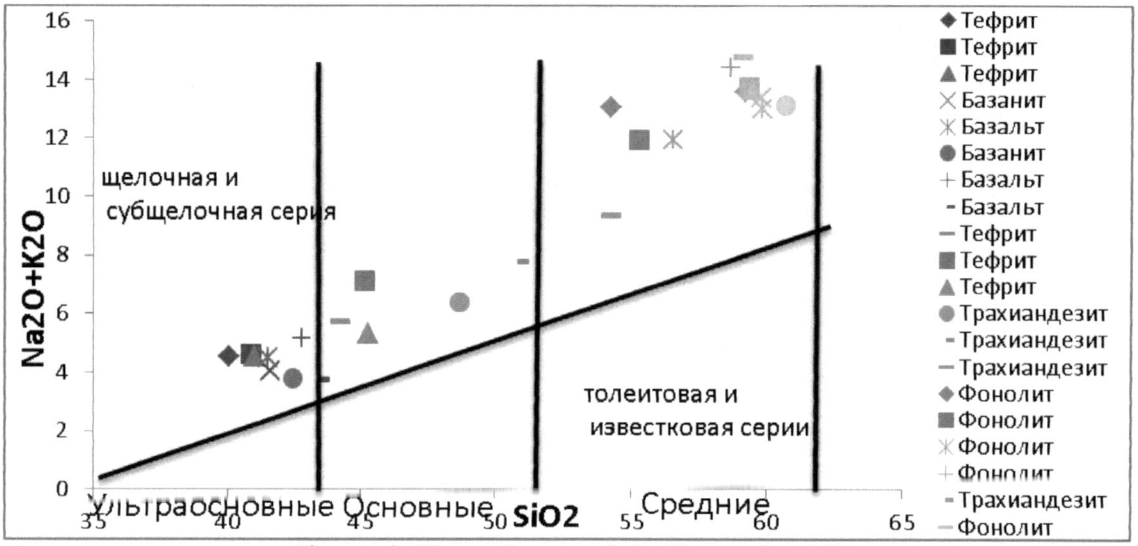

Figure. 4. Binary diagram of Na2O + K2O - SiO2.

In this diagram (Fig. 4), it can be observed that the rocks correspond to families of ultrabasic, basic and medium breeds. Figurative points, located on the interval of SiO2 content of 40-45%, belong to ultrabasic rocks. The SiO2 content (45-53%) corresponds to the composition of the main rocks. The SiO2 content (53-61%) belongs to the middle rocks. According to the content of alkalis, all rocks vary from 3 to 15%, where the most alkaline are phonolites, and the least alkaline - subalkaline basalts and basanites. The ratio of iron and magnesium oxides allows them to be diagnosed as products of the tholeiitic series (Fig. 5).

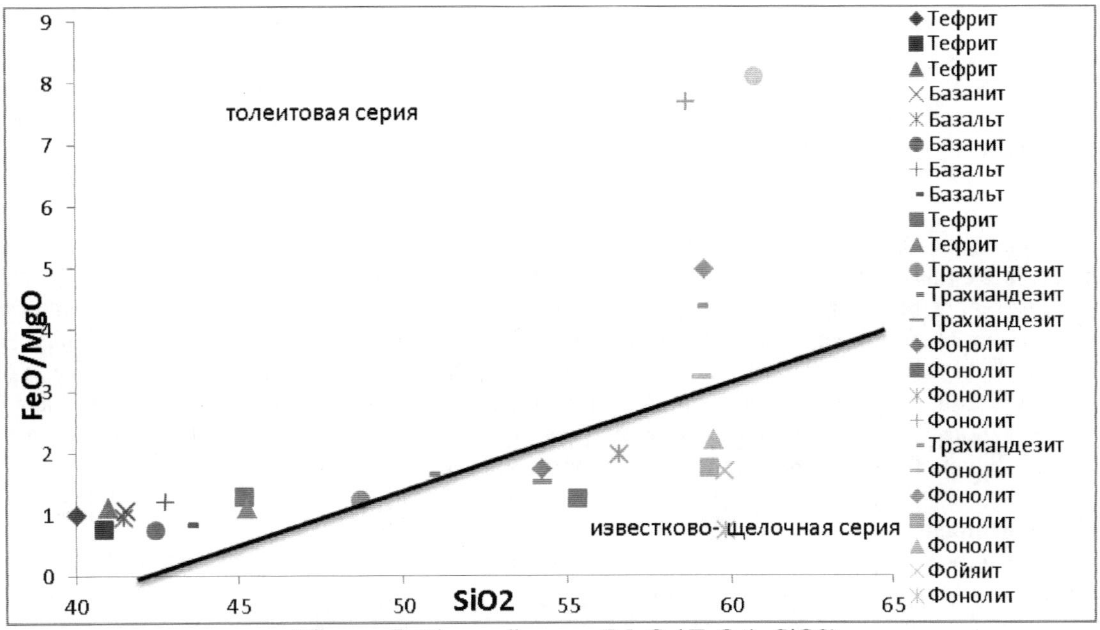

Figure. 5. Binary diagram (MgO / FeO * -SiO2).

However, it should be noted, that the products of high-alkaline magmas of K-Na specialization are characterized by increased iron content and, as a rule, correspond to derivatives of the thallium trend, whereas their varieties of increased potassium can correspond to calc-alkaline specialization.

According to the results of the analysis of the diagram "K2O / Na2O) -SiO2" (Figure 6), we see that most rocks belong to the potassium-sodium series, and only individual figurative points of tephrite fall into the region of sub-potassium and potassium derivatives proper. These deviations may be associated with late metasomatic changes in these rocks or with the process of interaction with groundwater enriched with potassium.

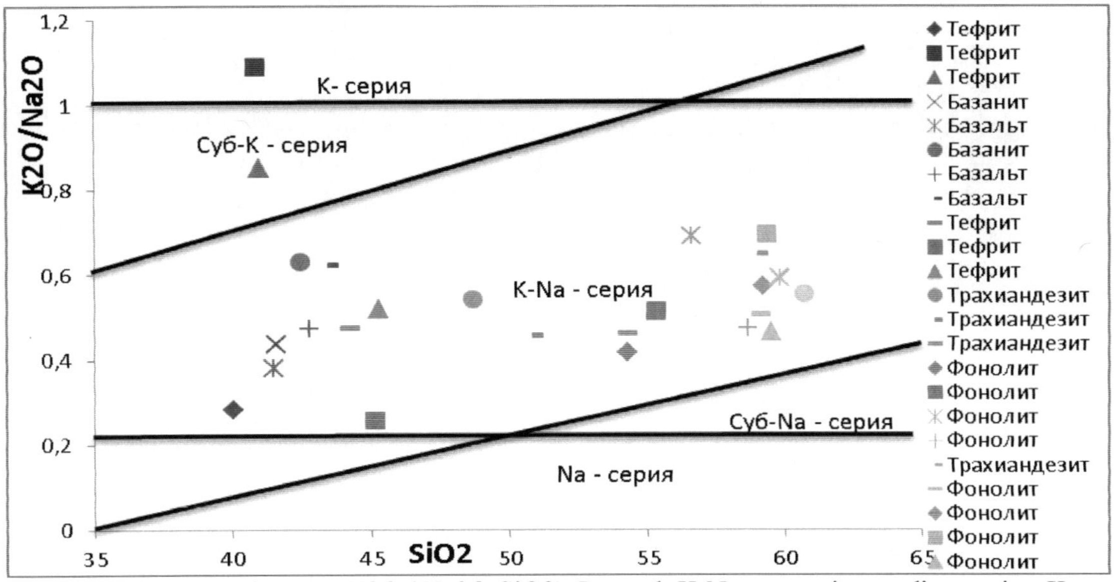

Figure. 6. Binary diagram (K2O / Na2O-SiO2). Legend: K-Na - potassium-sodium series; K - potassium series; Na is a series.

As a result of petrochemical analyzes, it can be concluded that the volcanic rocks of the Canary Archipelago and the Tenerife Islands in particular belong to the products of the alkaline and subalkaline K-Na series of specialization, which is a typical example of plume magmatism under oceanic crust conditions. Further more detailed geochemical and isotopic studies will make it possible to clarify the sources of matter of mantle substrates and their interaction with the oceanic crust.

5. Conclusions

The conducted researches allow to distinguish several stages of formation of the Canary Oceanic archipelago by the example of the island of Tenerife. In addition to the volcanic eruptions proper, which are responsible for the growth of the relief above the ocean level, the processes of weathering play an important role, creating the formation of a peculiar soil cover and unique flora and fauna. The analysis of the material composition of basaltoids and their probable differentiates, as the most representative species of the Tenerife rocks, makes it possible to compare them with the products of magmatic activity of increased alkalinity in the potassium-sodium specialization, which is typical for the manifestation of plumes on the structures of modern oceans.

Acknowledgements

This work was conducted as a government task of the Ministry of Education and Science of the Russian Federation, Project Number 5.2352.2017/4.6.

References

[1] Canary Islands, resource [http://priroda36.ru/priroda-ostrova-tenerife.html]
[2] Bether OV Vologdina IV 2012 Sedimentary rocks. Systematics and classification. (Examples of description: Textbook. - Tomsk: Publishing house Tom. Un-ta) 170 p (in Russian)
[3] Utkin Yu V 2002 Volcanogenic clastic rocks (taxonomy, structure, genetic types). Tutorial. (Tomsk: TsNTI) 148 p (in Russian)
[4] Sharkov E V [and others] 1985 Magmatic rocks / volume 3 / (Moscow: Nauka) 15p (in Russian)
[5] Vrublevsky VA Kortusov MP Utkin Yu V 2015 Optical properties of rock-forming minerals: textbook / comp. (Tom. state. Un-t, cafe. petrography – Tomsk) 53 p (in Russian)
[6] Kuzovatov NI Utkin YuV Chernyshov AI Gertner IF Bether OV 2012 Systematics and classification of igneous rocks (Textbook - Tomsk: Tomsk State University) 97 p (in Russian)
[7] Chernyshov AI Vologdina IV 2014 Structures and textures of igneous and metamorphic rocks: educational-methodical manual. (Tomsk: TSU Publishing House) 36p (in Russian)

Geological structure and prospects of noble metal ore mineralization of the Khayrkhan gabbroid massif (Western Mongolia)

K R Kurumshieva, I F Gertner, P A Tishin

Department of Geology and Geography, Tomsk State University, Lenin Avenue 36, Tomsk, 634050, Russia

E-mail: kkurumsieva@gmail.com

Abstract. An analysis of the distribution of noble metals in zones of sulfide mineralization makes it possible to justify the isolation of four ore-bearing horizons with a specific geochemical zonation. A rise in the gold content relative to palladium and platinum is observed from the bottom upwards along the section of the stratified series of gabbroids. The study of the mineral phases of sulphides and the noble minerals itself indicates the evolution of hydrothermal solutions, which determines the different activity and mobility of the fluid (mercury, tellurium, sulfur) and ore (copper, nickel, iron, platinum, gold and silver) components.

The Khairkhan dunite-troctolite-anorthosite-gabbro massif is a part of the Bayantsagan range of peridotite-pyroxenite-gabbronorit intrusions of the Hirgiznur Early Paleozoic complex [1]. It is located in the central part of the Lake Zone of Western Mongolia and forms a large lenticular body (18×5 km^2), stretched in the north-west direction. This massif is one of the largest layered gabbroic plutons and composes the zone of bald mountains of the Bumbat-Khairkhan ridge (Jargalantu) (elevations of about 3000 m). In the northern part of the intrusion in the upper reaches of the Urtuin-Gol river gabbronorites break through dislocated schists, sandstones, siltstones and limestones, which on regional maps belong to the Lower Cambrian burgasutai series. The host rocks in contact experience hornfreezing and, occasionally, contact melting. Intrusive contact of gabbroids with a conglomerate layer is also observed near of the Shara-Bulak spring in the southern part of the massif. The eastern part of the Khairkhan massif is destroyed by a fault of the northwestern strike, through which the deposits of the Burgasutai suite are brought into contact with the various facies of the intrusion, broken by the multiphase intrusion of quartz diorites, granodiorites and granites of the Tokhogenshil Early Paleozoic complex. The western part of the intrusion is destroyed by a large massif of the same complex. Intrusive contacts of quartz diorites of the Tohtogenshil complex and gabbronorites of the endocontact facies of the Khairkhan massif were observed northwest of the Ulin-Daba pass. The gabbroids in contact are hornfels and uralitized, which strongly obscures the primary magmatic structures of gabbroids and makes their field diagnostics more difficult. This led some researchers to erroneous ideas about gradual transitions between the diorites of the Tokhtoghenshil complex and differentiated gabbroids [2-3]. Further researches of this massif indicate the intrusive nature of both gabbroids and quartz diorites, while the latter break through the rocks of the marginal facies and the stratified series of the Khairkhan massif.

The primary structure of the Khairkhan massif was disturbed by later folded processes and the introduction of the intrusive bodies of the Tokhogenshil diorite-granodiorite complex. Most of the massif on its northeastern flank is tectonically disintegrated. The dynamometamorphic transformation is expressed in the tectonic character of the northeastern contact of the pluton with host Vendian-Early Cambrian volcanic-sedimentary deposits and in the development of isolated fragments of igneous rocks of a similar petrographic composition on the extension both along and across the main body. The intrusive bodies of the Tokhogenshil diorite-granodiorite complex on the southwestern flank cut the primary magmatic contact of the gabbroid series of the Khairkhan massif, making it difficult to estimate the true dimensions of the latter.

In the structure of the Khairkhan massif participate gabbroids of the stratified series and the marginal facies. The stratified series composes most of the pluton, mainly its central part. It is characterized by peri-magmatic stratification, differentiation and rhythmicity. The elements of the stratification are characterized by a northwest strike, consistent with the extent of the mass itself, and steep incidence angles. The stratified series is composed of plagioclase-containing dunites, troctolites, olivine gabbros, olivine gabbronorites, and anorthosites, with the most common troctolites and olivine gabbros. Ultrabasic rocks are widely distributed and occur at the base of the section, while anorthosites and leucocratic olivine gabbronorites predominate in its upper part, among which there are individual thin horizons of titanomagnetite ores. In the stratified series, low concentrations of Cu, Ni and Cr are recorded, which is linked to high alumina content and low magnesia of the parent melt. Higher copper contents, at low nickel concentrations, are established only in sulphide-containing gabbroids and sometimes reach industrial concentrations. Stratified and trachytoidal textures are characteristics of the stratified series, fixing the elements of the primary magmatic stratification. The marginal facies of the Khayrkhan massif has an apparent thickness of about 0.5-1 km and is developed predominantly on its northeastern flank. The marginal facies are characterized by a relatively homogeneous structure with gradual transitions between the varieties. Nevertheless, especially in the mesotypic and leucocratic differences, the elements of gneiss and trachytoids are fixed. The main petrographic varieties are the harmless gabbro-norites with local development of plagiovebsterites and pyroxene anorthosites, which are gradually replaced by amphibole-quartz-bearing gabbro-norites towards the enclosing stratum. The transition between the rocks of the marginal facies and the stratified series is rather gradual and, as a rule, is traced by the development of veins of pegmatoid gabbro-norites and leuco-gabbro-norites. The rocks of the marginal facies have lower Co, Ni, and Cr contents than those of the stratified series, but higher copper grades, which sometimes reach industrial values. Analysis of the patterns of spatial orientation of elements of internal anisotropy of gabbroids (banding, trachytoidity, mineral flattening) allows us to assume the general anticlinal structure of the differentiated series of the array with the southwestern vertex of the axial surface. The northeastern flank is dominated by a fall in the eastern rumbles at angles of 50-70°, and in the southwestern direction, respectively, in the western at angles of 70-90°. The generalized anti-form of the intrusive body is complicated by the development of lower-order kinematic folds and tectonic disturbances of the up-and-over or shear kinematic nature [5].

Prospects of ore content of the Khayrkhan gabbroid massif were previously associated with the development of dispersed chalcopyrite-pyrrhotite mineralization, which reveals an increase in the concentrations of platinoids (up to 0.3 g / t according to scintillation analysis). Mineralogical researches of the richest PGE samples allowed to determine the presence sperrylite, mertiit, osarsit and stibiopalladit [5]. Among the gabbroic blocks outside the main body, industrial gold contents (up to 5 g / t) were also found in the local copper sulfide accumulation zones. These data served as the basis for carrying out prospecting for precious metals that were performed by employees of Tomsk State University with the support of the Australian mining company Elayne Nominees P TY.LTD. In the course of field and laboratory studies, about 200 large-volume (3-7 kg samples) samples from sulfidated varieties of gabbroids of the Khairkhan Massif were selected and analyzed. The obtained results made it possible to establish that elevated concentrations of noble metals are fixed only in the form of local horizons (with a thickness of about 3-5 m), sustained along a strike up to 10-15 km. Such a character of ore deposits corresponds to the concept of "reef", which is used when characterizing many stratified ultramafic-mafic intrusions. In total, four such reefs with a total content of gold, palladium and platinum from 0.03 to 1.6 g / t were found within the Hayrkhan massif (table 1) [5].

The first horizon is confined to the contact of the marginal facies and the stratified series. It occupies the extreme north-eastern position in the structure of the massif. The maximum concentrations of noble metals (140-430 mg / t), as a rule, are tied to the veins of pegmatoid leucocratic gabbroids, revealing a pronounced gold specialization with an average value (Pd + Pt) / Au of about 0.2.

The second and third horizons are localized within the stratified series of mafites and are associated with melanocratic differences of the gabbro-peridotite type, plagiuverlites and olivine plagiovebsterites. These horizons show increased concentrations of gold and platinoids. Moderate accumulation of precious metals (an average of 90 mg / t) with a PGE / Au ratio of about 0.75 fixed for the second horizon. In the third "reef", the average content of useful components is slightly higher (up to 320 mg / t, and in some samples approaches industrial concentrations to 1600 mg / t) with a more significant role of platinoids (PGE / Au = 0.80). Both these horizons are traced along the entire massif, over a distance of about 10 kilometers. The estimated thickness of these horizons varies from 2 to 10 meters [5].

Table 1. The content of noble metals (ppb) in the ore-bearing horizons of the Khairkhan massif[a] [5].

Metal	Reef 1 (11)[b]			Reef 2 (9)		
	min	max	av	min	max	av
Au	29	353	134	7	90	53
Pd	5	60	23	13	129	33
Pt	5	13	36	5	18	7
Sum	40	432	163	35	234	93
(Pd+Pt)/Au	0,06	0,62	0,22	0,21	4	0,75
Metal	Reef 3 (9)			Reef 4 (6)		
	min	max	av	min	max	av
Au	9	972	178	3	36	14
Pd	5	340	98	13	70	35
Pt	5	296	44	5	24	8
Sum	51	1591	320	28	111	57
(Pd+Pt)/Au	0,29	4,67	0,8	1,1	28	3,07

[a] Analyzes were performed by assay method with ICP-MS termination in the analytical center of Alex Stuart Assayers Kyrgyzstan LTD (Bishkek, Kyrgyzstan).
[b] The number of samples taken from each horizon is indicated in parentheses.

The fourth ore horizon is located in the axial part of the massif among the olivine gabbroids at a considerable distance from the previous ones (up to 1000 meters). It is tracked in fragmentary single samples for a distance of up to 8 km. This reef is characterized by a sharp predominance of platinoids over gold (PGE / Au = 3.0) with a relatively low level of accumulation of about 60 mg / t [5].

The richest noble metals were the first and third reefs. The study of the material composition of the ore minerals of these reefs gave the following results. Pyrite, chalcopyrite and bornite are most common among sulfides. In most cases, their composition is close to stoichiometric, but reveals, nevertheless, a certain discreteness. In particular, the sulfides of the first (predominantly gold-bearing) horizon are characterized by elevated sulfur concentrations that determine the excess of this anion in the calculated formula of minerals. Electrum and native gold are also used as the actual Au mineral phase, and higher (from 3.2 to 6.8% by weight) silver and a constant admixture of platinum (0.02-0.06% by weight) are found in the bornite. This pattern suggests an increased acidity of ore-forming solutions. In the sulfides of the third (gold platinum) horizon, as a rule, a weak but constant sulfur deficit is observed against the background of increased concentrations of mercury and tellurium. The detection of a fairly rare palladium telluride temagamite Pd3HgTe3, completely confirms the general evolution of the ore-bearing hydrothermal column, where sulfur is the most mobile component [5].

In general, the ore paragenesis of the gold-platinum-bearing reefs of the Khairkhan Massif indicates a significant role for volatile components in the formation of noble metal mineralization at relatively low temperatures, i.e. real participation in their secondary enrichment of hydrothermal solutions. The observed zoning in the deposition of gold and platinoids seems to reflect their mobility at the magmatic and hydrothermal stages of the evolution of the initial basite magma under conditions of variations in the partial pressure of mercury, tellurium, sulfur and oxygen.

Thus, the object studied by us can be regarded as quite promising for noble metal mineralization. Its considerable size, taking into account the continuity of the ore-bearing horizons over considerable strike distances, allows one to assume the industrially significant reserves of platinum and gold. According to its structural-genetic type, the mineralization of the Khairkhan massif corresponds more to the low-sulfide platinum-bearing formation of the JM reef in the Styllurian massif, however its purely copper specialization resembles the Duluth complex in the state of Minnesota [5]. Considering that in most differentiated ultramafic-mafic intrusions, platinum-bearing reefs have an average thickness of about 10-30 cm, and in our case, wider ranges of the stratified gabbroide series were estimated (up to 5-10 m), while further prospecting should allow for much higher concentrations of precious metals in the rocks of this pluton.

Acknowledgements
This work was conducted as a government task of the Ministry of Education and Science of the Russian Federation, Project Number 5.2352.2017/4.6. The examinations of samples were performed on the equipment of Analytic Center of Natural Systems Geochemistry of Tomsk State University.

References
[1] Izokh A E, Polyakov G V, Krivenko A P, Bognibov V I, Bayarbiyeg L 1990 *The Gabbroid Formations of Western Mongolia* (Novosibirsk: Science, Sib. Dep.) p 269
[2] Pavlenko A C, Fedorova M E, Kovalenko V I, Kuzmin M I, Louvsandanzan B, Orlova L P, Pavlenko V C, Filippov L V 1974 Granitoid formations *Tect. of the Mong. Rep.* 210 – 34
[3] Dergunov A B, Louvsandanzan B, Pavlenko V C 1980 *Geology of Western Mongolia* (Moscow: Science) p 195
[4] Izokh A E, Polyakov G V, Anoshin G N, Golovanova N P 1991 Geochemistry of platinum metals, gold and silver in the Nomgon troctolite-anorthosite-gabbro massif (MNR) *Geochemistry* 10 114 – 17
[5] Gertner I F, Tishin P A, Bazhenov R S, Voitenko D N 2003 Prospects for the noble mineralization of the Khayrkhan gabbroid massif (Western Mongolia) *Tomsk State University Journal* 3 223 – 25

IOP Conf. Series: Earth and Environmental Science **110** (2017) 012014 doi:10.1088/1755-1315/110/1/012014

Thermodynamic modelling of metamorphic processes: state of the art in pseudosection approach

A. López-Carmona[1,2], G. Gutiérrez-Alonso[1,2], P. A. Tishin[2], I. F. Gertner[2]

[1]Department of Geology, University of Salamanca, Plaza de los Caídos s/n, 37008, Salamanca, Spain

[2]Tomsk State University, Tomsk 634050, Russia

E-mail:alioli@usal.es,gabi@usal.es,labspm@ggf.tsu.ru

Abstract. Understanding global-scale orogenic processes related to supercontinents, and their relationship to the secular evolution of the Earth's lithosphere, represent important challenges for Earth scientists today. The record of these processes is preserved in the microstructures, mineral assemblages and mineral compositions of lithospheric rocks exhumed to the Earth's surface. Given a well-characterized microstructural evolution, thermodynamic modelling is the key to quantifying changes in pressure and temperature, with the rate of these changes being provided by rock-forming and accessory mineral-based geochronology. Thus, metamorphic rocks provide Pressure–Temperature–time–deformation (P–T–t–d) data that help to parameterize orogenic processes.

1. Introduction

Equilibrium thermodynamics provides the formal mathematical framework that links measurable variables (whole rock or domainal compositions, mineral and fluid/melt inclusion compositions) with intensive variables (pressure, temperature and chemical potentials). This approach comprisesthe field of thermobarometry [1]. Thermobarometry has evolved significantly in the last twenty-five years with the development of internally consistent thermodynamic datasets and sophisticated thermodynamic mixing models for phases of interest. These advances have enabled the calculation of equilibrium phase diagrams for complex chemical systems, close to those in nature, for a diversity of variables. Such diagrams are called pseudosectionsfor the reasons given by Powell *et al.* [2]. Critically, inverse modelling using pseudosections allows using all available compositional informationfor both, the system and the individual minerals, which is not the case for conventional thermobarometry, where the chemical system is largely ignored. This development represents one of the most important advances of modern metamorphic petrology. The numerical modelling of phase relations requires a set of thermodynamic data for end-members, a description of the activity-composition (a-x) relations (the solid- or liquid-solution models) and a software to perform calculations.

The utility of pseudosectionis that they permit natural observations, and interpretations from thin sections, to be explained in terms of the evolving P–T conditions, within the context of a structural evolution, and from place to place. When combined with petrologically-linked age data [3], the resulting P–T–t–d paths provide essential input into constraining the processes involved in mountain-building and the evolution of continental lithosphere ([4]; *Figure. 1*).

Despite these advances, there remain considerable uncertainties in both, the thermodynamic framework used (end-member thermodynamics and a-x models) and more importantly, in estimating the volume of equilibration and changes in rock composition during metamorphism. These issues require the approach to be used with care. Thus, the results of any calculations are best viewed as an independent theoretical framework in which to interpret a rock rather than an exact reproduction of its evolution [5].

Thermodynamic modelling of petrological systems is only a tool, but because the pseudosection approach uses the maximum amount of available information, it is extremely useful and, today is an essential tool in the parameterization of metamorphic histories.

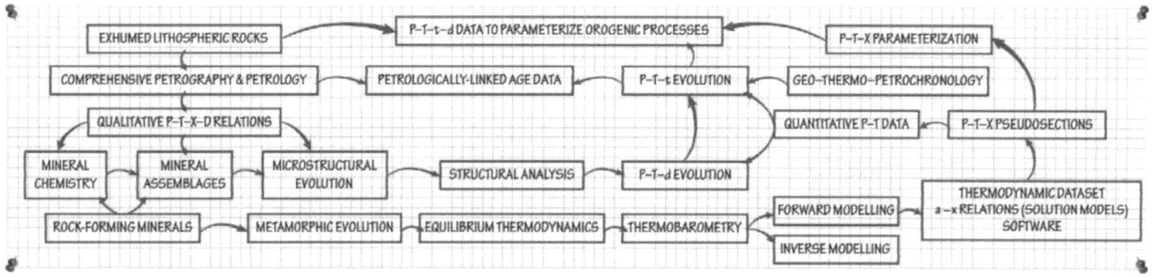

Figure 1.Flow chart for P–T–t–d path calculations

2. From inverse to forward modelling: a short review

Knowledge on thermodynamics is a key to understand physicochemical changes occurring in Earth's lithosphere, producing metamorphic rocks. All systems naturally tend to the minimum energetic conFigureuration, i.e. equilibrium. Chemical equilibrium implies no change over time. Thus, equilibration conditions attained by a mineral assemblage would preserve the characteristics of the phases involved in the system at a certain moment, e.g. the P–T formation conditions of a metamorphic rock. As soon as the conditions (e.g. P or T) change, the system will immediately tend to a new equilibrium state. However, since diffusion is involved, the scale at which the equilibrium will be achieved depends on the particular conditions (especially the temperature), on the character of the phases involved (e.g. diffusion in fluids is much faster than in solids), and on the available time. If conditions change rapidly, the scale of equilibration will be small. However, it can be argued that even in this case the system will develop local equilibria – small domains where equilibrium is achieved or approached [6, 7]. The equilibrium phase (mineral) assemblages and compositions can then be identified through a rigorous petrological analysis, and the principles of equilibrium thermodynamics may be applied in order to understand the rock, and determine its conditions of formation [8, 9].

The first to apply the concept of chemical equilibrium in metamorphic rocks was Goldschmidt [10], suggesting through his research that thermodynamics was an essential tool for the study of metamorphism. He also pointed that the development of distinctive minerals in each type of metamorphic rock could be influenced by variations in depth (pressure), temperature and changes in the bulk rock chemical composition. For long, it has only been possible to infer qualitative comparisons on the P–T formation conditions of metamorphic rocks through the presence of index minerals, or based on the assemblages resulting by the metamorphism of a wide range of protoliths under a particular facies.

Chemical thermodynamics is the formal mathematical framework that links measurable variables (i.e. mineral composition) with those that cannot be directly measured [2, 11], using mathematical expressions based on the application of three laws: the ideal gas equation, and the first two laws of thermodynamics. This enables deciphering the P–T history of metamorphic rocks, which is the main aim of geothermobarometry.

Experimentally determined phase equilibrium lie at the heart of nearly all thermobarometric methods. The experimental calibration of metamorphic reactions is the so-called***conventional or***

classical inverse modelling thermobarometry, and was the only numerical method used until the late 80s for the quantification of the P–T formation conditions in a metamorphic rock [12, 13]. It involves the use of two (end-members) mineral reactions, ideally one highly pressure-dependent (geobarometer) and another strongly dependent on temperature (geothermometer). The simultaneous solution of both reactions in the P–T space provides a quantitative estimation to be valid only if minerals are in equilibrium.

Although this technique may provide reasonable results, likewise requires important simplifications and show several limitations (errors derived from thermodynamic consistency, from the experimental calibrations or from the assumed equilibrium condition; [14, 15]). One of the most important limitations is the use of single sets of equilibria to calculate the P–T conditions under which a rock equilibrated.

A rock contains multiple minerals and is affected by several reactions. Then, increasing these information increases the accuracy on the estimations. Thus, to overcome these issues, and allow calculation of all equilibriums between a set of not only mineral end-member data, but also to activity-composition data (together with calorimetric or heat capacity and volume data among others), thermodynamic databases were created [16].

The so-called***multi-equilibrium thermobarometry*** begins to develop during the 90s for this purpose. It involves the use of internally consistent thermodynamic datasets for a wide range of minerals and chemical systems, implying a breakthrough on the research of phase relationship for mineral assemblages. These datasets are known as internally consistent because the thermodynamic parameters regressed for each phase depend on the others to produce equilibria which best fit the experimental determinations. The two most commonly used internally consistent databases in metamorphic petrology are those compiled by Holland and Powell [1, 14, 15, 17, 18] and Berman [15, 19, 20]. Currently, several thermodynamic calculation software use internally consistent datasets for modelling phase equilibrium in petrological systems. THERMOCALC [17], Perple_X [21] and Theriak-Domino [22] are most common in metamorphic petrology.

3. Thermodynamic calculation software

At their heart, these computer programs are an internally consistent thermodynamic datasets that allows using thermodynamic data in a variety of ways, solving how to handle multiple dimensions represented in two.

Perple_X is a command-line-driven software package that performs Gibbs energy minimization to create phase diagrams and pseudosections.

Theriak-Domino is a suite of programs to calculate and plot thermodynamic functions, equilibrium assemblages and rock-specific equilibrium assemblage diagrams (i.e. pseudosections).

Straightforward application is granted by both Perple_X and Theriak-Domino, which are based on fast computing algorithms and has the ability to calculate and plot "ready to use" phase diagrams in reasonably short times without user intervention, performing calculations with a wide range of thermodynamic databases.

On the other hand, THERMOCALCuses a nonlinear equation solver to calculate user-specified equilibria via only its own internally consistent thermodynamic dataset [14, 16, 17, 18]. It handles mineral equilibria through inverse modelling (geobarometry/thermometry using average P–T), and forward modelling (calculating phase diagrams for model systems, i.e. P–T–X–μpseudosections). The average P–T method of THERMOCALCconsiders an independent set of equilibria (from which all the other possible equilibria can be calculated) and a robust statistical treatment to indicate quantitatively the uncertainty on the result [1]. Create a diagram in THERMOCALCis laborious as curves must be built up one by one and manually combined, and can be extremely time-consuming without granting success. On the other hand, it allows understanding how phase diagrams are calculated (user monitors the *Phase Rule*, *Schreinemaker's analysis*, and how variance changes across field boundaries).

Summarizing, numerical modelling of phase relations requires a set of thermodynamic data for mineral end-members, a description of the activity-composition relations and a software to perform the calculations.

IOP Conf. Series: Earth and Environmental Science **110** (2017) 012014 doi:10.1088/1755-1315/110/1/012014

Among the most frequent softwareimplemented in metamorphic petrology, the rapidity and relative easiness of use, together with the possibility of using multiple datasets, make Perple_X and Theriak-Domino interesting alternatives for calculating phase diagrams in relatively simple chemical systems. Calculating diagrams in complex chemical systems requires considerable understanding on the operation of both.THERMOCALCmay be the best choice for calculating accurate phase diagrams in complex chemical systems, or with phases requiring complex mixing models (see example in *Figure.* 2).

4. Pseudosection approach

The advantage of multi-equilibrium thermodynamic calculation programs resides in their ability to generate pseudosections.

After the pioneering work of Hensen [24], and subsequent inquiries by some isolated workers [25], it was the development of the internally-consistent thermodynamic datasets and the software THERMOCALCat the end of the 1980s, what allowed the widespread use of pseudosections for metamorphic rocks[14, 17, 26].

Pseudosectionsare phase diagrams that show stability fields of different equilibrium mineral assemblages for a particular bulk-rock composition [1]. Considering the rock chemical composition (X) provides added constraints unavailable in conventional thermobarometry, such as phase's coexistence, when phases (minerals, fluids or liquids) in the mineral assemblage are no longer stable, or when additional phases join the mineral assemblage. Through the comparison between the natural observations and the results of the numerical model, pseudosections allow understanding the metamorphic evolution of a specific rock as P, T and X change. Pseudosections have the power to provide valuable additional thermobarometric information and do not depend on measurements of mineral composition. Likewise, even partly preserved assemblages (e.g. inclusions or partly pseudomorphed minerals) may be useful to infer segments of the P–T path of a rock, before and after the main equilibration stage [27, 28, 29].

Moreover, at each point of a pseudosection, the composition and proportion of all phases may be calculated. Then, pseudosections allow estimating specific equilibria by calculating the P–T stability of particular mineral compositions with respect to the bulk rock composition, i.e. allows calculating mineral composition (and proportion, expressed in relative molar abundance) isopleths. The value set for the isopleth compositional variable derives from the EPMA analysis of that particular mineral in a specific sample [32, 33]. Isopleths then are used to further constrain equilibration conditions. Isopleths are particularly useful when applied on understanding chemical zoning of minerals. These isolines allow to quantitatively investigate the effects of crystal fractionation on phase stability. Therefore, applying pseudosections to real rocks provides P–T estimates at which particular minerals grew within a rock. However, pseudosections are not fool proof and show certain drawbacks that are shortly addressed below.

When calculated using a thermodynamic dataset, pseudosections represent the result of a forward numerical modelling, and are independent on the petrographic observations in the specific rock. An important restriction when using multi-equilibrium tools is implicit in the internally consistent thermodynamic dataset. Knowledge on the thermodynamic parameters for mineral phases used in thermobarometry has largely improved since databases appeared, but there is still much to be done. Occasionally, the use of provisional data in the internally consistent database is subject to unknown errors that can slightly alter phase's stability, and thus the P–T estimates. Nevertheless, even the best software cannot overcome problems arising from missing or unconfident thermodynamic data.

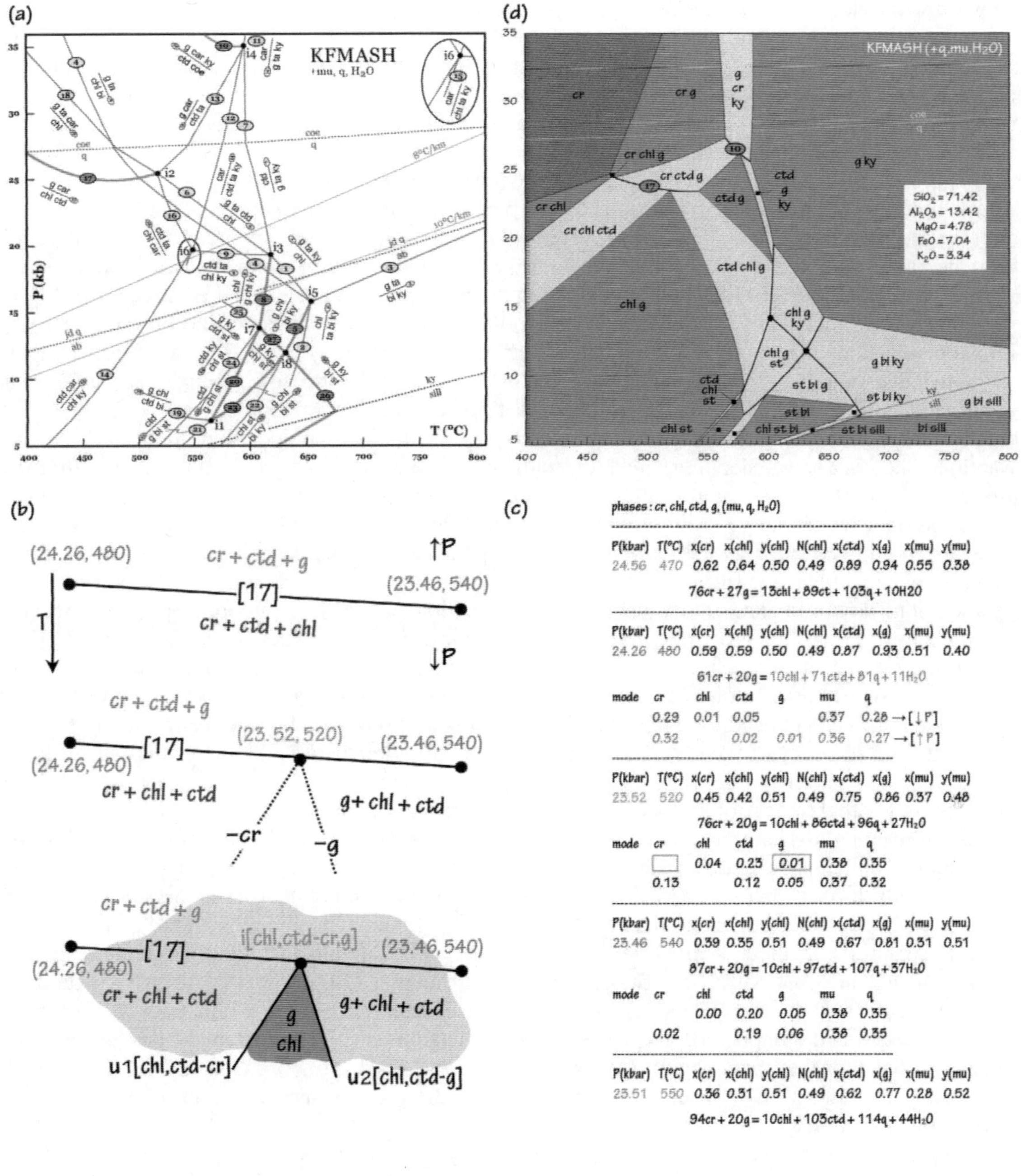

Figure 2. (a)Petrogenetic grid for various high-pressure metapelites. Black dots and blue lines represent the invariant points and univariant reactions "seen" by sample BS, used as an example composition. Modified after Wei and Powell (2003); (b) user's draft for reaction [23]; (c)THERMOCALC "output" for reaction [17]; (d) P–T pseudosection in the KFMASH system for sample BS. Black dots and highlighted black lines represent the invariant points and univariant reactions "seen" by the BS composition. Modified after López-Carmona *et al.* [31]

Two crucial elections have to be carefully considered before start working with THERMOCALC: choosing the chemical system, and the right bulk rock composition. The chosen***chemical system*** must be as close as possible to nature. Ignoring a phase component (such as e.g. apatite) can artificially modify the bulk rock composition. This can be solved applying the appropriate corrections (e.g. Ca in apatitecorrection $CaO=CaO-3.33*P_2O_5$). The available chemical system usually depends on the accessible solid solution models. In general, simpler systems may be used to forward modelling theoretical scenarios, whereas inverse modelling requires larger systems. However, complex systems imply complex mixing models for solid-solution phases that may be a source of additional errors. Choosing the right ***bulk rock composition*** (expressed in mol %) is decisive since the pseudosection would be only valid for such election. Nevertheless, as many minor elements are systematically ignored (e.g. P_2O_5 or BO_3), bulk rock compositions in THERMOCALCare always (reliable) approximations to real rocks. The effects of overlooking these components depend on the sensitivity with which the adjustments on the bulk composition have been done, which relies on the petrographic user-knowledge of the sample considered. THERMOCALCmanages the so-called***effective bulk rock composition***, which is that available to the reacting assemblage of phases (see Stüwe, 1997). Metamorphic rocks are commonly heterogeneous; this effective composition then may change throughout the history of metamorphic recrystallization, and largely depends on the ***equilibration volumes***. Two mechanisms can alter the effective bulk-composition of a rock during metamorphism, the open system behaviour (e.g. deformation or fluid effects) and the successive (re)equilibrations. The occurrence of ***chemically zoned phases***(typically garnet) in a sample implies crystal fractionation. Thus, the chemical reservoir from which minerals are growing (whole-rock composition before garnet crystallization) would change into a new rock composition devoid of the components that reacted while the zoned crystal was growing (effective rock composition after garnet crystallization; [34, 35]).

Accordingly, the effective bulk-composition of a rock through its metamorphic history may be characterised by a multitude of relatively small equilibration volumes [9, 34, 36, 37, 38]. Currently two main approaches are used to obtain the effective-bulk composition, but the method used depends on the scale and accuracy of the problem to be solved. Small compositional domains may be analysed by quantitative X-ray mapping with an EPMA over a selected area of the sample [39; 40]. This method allows both, the possibility to selectively eliminate individual phases that are interpreted not to have been reacting during recrystallization of surrounding phases, and the chance to eliminate compositional domains that are distinctly different from that of the region of primary interest in compositionally layered samples [38]. Equilibrium is generally thought to be attained at the hand-sample scale if no clear evidence to the contrary is found [41; 42]. Theoretically, if the compositional domains of interest are about 1 cm wide or greater, up to hand-sample scale, individual domains can be separated and analysed with X-ray fluorescence spectroscopy (XRF). THERMOCALCis provided of a facility (the rbi "read bulk info" code) that allows defining or changing a bulk rock composition through the mode and composition of phases. This application can be used to investigate the internal buffering of fluids or the fractionation of bulk rock compositions due to porphyroblast growth, as well as to assess the loss of melt from a bulk rock composition or to make an approximate protolith composition from a melt depleted rock.

Finally, the difficulty to determine the proportion of significant components such as H_2O and Fe_2O_3 ("O" in THERMOCALC) in the bulk composition, by routine analytical techniques, raises important uncertainties when modelling phase equilibria, and can largely influence phase relations and hence,P–T estimates. The amount of H_2O has to generally be guessed, if not in excess, whereas the Fe_2O_3 can be determined. Total iron (expressed as $Fe_2O_{3(T)}$) can be measured by XRF and FeO can be analysed by wet chemical titration. Then, the amount of Fe_2O_3 is calculated stoichiometricazlly.

Acknowledgments

This work was conducted as a government task of the Ministry of Education and Science of the Russian Federation, Project Number 5.2352.2017/4.6

References

[1] Powell R & Holland T J B 2008 Onthermobarometry *Journal of Metamorphic Geology* **26** 155–179

[2] Powell R Holland T J B & Worley B 1998 Calculating phase diagrams involving solid solutions via non-linear equations, with examples using THERMOCALC *Journal of Metamorphic Geology* **16** 577–588

[3] Kohn M.J Engi M &Lanari P 2017 Significant Ages: an introduction to Petrochronology *Reviews in Mineralogy and Geochemistry* **83(1)** p 1–12

[4] Brown M 2014 The contribution of metamorphic petrology to understanding lithosphere evolution and geodynamics *Geoscience Frontiers* **5** p 553–569

[5] White RW Powell R & Johnson T E 2014 The effect of Mn on mineral stability in metapelites revisited: new a–x relations for manganese-bearing minerals *Journal of Metamorphic Geology* **32** p 809–828

[6] Korzhinskii S 1959 *Physico-chemical basis of the analysis of the paragenesis of minerals* Consultants Bureau, New York 142 p.

[7] Thompson J B 1959 *Local equilibrium in metasomatic processes* In: Abelson, P.H. (ed.) Researches in Geochemistry Wiley New York p 427– 457

[8] Powell R 1978 *Equilibrium thermodynamics in petrology* (An Introduction) Harper & Row London 284 p

[9] Pitra P 2011 Numerical modelling of phase equilibria and its contribution to understanding orogenic processes. Mémoire présenté en vue d´obtenir l´Habilitation à diriger des recherches66 p Unpublished

[10] Goldschmidt V M 1911 *Die Kontakt metamorphism Kristianiagebiet Vidensk Skrifter* L Mat.-Naturv K 11

[11] Wood B J & Fraser D G 1977 *Elementary Thermodynamics for Geologists* Oxford University Press, Cary, North Carolina 318 p

[12] Hodges K W 1991 Pressure–temperature–time paths Annual *Review of Earth and Planetary Science* **19** p 207–236

[13] Spear F S 1993 *Metamorphic Phase Equilibria and Pressure–Temperature–Time* Paths. 1st edition Mineralogical Society of America Washington D.C 799 p

[14] Holland T J B & Powell R 1985 An internally consistent thermodynamic dataset with uncertainties and correlations 2. Data and results *Journal of Metamorphic Geology* **3** p 343–370

[15] Berman R G 1991 Thermobarometry using multi–equilibrium calculations: a new technique, with petrological applications In: Gordon, T.M. & Martin, R.F. (eds). Quantitative methods in petrology: an issue in honor of Hugh J. Greenwood *Canadian Mineralogist* **29** p 833–855

[16] Holland T J B & Powell R 1990 An enlarged and updated internally consistent ther–modynamic dataset with uncertainties and correlations: the system K_2O–Na_2O–CaO–MgO–MnO–FeO–Fe_2O_3–Al_2O_3–TiO_2–SiO_2–C–H_2–O_2. *Journal of Metamorphic Geology* **8** p 89–124

[17] Holland T J B & Powell R 1998 An internally consistent thermodynamic data set for phases of petrological interest *Journal of Metamorphic Geology* **16** p 309–343

[18] Holland T J B & Powell R 2011 An improved and extended internally consistent thermodynamic dataset for phases of petrological interest, involving a new equation of state for solids *Journal of Metamorphic Geology* **29** p 333–383

[19] Berman R G 1988 Internally–consistent thermodynamic data for minerals in the system Na_2O–K_2O–CaO–MgO–FeO–Fe_2O_3–Al_2O_3–SiO_2–TiO_2–H_2O–CO_2 *Journal of Petrology* **29** p 445–522

[20] Berman R G 2007 WinTWQ (version 2.3) a software package for performing internally–consistent thermobarometric calculations Geological Survey of Canada Open File 5462 (ed. 2.32) 41 p

[21] Connolly J A D 1990 Multivariable phase diagrams; an algorithm based on generalized thermodynamics *American Journal of Sciences* **290** p 666–718

[22] De Capitani C & Brown T H 1987 The computation of chemical equilibrium in complex systems containing non–ideal solutions *Geochimica et Cosmochimica Acta***52** p 639–2652

[23] Kohn M 2016 Metamorphic chronology–a tool for all ages: past achievements and future prospects *American Mineralogist* **101** p 25–42

[24] Hensen B J 1971 Theoretical phase relations involving garnet and cordierite in the system MgO–FeO–Al_2O_3 –SiO_2 *Contributions to Mineralogy and Petrology* **33**, p 191–214

[25] Hudson N F C 1980 Regional metamorphism of some Dalradianpelites in the Buchan Area NE Scotland *Contributions to Mineralogy and Petrology* **73** p 39–51

[26] Guiraud M Holland T & Powell R 1990 Calculated mineral equilibria in the greenschist–blueschist–eclogitefacies in Na_2O–FeO–MgO–Al_2O_3–SiO_2–H_2O: methods, results and geological applications *Contributions to Mineralogy and Petrology* **104** p 85–98

[27] López–Carmona, AKusky T M Santosh M &Abati J 2011 P–T and structural constraints of lawsonite and epidoteblueschists from Liberty Creek and Seldovia: Tectonic implications for early stages of subduction along the southern Alaska convergent margin *Lithos* **121**, p 100–116

[28] López–Carmona A Pitra P &Abati J 2013Blueschist–faciesmetapelites from the Malpica–Tui Unit (NW Iberian Massif): phase equilibria modelling and H_2O and Fe_2O_3 influence in high–pressure assembalges *Journal of Metamorphic Geology* **31** p 263–280

[29] López-Carmona A Abati J Pitra P & Lee J K W 2014 Retrogressed lawsoniteblueschists from the NW Iberian Massif: P–T–t constraints from thermodynamic modelling and $^{40}Ar/^{39}Ar$ geochronology *Contributions to Mineralogy and Petrology* **167(3)** p 1–20

[30] Powell R Guiraud M & White R W 2005 Truth and beauty in metamorphic mineral equilibria: conjugate variables and phase diagrams *Canadian Mineralogist* **43** p 21–33

[31] López–Carmona A Abati J &Reche J 2010Petrologic modeling of chloritoid–glaucophane schists from the NW Iberian Massif *Gondwana Research* **17** p 377–391

[32] Evans T P 2004[a]A method for calculating effective bulk composition modification due to crystal fractionation in garnet–bearing schist: implications for isopleththermobarometry *Journal of Metamorphic Geology* **22** p 547–557

[33] Evans T P 2004[b]Reconciling the structural and metamorphic record of orogeny in central western New Hampshire through microstructure and garnet isopleththermobarometry Unpublished PhD Thesis, James Cook University 89 p

[34] Stüwe K 1997 Effective bulk composition changes due to cooling: a model predicting complexities in retrograde reaction textures *Contributions to Mineralogy and Petrology* **129** p 43–52

[35] Hollister L S 1966 *Garnet zoning: an interpretation based on the Rayleigh fractionation model Science* **154** p 1647–1651

[36] Atherton M P 1968The variation in garnet, biotite, and chlorite composition in medium grade pelitic rocks from the Dalradian, Scotland, with particular reference to the zonation in garnet *Contributions to Mineralogy and Petrology* 18 p 347–371

[37] Spear F S 1991 On the interpretation of peak metamorphic temperatures in light of garnet diffusion during cooling *Journal of Metamorphic Geology* **9** p 379–388

[38] Tinkham D K & Ghent E D 2005 Estimating P–T conditions of garnet growth with isochemical phase diagram sections and the problem of effective bulk–composition *Canadian Mineralogist* **43** p 35–50

[39] Clarke G L Daczko N R &Nockolds C 2001 A method for applying matrix corrections to X–ray intensity maps using the Bence–Albee algorithm and Matlab *Journal of Metamorphic Geology* **19** p 635–644

[40] Marmo B.A Clarke G L & Powell R 2002 Fractionation of bulk rock composition due to porphyroblast growth: effects on eclogitefacies mineral equilibria, Pam Peninsula, New Caledonia *Journal of Metamorphic Geology* **20** p 151–165

[41] Chernoff C B & Carlson W D 1997 Disequilibrium for Ca during growth of pelitic garnet *Journal of Metamorphic Geology* **15(4)** p 421–438

[42] Hirsch D M Prior D J & Carlson W D 2003 An overgrowth model to explain multiple, dispersed high-Mn regions in the cores of garnet porphyroblasts *American Mineralogist* **88** p 131–141

IOP Conf. Series: Earth and Environmental Science **110** (2017) 012015 doi:10.1088/1755-1315/110/1/012015

Magma storage constrains by compositional zoning of plagioclase from dacites of the caldera forming eruptions of Vetrovoy Isthmus and Lvinaya Past' Bay (Iturup Island, Kurile Islands)

I A Maksimovich[1,2], S Z Smirnov[2,3], A A Kotov[1,2], T Yu Timina[2], A V Shevko[2]

[1]Novosibirsk State University, Pirogova, 1/2, 630090, Novosibirsk, Russia
[2]V S Sobolev Institute of Geology and Mineralogy SB RAS, pr. ak. Koptyuga, 3, 630090, Novosibirsk, Russia
[3]Tomsk State University, pr. Lenina, 36, 634050, Tomsk, Russia

Abstract. The Vetrovoy Isthmus and the Lvinaya Past' Bay on the Iturup island (Kuril island arc) are the results of large Plinian eruptions of compositionally similar dacitic magmas. This study is devoted to a comparative analysis of the storage and crystallization conditions for magma reservoirs, which were a source of large-scale explosive eruptions. The plagioclase is most informative mineral in studying of the melt evolution. The studied plagioclases possess a complex zoning patterns, which are not typical for silicic rocks in island-arc systems. It was shown that increase of Ca in the plagioclase up to unusually high An_{95} is related to increase of H_2O pressure in both volcanic magma chambers. The study revealed that minerals of the Vetrovoy Isthmus and Lvinaya Past' crystallized from compositionally similar melts. Despite the compositional similarity of the melts, the phenocryst assemblage of the Lvinaya Past' differs from the Vetrovoy Isthmus by the presence of the amphibole, which indicates that the pressure in the magmatic chamber exceeded 1-2 kbar at a 4-6 wt. % of H_2O in the melt. The rocks of the Vetrovoy Isthmus do not contain amphibole phenocrysts, but melt and fluid inclusions assemblages in plagioclase demonstrate that the magma degassed in the course of evolution. This is an indication that the pressure did not exceed significantly 1-2 kbar.

1. Introduction

The Kurile-Kamchatka island-arc system stretching from the Kamchatka peninsula to the Hokkaido Island is an area of modern active island-arc volcanism. A wide spectrum of volcanic rocks: from basalts to rhyolites is represented on the islands of the Great Kurile island chain. The most common rocks are andesites. Silicic rocks are subordinate to the mafic and intermediate rocks. Their proportion increases southward. Nevertheless, the silicic rocks of the Kurile Islands are poorly studied and are of great importance for understanding the evolution of island-arc magmas and the formation of the Kurile island-arc crust.

Lvinaya Past' (LP) caldera is located in the southern part of Iturup Island. The northwestern part of the caldera ridge is absent. Caldera's interior is filled with the waters of the Sea of Okhotsk. The caldera has the shape of an ellipse. Different erupted volume estimates vary from 20 to 170 km^3 [1-2]; The maximum diameter of the caldera is 8 km. Fragments of the somma form the Bezvodnyi (Waterless) Ridge, which raises up to 350-400 meters above the sea level. The inner walls of the caldera are composed of alternating lava and pyroclastic material of the pre-caldera stratovolcano [3].

Dacitic pumaceous tuffs are developed on a periphery of the broken stratovolcano and cover a wide area between Breutarube volcano on the south and the Golubka Mountain on the north. The size of Lvinaya Past' caldera is comparable to the caldera of Kurilskoe Lake, which is associated with the most powerful explosive eruption in the Holocene on the Kamchatka peninsula [4].

The Vetrovoy Isthmus (VI) is located in the northern part of the island and separates the Medvezhiy Peninsula from the rest of the island. The entire area of the VI is covered by pumice deposits with a a thickness more than 264 meters. Formation of these rocks is associated with a caldera eruption that occurred in the Late Pleistocene. According to the state geological mapping [5], the pyroclastic rocks extend for 10 km to the Sea of Okhotsk bottom in the north-west direction. Remnants of large caldera have are more than 7 km in diameter. with a maximum height of 264 m. The erupted material volume was estimated by I.V. Melekestsev [6] at about 100 km^3.

The aim of this article is to determine the magma storage conditions that preceded catastrophic eruptions of VI and LP on the basis of comparative analysis of chemical zoning and inclusions of fluids and melts in the plagioclase phenocrysts.

2. Research methods

The bulk composition of the rocks was determined by X-ray fluorescence (XRF) and ICP-MS analyzes at the Center for Collective Use of Scientific Equipment for Multielement and Isotopic Studies in IGM SB RAS (Novosibirsk).

2.1. X-Ray spectral analysis.

Energy-dispersive electron-microprobe analysis (EDS) on TESCAN MIRA-3 LMU SEM equipped with a microanalysis system, including the INCA Energy 450+ software and X-Max 80 analyzer, was used for determination of the compositions of minerals and melt inclusion glasses in phenocrysts of pumices. The accelerating voltage was 20 kV, probe current 1.4 - 1.6 nA, the diameter of the focused electron beam was 10 nm. The counting time was 60 seconds. With these parameters, the detection limits of different elements vary within 0.1-0.3 wt. %. The analysis of the melt inclusion glass was carried out by scanning the 10x10 μm area in order to reduce the effect of sodium loss [7-9]. For each inclusion, three glass analyzes and one analysis of the host mineral near the inclusion were performed. The analysis of the host mineral was carried out to verify the correctness. If necessary, a correction was applied.

3. Mineralogy and petrography

The rocks of the LP and VI are presented by white or light gray strongly porous pumice chunks and volcanic sands. The pumices contain about 25-30% phenocrysts. Porphyritic phenocrysts are immersed in the vitreous silicic (SiO_2 73-76 wt.%) groundmass with a fibrous texture. The phenocrysts are presented by plagioclase, quartz, augite (Wol 0.43-0.44, Mg # 0.73-0.74), hypersthene (Mg # 0.62-0.64) and Fe-Ti oxides [10]. Major feature that differs LP phenocryst assemblage from that of VI is a presence of amphibole. Apatite does not occur as phenocrysts, but is abundant in as mineral inclusions in all minerals. Sulfides of Fe and Cu were found also as mineral inclusions in mafic minerals and Fe-Ti oxides.

The composition of the plagioclase is sensitive to changes in the PTX conditions. This is manifested in compositional zoning. The phenocrysts of the plagioclase of both volcanic centers usually form well-shaped transparent crystals with abundant inclusions. Plagioclase dominates the both phenocryst assemblages (more than 50-60 % of all phenocrysts). The K_2O content does not exceed 0.17 wt. %. Compositions of phenocrysts vary from andesine to labrador and, less often, to anorthite (An_{41-95}) in VI rocks and from andesine to labrador (An_{40-85}) in the LP rocks.

The plagioclases possess complex patchy and concentrically zoned patterns [11-12] (figure1). Patchy plagioclase consists of domains with contrast compositions. In VI plagioclase they belong to An_{84}-An_{95} and An_{41}-An_{55} (figure1 b). In the LP plagioclases – to An_{79}-An_{85} and An_{40}-An_{55} (figure1 a). There is a gap between An_{70-80} and An_{50} in the domain compositions. There are no obvious signs of skeletal growth in the both LP and VI patchy plagioclase.

Oppositely to patchy plagioclase compositions gradually change in the concentrically zoned one from An_{95} to An_{41} on VI plagioclase and from An_{85} to An_{41} in the LP plagioclase. When both patterns are presented in the same crystal the patchy one composes the core and the concentrically zoned form a periphery. The highest Ca contents are confined to the boundary separating the patchy and concentric-zoned parts (figure 1 a, b).

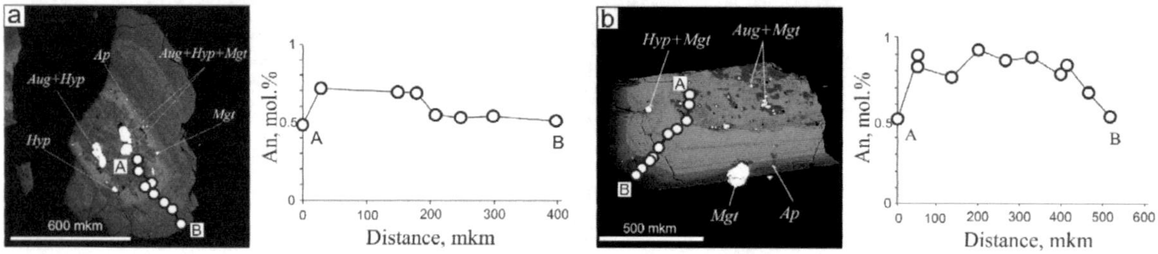

Figure 1. BSE-images and EDS profiles of plagioclase phenocrysts from LP (a) and VI (b).

4. Mineral-forming media

In the phenocrysts of plagioclase, a large number of melt inclusions (MI) were found. They are mostly distributed along the growth zones or form irregular "clouds". By the phase compositions at room temperature, MI in VI plagioclases can be divided into several groups: single-phase, containing only glass (figure 2 a); two-phase, containing glass and shrinkage or gas bubbles (figure 2 b); multiphase, containing glass, crystals and a gas bubble (figure 2 c). Mineral phases in the multiphase inclusions are represented by apatite, magnetite, ilmenite, pyrrhotite, hypersthene and augite. All these minerals occur as crystalline inclusions and are not daughter phases of MIs. Apatite, magnetite, ilmenite and augite are found in patchy parts of the VI plagioclase as crystalline inclusions. In the vast majority of cases, the inclusion of hypersthene occurs only in the concentrically-zoned parts of the plagioclase.

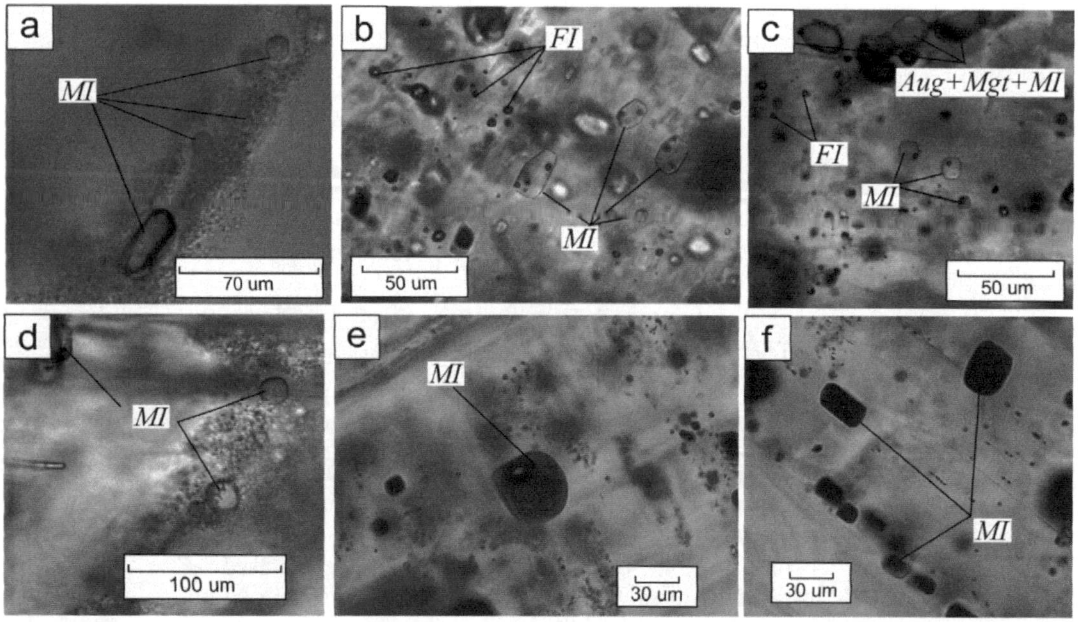

Figure 2. Melt inclusions in transmitted light at room temperature; a-c - MI in plagioclases of the Vetrovoy Isthmus; d-f - MI in plagioclases of Lvinaya Past'; a, d - single-phase vitreous MIs, b, c - two-phase MIs, e, f – devitrified.

The LP plagioclases contain all the types of MIs (figure 2 d, e) reported above along with finely crystallized inclusions (figure 2 f), in which the crystalline phases appear as a result of the devitrification of glass. Crystalline inclusions in the LP plagioclases are represented by apatite, magnetite, ilmenite, pyrrhotine, hypersthene, augite, amphibole and quartz. Unlike VI, hypersthene occurs in association with augite in patchy cores of the LP plagioclase. Inclusions of the amphibole are extremely rare. Amphibole in an intergrowths with pyroxenes was found on the boundary between the patchy core and the concentrically zoned periphery of one of the phenocrysts of plagioclase with An_{70}. Inclusions of quartz were found only in the low-calcium outer zones of phenocrysts.

Fluid inclusions (FI) are found only in the VI plagioclases only in areas with a high Ca contents. At room temperature, FIs contain two phases: low-density CO_2 with a thin rim of liquid water (figure 3 a). FIs always appear in the association with MIs (figure 3 b).

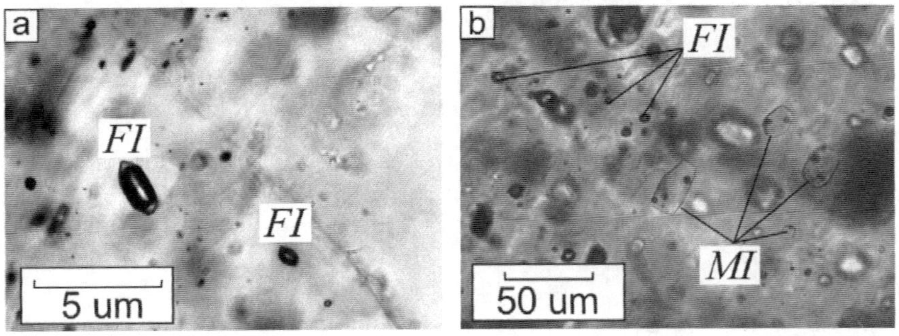

FIgure 3. Fluid inclusions in plagioclase phenocrysts from the pumice of the Vetrovoy Isthmus at room temperature (a, b); fluid and melt inclusion assemblage (b).

4.1. Melt inclusions composition

The rocks of both eruptions correspond to dacites with normal alkalinity (figure 4). The MIs compositions in the LP and VI plagioclases are similar to groundmass glass and correspond to the high-alumina (A / CNK 1.0-1.2 mol.%) rhyolite (SiO_2 - 74-75 wt.%) with normal alkalinity, low potassium (K_2O 1.6-2.0 wt.%) (figure 4) and high Cl content (0.2-0.4 wt. %).

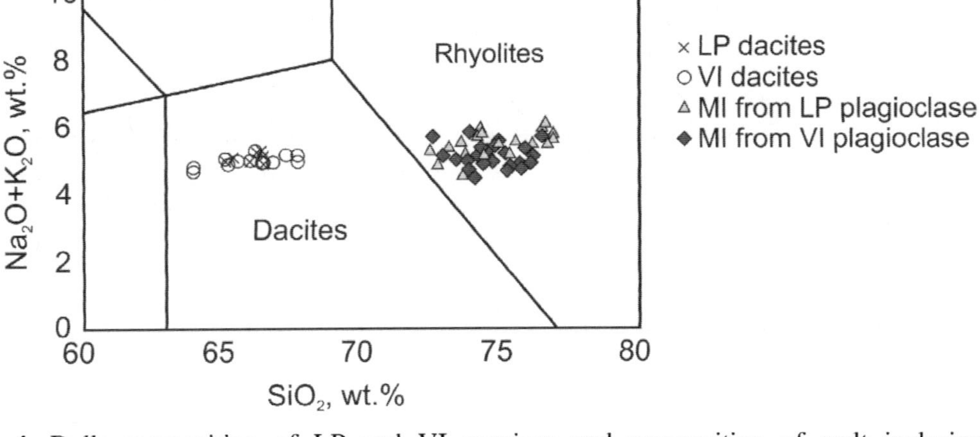

Figure 4. Bulk composition of LP and VI pumices and composition of melt inclusions from plagioclase phenocrysts.

Compositions of the plagioclase MIs shown in (figure5) demonstrate positive correlation between CaO and Al_2O_3 within narrow concentration range and slight positive or no correlation between FeO and Al_2O_3 contents. These regularities are similar to MIs in other minerals of phenocryst assemblages both from LP and VI [10]. This means that the trapped melts changed their compositions due to crystallization of Ca-Al rich mineral, e.g. plagioclase, and were only slightly or not affected by crystallization of mafic minerals. Compositional similarity with MIs in other minerals allows conclusion that this regularity reflects plagioclase fractionation in the reservoir rather than crystallization of plagioclase on inclusion walls after entrapment. Thus MI glass compositions could be considered as compositions of the parent melt.

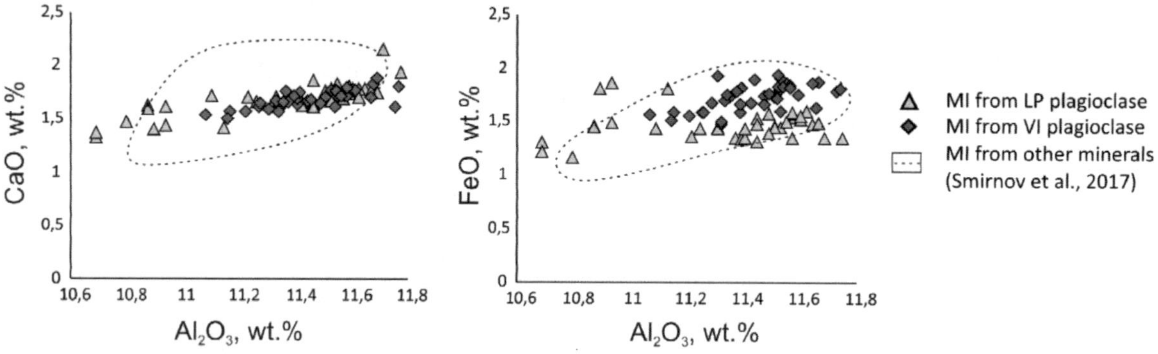

Figure 5. Composition variations of the melt inclusions glasses in plagioclases for the Lvinaya Past' caldera and Vetrovoy Isthmus. Dotted line - compositions of glasses of melt inclusions from other minerals according to Smirnov et al. (2017).

The water content in the VI and LP melts was estimated by secondary-ion mass-spectrometry (SIMS). The water content of MIs in the VI plagioclase varies from 2.3 to 5.3 wt. %, while for MIs in the LP plagioclase – within 4.5 – 4.9 wt. %. Raman spectroscopy demonstrated that water contents of MIs in the VI plagioclase varies from 3.5 to 6.0 wt.%, which agrees well with the data of SIMS.

5. Mineral thermometry

Crystallization temperatures were estimated using plagioclase-liquid [13-14] and two-pyroxene geothermometers [15]. In order to estimate the temperature by the plagioclase-liquid geothermometer, the compositions of melt inclusion and plagioclase from the zone containing the inclusion were taken into account. The calculated temperatures for both LP and VI plagioclases at 1 kbar and 4 wt.% H_2O were 843 to 868°C. For 5 wt.% H_2O – 825 - 845°C, and for 6 wt.% H_2O – 800-823°C.

Two-pyroxene temperatures at 1 kbar were 832 - 848°C for VI and 845-860°C for LP, which agrees well with data obtained from the plagioclase-liquid geothermometer.

6. Discussion

The earliest mineral paragenesis of VI is represented by the assemblage of the intermediate plagioclase (An_{41-55}) with augite and Fe-Ti oxides. The patchy zones occur due to the replacement of the intermediate plagioclase by the Ca-rich one (An_{84-95}). This process coincides with the beginning of the melt degassing and release of water-carbon dioxide fluid that was trapped by plagioclase as fluid inclusions [10]. The domains and zones of Ca-rich plagioclase contain less augite inclusions, while inclusions of hypersthene and apatite appear. An increase in the calcium content is also associated with a change of the pattern of zoning in plagioclase. It becomes concentrically zoned and demonstrates gradual outward decrease in Ca-content.

Variations in the melt compositions during the evolution of both magma chambers were insignificant and were related to plagioclase crystallization. The absence of compositionally contrasting MIs implies that no mixing or mingling with andesitic or basaltic magmas took place.

The beginning of relatively monotonous outward decrease in the calcium content in the VI plagioclase from An_{95} to An_{41} coincides with the disappearance of fluid inclusions, i.e. with the end of the degassing process. The outer zones contain less crystalline inclusions. The latter are mainly represented by rare apatite needles. The quartz joins the plagioclase at the final stages immediately before the catastrophic eruption, as is evidenced by the absence of quartz in the form of inclusions in plagioclase and the presence of rare intergrowths of quartz and plagioclase.

The nature of the zoning of plagioclases in the pumice of the LP (figure 1) allows us to conclude that the crystallization process was similar in general to the VI. The earliest plagioclase had an intermediate composition (An_{40-55}) and was associated with hypersthene, augite, and Fe – Ti oxides. The patchy structure of cores of some phenocrysts resembles the substitution of the low-Ca plagioclase by the high-Ca one, similarly to VI. This similarity along with similarity in mineral and bulk rock compositions implies that this process also was controlled by increase in H_2O pressure. However, unlike VI, pressure increase did not trigger the degassing. Oppositely it favored crystallization of of magnesian hornblende amphibole phenocrysts. Apparently, this it could be a reason why Ca contents in the Ca-rich zones of LP plagioclases are lower (less than An_{85}) than in those of VI plagioclases. A monotonic decrease in the Ca contents towards the outermost zones suggests a decrease in fluid pressure, the cause of which remains to be established. Quartz begins to crystallize at later stages, but unlike VI, it occurs as inclusions in concentrically zoned plagioclase.

The above analysis indicates that VI magma storage developed at pressure less than 1 - 2 kbar. Low total pressure impeded the formation of phenocrysts of amphibole but favored fluid degassing. The presence of amphibole in assemblage with plagioclase and two pyroxenes in LP rocks indicated that pressures were more than 1 – 2 kbar as confirmed by experiments on the dehydration melting of andesibasaltic metabasites [16]. According to experimental data on the melting and crystallization of the Mt. St. Helens rhyodacites (USA) amphibole becomes stable at 885 ° C [17] and pressures that exceed 1.5-2.5 kbar, at the water contents similar to the melts of the LP and VI.

7. Conclusion

The appearance of high-calcium zones in plagioclase crystallizing from rhyolitic melts of VI and LP is explained by an increase in the partial pressure of H_2O [18], [12], and not by the influence of more mafic melts intruding the dacite magma [19], [12]. At 5 kbar, the liquidus and solidus temperatures in the binary albite-anorthite system decrease by ~ 300°C and plagioclase with higher Ca contents become more stable [20].The results of melting experiments on mafic and intermediate rocks in water-saturated systems [24-26] showed that the addition of water to the melt of the basaltic and andesitic composition can increase the anorthite content in plagioclase by 10 mol.%, and when the melt is saturated with H_2O - by more than 20 mol.%. The authors of [21], [24] emphasize that for island arc magmas H_2O affects composition of plagioclase during magmatic crystallization more than the bulk magma composition.

On the basis of the data obtained, it can be concluded that the rocks of the Lvinaya Past' caldera and the Vetrovoy Isthmus were formed as a result of crystallization of low-potassium plagiorhyolite-rhyodacite melts with a high water content at temperatures close to 850°C. Despite the same magma chemistry and similar crystallization temperatures, the assemblages of porphyritic phenocrysts are different because the evolution of magma reservoirs occurred at different pressures.

Aknowledgements

The work was carried out within the framework of projects 0330-2016-0005 and 0330-2016-0001 of the state assignment of the IGM SB RAS and with the support of the RFBR grant 16-05-00894. Analytical studies were carried out in the Center for Multielement and Isotope Studies of the SB RAS.

References

[1] Degterev A V Rybin A V Arslanov H A 2014 Calderaforming eruption of Lvinaya Past' (Iturup island, Kurile Islands): stratigraphy and age *VII Sibirean scientific and practical conference materials of young scientists of Earth sciences* VS Sobolev IGM SB RAS p. 14-15 (In Russian).

[2] Laverov N P 2005 Modern and holocene volcanism in Russia (Moscow: Nauka) 604 p.

[3] Lomtev V L 2008 Extrusion of the south of the Okhotsk outskirts of the Kuril arch near theLvinaya Past' caldera (Iturup Island) *Geologiya i razvedka* (short articles) **4** pp 72-74 (In Russian)

[4] Bazanova L I Melekescev I V Ponomareva V V Dirksen O V Dirksen V G 2016 Late Pleistocene and Holocene volcanic catastrophies in Kamchatka and in the Kuril islands. Part 1. Types and classes of catastrophic eruptions as the leading components of volcanic catastrophism *Journal of volcanology and seismology* **10(3)** pp 151-169

[5] Kovtunovich P Yu Safonov A D Udodov V V Raschepkina E V 2002 State geological map of Russian Federation, 2nd edition Page L-55-XXXIII Explanatory letter Saint-Petersburg: VSEGEI, 269 p. (In Russian)

[6] Melekescev I V Braitseva O A Sulerzhitsky L D 1988 Catastrophic explosive eruptions of volcanoes of the Kurile-Kamchatka region at the end of the Pleistocene-Early Holocene DAS USSR, , **300(1)** pp 175-181 (In Russian)

[7] Lineweaver, J L 1963 Oxygen outgasing caused by the electron bombardment of glass *Journal of Applied Physics* **34** pp 1796-1791

[8] Morgan G B London D 1996 Optimizing the electron microprobe analysis of hydrous alkali aluminosilicate glasses *American Mineralogist* **8(9-10)** pp 1176-1185

[9] Morgan G B London D 2005 Effect of current density on the electron microprobe analysis of alkali aluminosilicate glasses *American Mineralogist* **90(7)** pp 1131-1138

[10] Smirnov S Z Rybin A V Sokolova E N Kuzmin D V Degterev A V Timina T Yu 2017 Felsic magmas of the caldera-forming eruptions on the Iturup Island: the first results of studies of melt inclusions in phenocrysts from pumices of the Lvinaya Past and Vetrovoy Isthmus calderas *Russian Journal of Pacific Geology* **11(1)** pp 46-63

[11] Kawamoto T 1992 Dusty and honeycomb plagioclase – indicators of processes in the Uchino stratified magma chamber, Izu Peninsula, Japan *Journal of Volcanology and Geothermal Research* **49(3-4)** pp 191-208

[12] Shcherbakov V D Plechov P Y Izbekov P E Shipman J S 2011 Plagioclase zoning as an indicator of magma processes at Bezymianny Volcano, Kamchatka *Contributions to Mineralogy and Petrology* **162(1)** pp 83-99

[13] Putirka K A 2005 Igneous thermometers and barometers based on plagioclase plus liquid equilibria: Tests of some existing models and new calibrations *American Mineralogist* **90(2-3)** pp 336-346

[14] Laura E Waters Rebecca A Lange 2015 An updated calibration of the plagioclase-liquid hygrometer-thermometer applicable to basalts through rhyolites *American mineralogist* **100** pp 2172–2184

[15] Putirka K D 2008 Thermometers and Barometers for Volcanic Systems *Minerals, Inclusions and Volcanic Processes* **69** pp 61-120

[16] Beard J S Lofgren G E 1991 Dehydration melting and water-saturated melting of basaltic and andesitic greenstones and amphibolites at 1.3 and 6.9 *Journal of Petrology* **32(2)** pp 365-401

[17] Riker J M Blundy J D Rust A C Botcharnikov R E Humphreys M C S 2015 Experimental phase equilibria of a Mount St. Helens rhyodacite: A framework for interpreting crystallization paths in degassing silicic magmas *Contributions to Mineralogy and Petrology* **170(6)** p 22

[18] Hattori K Sato H 1996 Magma evolution recorded in plagioclase zoning in 1991 Pinatubo eruption products *American Mineralogist* **81(7-8)** pp 982-994

[19] Snyder D Tait S 1996 Magma mixing by convective entrainment *Nature* **379(6565)** pp 529-531

[20] Yoder H S Stewart D B Smith J R 1957 Ternary feldspars *Carnegie Institution of Washington Year Book* **56** pp 206-214

[21] Pletchov P Y Gerya T V 1998 Effect of H_2O on plagioclase-melt equilibrium *Experiment in Geosciences* pp 7-9

[22] Housh T B Luhr J F 1994 Plagioclase-melt equilibria in hydrous systems *American Mineralogist* **79(3-4)** pp 352-352

[23] Sisson T W Grove T L 1993 Experimental investigations of the role of H2O in calc-alkaline differentiation and subduction zone magmatism *Contributions to Mineralogy and Petrology* **113(2)** pp 143-166

[24] Panjasawatwong Y Danyushevsky L V Crawford A J Harris K L 1995 An experimental-study of the effects of the effects of melt composition on plagioclase-melt equilibria at 5-kbar and 10-kbar – implications for the origin of magmatic high-An plagioclase *Contributions to Mineralogy and Petrology* **118(4)** pp 420-432

[25] Almeev R R Ariskin A A 1996 Mineral-melt equilibria in a hydrous basaltic system: Computer modeling *Geochemistry International* **7** pp 624-636

[26] Bachmann O Bergantz G W 2003 Rejuvenation of the Fish Canyon magma body: A window into the evolution of large-volume silicic magma systems *Geology* **31(9)** pp 789-792

IOP Conf. Series: Earth and Environmental Science **110** (2017) 012016 doi:10.1088/1755-1315/110/1/012016

Features of geology and composition of rocks from the alkaline-gabbroic University massif (N-E Kuznetsky Alatau ridge, Siberia)

A A Mustafayev[1], I F Gertner[1], P A Serov[2]

[1] National research Tomsk State University, Lenin Avenue 36, Tomsk, 634050, Russia
E-mail: labspm@ggf.tsu.ru
[2] Kola scientific center of the Russian Academy of sciences, Fersman street 14, Apatity, 184209, Russia

Abstract. This paper discusses specifics of petrography and structural tectonic position of the alkaline-gabbroic University massif, which is a sort of analogue of the Kiya-Shaltyr deposit of nepheline ores in this area that was very slightly eroded. In this work, we will also present new precise analytical data on rare and trace element distribution in major intrusive phases of this geological object, and also isotope dating on the basis of method Sm-Nd.

1. Introduction
The University Massif is one of the least studied objects in alkaline-gabbroic association on the Northern slope of Kuznetsk Alatau also known as Mariinsk taiga (local term for this area). It was discovered during prospecting works by geologists of Martaiga geological prospecting expedition operation and employees of Tomsk State University in 1981-1987. Up until now, some researchers think that this geological object does not exist (verbal message from leading geologist of "WestSibgeologsurvey" A.N. Uvarov). However, during magnetometric survey, above ground and drilling works in 1987 they were able to outline and obtain samples of holocrystalline rocks correlating to main intrusive phases of Kiya Shaltyr and other adjacent alkaline-basic massifs in this region. In many ways, the University intrusive is unique due to its geotectonic location and geomorphological settings of the bedding formation.

2. Geotectonic location and formation ages of alkaline-gabbroid magmatism in Northern part of Kuznetsk Alatau
There are two structural compositional levels in geological composition of this territory. Lower level is composed by Vendian-Early Cambrian dislocated sediments, which are developed fragmentally in the areas of the most erosional cross section and are represented by carbonate sediments with layers of effusive-sedimentary rocks. They are grouped in narrow linear folds with N-E and submeridional (near N-S) trends, thus correlating to main structures of the region. Upper structural level is represented by volcanogenic formations usually dated as Middle Cambrian. It contains basaltic and andesite rocks, their tufas and high Mg lava breccia. Effusive distribution fields are enriched with alkaline and subalkaline dykes. It is typical for this level sediments to have quite flat inclination (falling angles) of 15-30°. The region has complex block-type composition caused by many disjunctive destructions of different type submeridional trend [13]. There are over 20 expressions of alkaline rocks in this region, which are of certain interest in exploring nepheline ore deposits of different categories (*Fig. 1*).

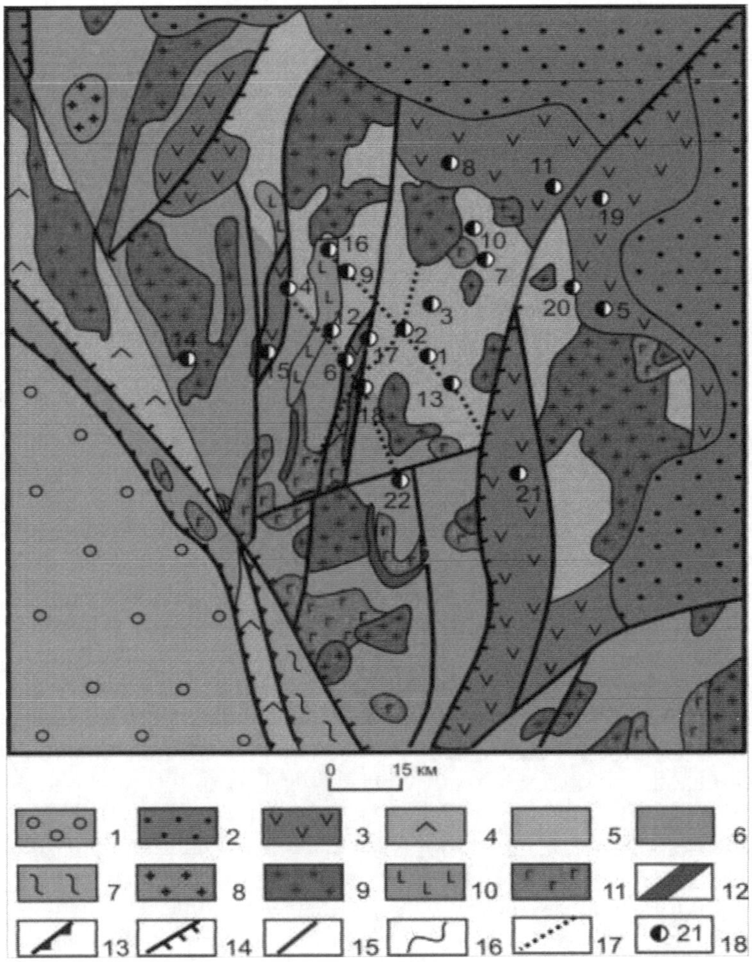

Figure 1. The scheme of placement alkaline basic massifs in geological structures of a northern part of Kuznetsk Alatau [1].

1 – deposits of the Kuznetsk carboniferous deflection; 2 – terrigenous deposits Devonian deyteroorogennic of hollows; 3 – volcanogenic deposits Devonian deyteroorogennic of hollows and grabens; 4 – Ordovician deposits of the Taydonsky graben; 5 – carbonaceous and volcanogenic deposits of the lower and middle Cambrian; 6 – siliceous and slate, volcanogenic and carbonaceous deposits of the top rifeya-lower Cambrian; 7 – average rifeya-high metamorphic complexes of the Tomsk ledge; 8 – superalkalinity granites; 9 – granites of a normal causeticity; 10 – moderate and alkaline gabbroic and syenite; 11 – basits of ophiolit association; 12 – ultramafita of ophiolites association; 13 – thrust; 14 – dumpings; 15 – other tectonic violations; 16 – geological borders; 17 – belts of milkings alkaline basites series; 18 – massifs of the alkaline and main breeds (1 – Kia-Shaltyrsky, 2 – University, 3 – Belogorsk, 4 – Svetlinsky, 5 – Podtayginsky, 6 – "Kiysky exits", 7 – Kurgusulsky, 8 – Batanayulsky, 9 – Verkhnepetropavlovsky, 10 – Cheremushinsky, 11 – Goryachegorsky, 12 – Zagorny, 13 – Dedovogorsky, 14 –. Motley, 15 – Malosemenovsky, 16 – Tuluyulsky and Medvedkinsky, 17 – Malokiyashaltyrsky, 18 – Velvet and Kiysky, 19 – Andryushkina the small river, 20 – Uchkuryupsky, 21 –. Bald, 22 – stream Dmitriyevsky).

One of the typical features of structural localization for these objects is dyke belts with certain trends, which connect real intrusive massifs. Generally speaking, the largest intrusions are located at the intersection of N-W and near N-S trended dyke belts or near them, which correlates to more permeable “magma plumbing systems and “magma concentration” systems or structures. Last ones are considered to be lithological and structural border between Vendian-Cambrian carbonate folded association and overlapping Middle Cambrian volcanogenic bedding thickness. This conclusion is also confirmed by concentration of majority of alkaline-gabbroic intrusives in carbonate substrate.

Only intrusive massifs located in central part of this region (adjacent to Kiya-Shaltyr deposit) correspond fully to criteria of Early Devonian magmatism in Mariinsk Taiga.

3. Specific features of geological composition and petrographic variety variance of the University Massif

In the modern erosional cross section, University Massif is barely exposed; it is located in lower part of relief and overlapped by a high thickness layer of mellow sediments (from 1 to 10 m). Under these conditions, it was extremely challenging to outline the massif and determine features of its geological composition. In order to do so, detailed above ground magnetometric survey was carried out along with major mining operations and core drilling. On a map (*Fig.2*), it is a body stretched in meridional (near N-S) direction (its length is 2.5 km, its width is from 0.2 to 0.6 km), whose structure is significantly complicated by tearing strains. Relationships with host rocks are ambiguous. It is thought that there are mostly tectonic contacts with gabbro and plagiogranites in the Voskresensk intrusion. With limestones from the Ust-Kundatsk Formation, there were fragments of intrusive contacts found, and both tectonic and active magmatic boundaries were seen with vulcanites of the Berikul Formation. In the active contact zone, marling and hornfelsing and propylitic alteration were noticed. Main intrusion phase of the massif is represented by a subalkaline gabbroic body, whose petrographic composition is similar to poikilitic basites of the Kiya-Shaltyr Massif. Other phases are feldspar and feldspar-free foydolites, subalkaline and alkaline gabbroic, nepheline and alkaline syenites that compose many small and sometimes thick (up to 10-30 m) dykes, as well as holocrystalline theralites and ijolites represented by "isometric" body on the eastern side of the massif.

Subalkaline gabbroic, which compose main body of the first intrusive phase, have quite uniform mineral composition and differ mostly in degree of crystallinity. Main rock-forming minerals are olivine, basic plagioclase (An_{50-54}), Ti-bearing augite, biotite and titanium magnetite; accessory minerals are usually apatite and sphene. It has hypidiomorphic-granular texture with elements of poikilo-ophytic and poikilitic textures.

Basic foydolites and theralites have smooth transitions from one to another with different variations in content of nepheline, plagioclase, olivine and pyroxene. Major difference in composition of foydolites and alkaline basites is that nepheline grains have higher degree of idiomorphism than feldspar. Medium-coarse grain theralites of the second intrusive body on the Eastern side demonstrate characteristics of secondary feldspathoids development in plagioclase, whose basicity lowers down to oligoclase-andesine in this paragenesis. This indirectly confirms hybrid nature of these formations. Taking into consideration that there are diluvial drops of almost completely feldspar-free medium-grain ijolites in association with these rocks, we can assume active interaction of later feldspatoid melt with early intrusion stage gabbroids. Additional prove of these rocks belonging to second phase of massif formation is presence of many small dykes if feldspar ijolites and nepheline camptonites (tamaraites) breaking through subalkaline gabbroic of main body and which can be considered as vein facies of feldspar ijolites.

Ultrabasic foydolites within the massif and its adjacent areas are represented by dykes with low thickness (from tens of cm to couple meters). They are mostly located in Western part of University Massif breaking through gabbroic and effusives of Berikulsk association. By compositional and textural characteristics, we can classify them in four main groups: a) phenocryst-free microijolites; b) ijolite-porphyrs with nepheline inclusions from 10 to 30%; c) urtite-porphyrs (with nepheline inclusions 50-70%); d) microijolites with large inclusions ("xenoliths") of holocrystalline urtites, rarely pegmatoid ijolites and microijolites. "Xenogenic" nature of last ones is suggested based on observations of similar variety of Kiya-Shaltyr deposit dykes, which have sharp contacts with ore body rocks and specific isotopic-geochemical characteristics [3].

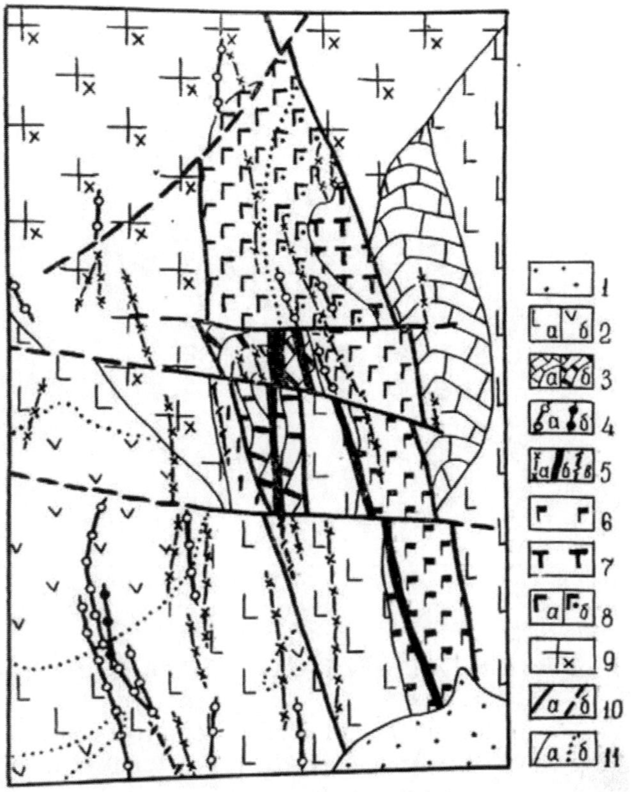

Figure 2. Scheme of a geological structure of the University massif [7].
1 – quarternary alluvial deposits; 2 – middle Cambrian, berekulsky suite: and – basalts, andesitobasalta, dacites, – lavobreccia, their tuffs; 3 – lower Cambrian, lips kundatsky suite: and – limestones, batts, interlay siliceous and batts, – marbles limestones; early Devonian kiysky (goryachegorsky) alkaline gabbro complex: 4 – 5 – high-alkaline dyke: 4 – and – mikroijolita, – urtit porphyries, 5 - feldspathic rocks: and – shallow and microgranular ijolita, - medium-grained ijolita, in – nepheline kamptonita; 6 – alkaline compact-grained porfirovidny gabbros of the facies of endocontact; 7 – leucoteralita with lenses the feldspathic of ijolit; 8 – the gabbro containing nepheline: and – to 5 and – up to 15%; 9 – early Paleozoic Voskresensky intrusion: gabbro, granodiorita, plagiogranita; 10 – tectonic violations: and – traced, - estimated; 11 – geological borders: and – traced, – gradual.

In general, composition patterns of University massif and its rock associations let us assume that this object can be considered as slightly eroded alkaline-gabbroid massif formed on lithological boundary of Early Cambrian carbonate and Middle Cambrian volcanogenic formations.

4. Geochemical characteristics of rocks and geodynamic setting for their formation

Main characteristic of petrogenic oxide distribution in high-alkaline rocks from Northern slope of Kunetsk Alatau is their discreteness demonstrated on «Na_2O+K_2O–SiO_2» diagram *(Fig. 3)*. Variation trends plotted for 13 massifs of this region correspond/correlate to at least three types of petrographic magmatic series:

a) Subalkaline gabbroids – moderate alkaline syenites;
b) Theralites – basic foydolites – nepheline syenites;
c) Feldspar-free melteigites – ijolites – urtites.

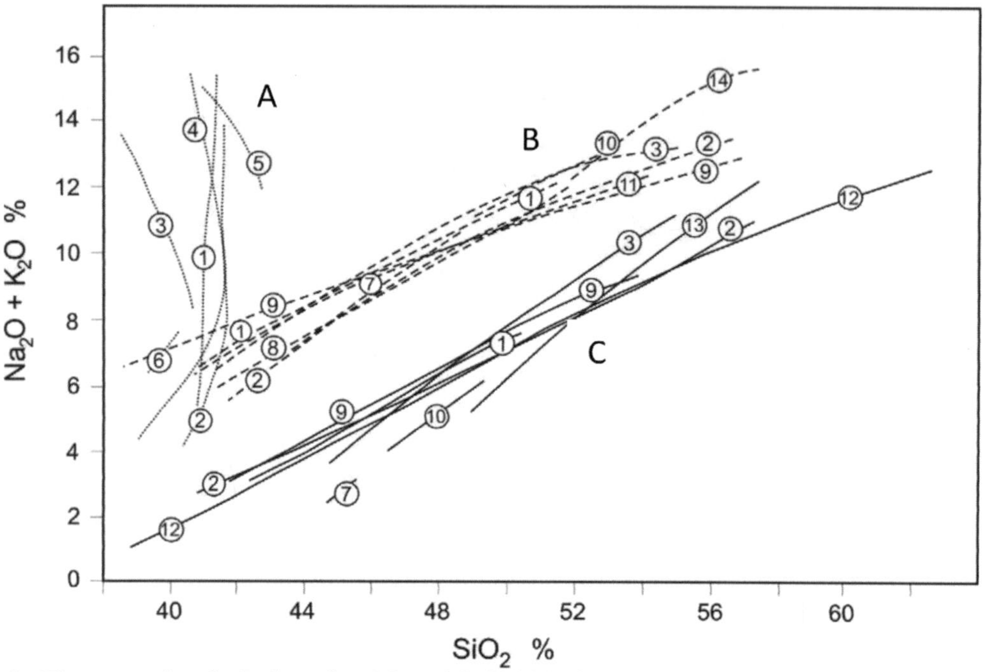

Figure 3. The petrochemical chart "Na2O+K2O-SiO2" (for breeds of alkaline massifs of Kuznetsk Alatau [3].

Legends: (A) variation curves of the ultramain foidolit; (B) variation curves of the main foidolit; (C) variation curves of alkaline gabbroic. Alkaline massifs: 1, Kia-Schaltyr; 2, University; 3, Belogorsk; 4, Svetlinsky; 5, Podtayga; 6, Kiysky exits; 7, Kurgusul-Listvenny; 8, Batanayula; 9, Verkhnepetropavlovsky; 10, Cheremushinsky; 11, Goryachegorsky; 12, Bolshe-Taskylsky; 13, Dedovogorsky; 14, mountain Pestraya.

Different combinations of mentioned series are observed in different geological objects. For example, only rocks of the first series (A) are typical in Barkhatno-Kiysk and Dedovogorsk massifs, only second series (B) in Goryachegorsk massif, and exclusively rocks from third series (C) are present in Svetlinsk massif. There are objects with contrast monolithic chemistry (like Kiya-Shaltyr, Upper Petropalovsk, Kurgusul-Listvenniy massifs) or more diversified massifs, which contain products of all three series. The University intrusive belongs to last ones; it has subalkaline gabbroic, theralites and basic foydolites, as well as ijolity-urtite dyke series. Chemical composition and concentrations of trace elements of representative rock varieties in the studied object are listed in Table 1. Petrogenic oxides were diagnosed analyzed by wet chemical methods and XRF, and rare and trace elements were studied by ICP MS method in Analytic center of geochemistry of nature systems in National Research Tomsk State University.

Table 1. Representative chemical compositions of magmatic breeds of the University massif

Oxide	1	2	3	4	5
	36/147	41/84.0	8A	UN-1	KC-7/1
SiO2	44,98	47,80	46,46	41,93	41,17
TiO2	0,95	1,055	1,27	0,375	0,49
Al2O3	15,11	19,58	14,71	25,96	28,5
Fe2O3	11,198	9,11	11,34	2,89	4,53
MgO	8,929	4,315	6,92	2,09	2,4
CaO	14,63	14,19	10,53	6,41	7,96

Na2O	2,96	2,8	4,23	10,35	10,24
K2O	0,95	0,39	2,43	9,26	3,28
P2O5	0,093	0,051	0,5	0,301	0,44
LOI	1,13	0,54	1,26	2,89	2,55
Sum	99,8	99,29	99,65	99,566	99,01
Infrequent elements					
Cr	224,1975	37,8801	55,7850	38,8518	19,8222
Ni	55,6799	29,2340	7,2418	6,2617	12,1023
V	155,2618	42,9117	11,7179	8,5373	63,7576
Co	48,4729	34,3551	12,1484	9,1979	14,2550
Cs	0,7782	0,4699	0,9851	0,6854	11,6256
Rb	23,4988	37,9099	41,1832	53,8740	100,1275
Ba	302,8811	345,7155	725,5941	395,4535	280,1494
Sr	537,59	1075,04	885,38	1063,35	35,6995
Nb	9,3328	10,4365	41,2114	12,6448	5,8020
Ta	0,5891	0,7326	2,4440	0,9502	0,4298
Zr	123,5396	95,0487	275,6794	81,0037	69,0379
Hf	2,6788	1,5559	3,9573	0,8246	2,2212
Y	22,3645	17,2153	45,2998	14,8354	8,2749
Th	2,7176	2,4501	7,0018	3,2981	4,1883
U	1,9170	2,2620	5,4251	3,1508	1,3037
Sc	23,6218	6,1947	0,8634	0,5078	5,7854
$\sum$ TR	111,3	94,37	245,97	103,58	79,185
La/Yb	9,691	9,588	10,277	12,419	17,76

Note. 1 – melanocratic gabbro; 2 – leucocratic gabbro; 3 – theralit; 4 – urtit porphyry; 5 – the urtit xenolith from urtit porphyry. Content of oxides (weight %) and infrequent elements (ppm) of RFA and ICP-MS. in "Analytical center of a geochemistry of natural systems" of NI TGU (Tomsk).

Transition elements (siderophile and chalcophile like Cr, Ni, Mn, V, Co, Sc, Cu, Zn). We observe quite low concentrations of these elements in the massif rocks. Their enrichment accumulation level comparing to contents in C1 (chondrite) standard [5]. is generally under one, except for Sc, Ti, and V (*Fig. 4*).

We can see on the diagram "sharp" negative anomalies for iron-group elements (Cr, Co, Ni)– their concentrations are 3 to 1000 times less than in chondritic component, and insignificant positive anomalies for (Ti, Sc, V), whose concentrations in average are 1,5 to 20 times higher than in C1 standard (*Table 2*).

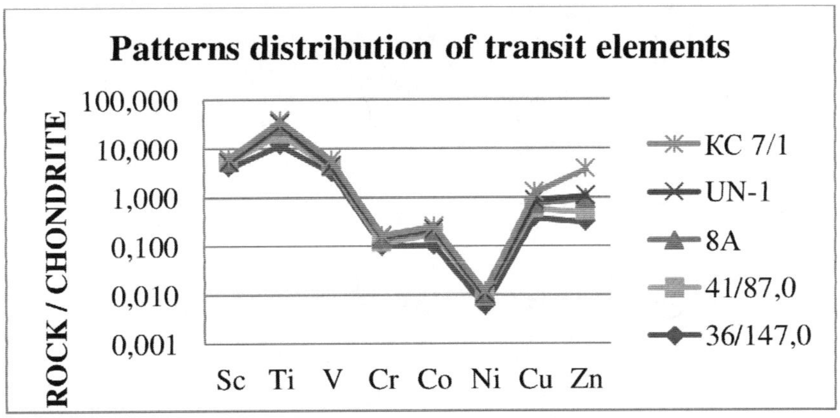

IOP Conf. Series: Earth and Environmental Science **110** (2017) 012016 doi:10.1088/1755-1315/110/1/012016

Figure 4. Patterns distribution of transit elements in breeds of the University massif
36/147,0 – melanocratic gabbro; 41/84.0 – leucocratic gabbro; 8A – leucotheralit; UN-1 – urtit porphyry; KC 7/1 – an urtit xenolith from urtit porphyry.
Concentration of a normalized on C1 [5].

Table 2. Distribution of infrequent and rare earths in breeds of the University massif

Rock	36/147.0	41/84.0	UN-1	8A	8A	KC-7/1
Elements	1	2	3	4	5	6
Be	1,503	1,615	2,819	5,217	4,907	0,752
Sc	23,622	6,195	0,508	0,863	0,861	5,785
Ti	4851,3	3369,1	1795,4	3176,9	3314,6	2094,2
V	155,3	42,912	8,537	11,718	10,830	63,758
Cr	224,2	37,880	38,852	55,785	55,505	19,822
Co	48,473	34,355	9,198	12,148	12,643	14,255
Ni	55,680	29,234	6,262	7,242	7,993	12,102
Cu	42,795	21,524	15,711	18,815	20,431	43,856
Zn	106,0	63,5	52,1	139,9	163,0	943,6
Ga	17,17	14,74	20,26	24,97	26,11	7,573
Rb	23,50	37,91	53,87	41,18	42,04	100,13
Sr	537,6	1075,0	1063,4	885,4	899,9	35,7
Y	22,365	17,215	14,835	45,300	44,745	8,275
Zr	123,5	95,05	81,00	275,7	281,5	69,04
Nb	9,333	10,436	12,645	41,211	42,890	5,802
Cs	0,778	0,470	0,685	0,985	1,015	11,626
Ba	302,9	345,7	395,5	725,6	796,6	280,1
La	20,98	17,91	21,76	48,93	47,96	18,97
Ce	44,48	37,76	44,58	101,63	101,74	31,46
Pr	5,205	4,514	4,919	9,161	9,104	3,737
Nd	21,23	17,56	17,85	44,12	43,30	14,33
Sm	4,545	3,395	3,097	8,532	8,541	2,746
Eu	1,321	1,385	1,039	2,717	2,630	0,721
Gd	4,412	3,191	2,732	8,088	8,041	2,144
Tb	0,700	0,516	0,432	1,334	1,316	0,320
Dy	4,267	3,124	2,619	8,311	8,095	1,926
Ho	0,881	0,663	0,578	1,788	1,761	0,379
Er	2,440	1,908	1,690	5,094	5,039	1,063
Tm	0,357	0,295	0,268	0,784	0,769	0,162
Yb	2,165	1,867	1,752	4,761	4,806	1,068
Lu	0,317	0,285	0,264	0,715	0,706	0,163
Hf	2,679	1,556	0,825	3,957	4,065	2,221
Ta	0,589	0,733	0,950	2,444	2,507	0,430
Th	2,718	2,450	3,298	7,002	6,717	4,188
U	1,917	2,262	3,151	5,425	5,692	1,304

Note. Breeds are presented: melanocratic gabbro – 36/147; leucocratic gabbro – 41/84.0; urtit porphyry – UN-1; leucotheralit – 8A; ijolit with "xenolith" urtit porphyry. Concentration of minerals (ppm) are determined by the ICP-MS method in "Analytical center of a geochemistry of natural systems" of NI TGU (Tomsk).

Rocks in the massif demonstrate the following pattern of transit element distributions. In genetic sequence "leucocratic - melanocratic gabbro", concentrations of all elements decrease from samples 41/84.0 to 36/147,0. Moreover, comparing with gabbroic in leucotheralite (sample 8A), urtite porphyry (sample UN-1) and xenolith (sample KC-7/1) have higher concentrations of transit elements. It should also be noted that there is high concentration of zinc in sample Kc-7/1 (ijolite with xenolith of urtite) comparing to other samples.

Specific characteristics of trace element behavior in rocks of University massif are shown on multi-elemental diagrams demonstrating patterns of REE, LILE and HFSE distribution. When plotted these graphs plots, primary analytical data was normalized by CI and OIB standards according to [9].

Spectrum graphs of lanthanide distributions for rocks from University massif are characterized by very tight ratios of light to heavy elements (*Fig. 5*). They are almost parallel to each other and correspond to La/Yb ratios of 10 to 18. Total REE enrichment accumulation level corresponds to the following. The least depleted ones are urtite inclusions ("xenoliths") from ijolite dykes (sample Kc-7/1) that correlates to the data on urtites from Kiya-Shaltyr intrusive [10]. In urtite porphyry, concentration of these elements is a little higher, and it keeps increasing in subalkaline gabbroic. At the same time, melanocratic gabbro is more enriched in rare earth elements than leucocratic gabbro. In last one, a week positive anomaly of Europium is observed. Such pattern can be explained by cumulative segregation of nepheline in foydolites and plagioclase in gabbroids as early crystallization stages of the melts.

Maximum level of enrichment accumulation is in leucotheralites, whose graph plot is kind of isolated. It could be the result of mixing of magmatic melts with different compositions.

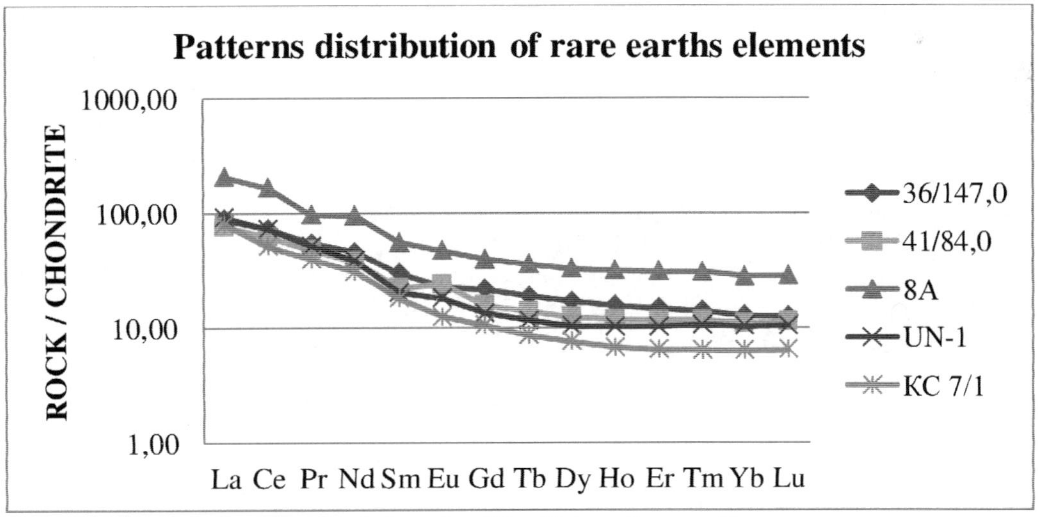

Figure 5. Distribution of rare earths in breeds of the University massif
36/147,0 – melanocratic gabbro; 41/84.0 – leucocratic gabbro; 8A – leucotheralit; UN-1 – urtit porphyry; KC 7/1 – an urtit xenolith from urtit porphyry.
The maintenance of REE is normalized on CI [9].

Multi-elemental graphs plots spectrums normalized by average OIB composition show following patterns. Foydolites (samples KC-7/1 and UN-1) demonstrate quite fast sharp depletion in most rare elements comparing to OIB standard (*Fig. 6a*). Higher concentrations are seen in LILE and U. Similar behavior is observed in multi-elemental graphs for subalkaline gabbroic, but at a slightly higher concentrations of rare elements (*Fig. 6b*).

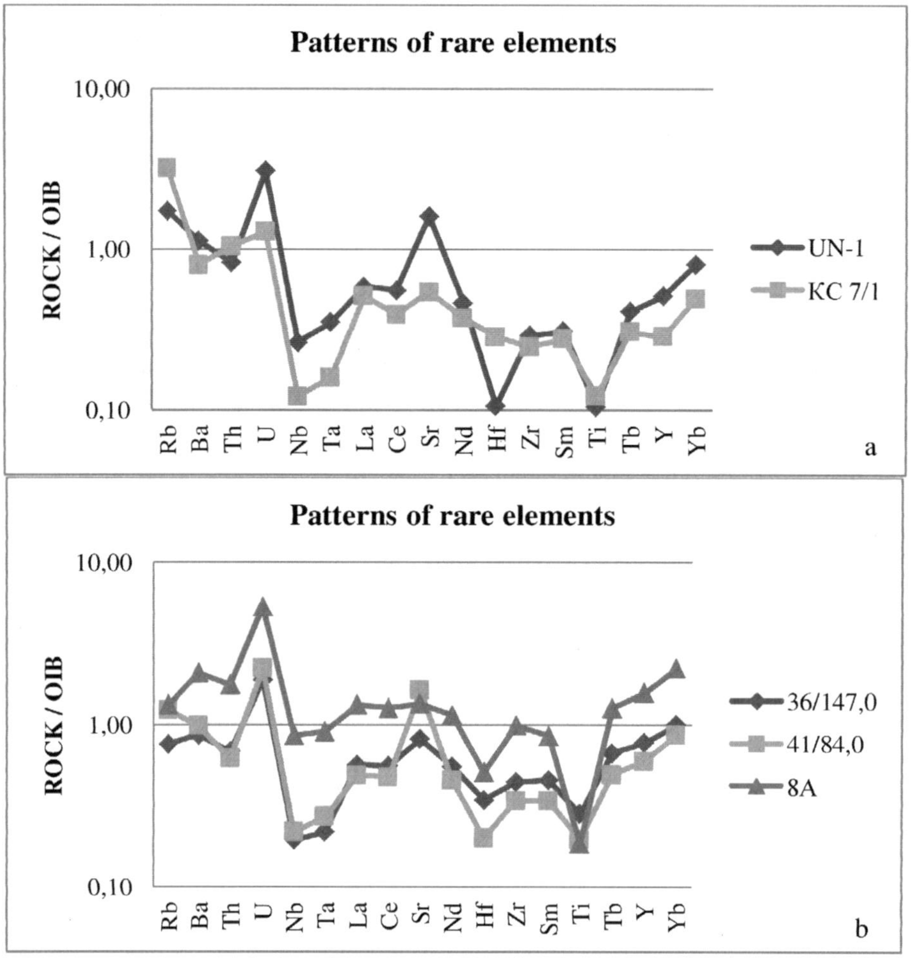

Figure 6. Patterns rare elements in breeds of the University massif
(a) Foidolita: UN-1 – urtit porphyry; KC 7/1 - "xenolith" of the urtit from ijolit porphyry.
(b) The gabbro, theralit: 36/147,0 – melanocratic gabbro; 41/84.0 – leucocratic gabbro; 8 A – leucoteralit;
Contents the rare of elements are normalized on OIB [9].

The closest to OIB graph is distribution in leucotheralites (sample 8A), where there is no fast enrichment in Sr. In general, presence of distinct negative anomalies of Nb, Ta, Hf and Ti with steep positive peaks of Rb, Sr and U indicate quite complex geodynamic setting for formation of Devonian alkaline rocks in Mariinsk taiga. It is very possible that generation of primary magmas took place in setting of active thermal interaction of mantle plumes and lithosphere mantle with further participation of crust matter and crust fluids [4].

Therefore, specifics of trace element distributions in rocks from University massif allow us to correlate them with analogue products from Kiya-Shaltyr and adjacent massifs (Dedogorsk and Belogorsk). Deficiency in niobium, tantalum, hafnium (appear as steep negative anomalies) along with generally lower mantle plum component normalized to OIB, as well as local enrichment in LILE (Sr, Rb, Ba) suggest complex geodynamic settings of alkaline-basic magmatism in central part of Kuznetsk

IOP Conf. Series: Earth and Environmental Science **110** (2017) 012016 doi:10.1088/1755-1315/110/1/012016

Alatau in Early Devonian. Similar conclusions were made by D.N. Voitenko while interpreting geochemistry of rocks from Kiya-Shaltyr massif [10].

5. Isotopic dating

Sm-Nd Method. Sm and Nd isotopic composition was measured on 7-channel solid-phase mass spectrometer Finnigan-MAT 262 (RPQ) in static double strand mode using rhenium and tantalum strands [8]. Average ratio of $^{143}Nd/^{144}Nd$ in La Jolla standard during period of measurements was 0.512081±13 (N = 11) *(Table. 3)*. Errors in values of (2 σ) $^{147}Sm/^{144}Nd$ do not exceed 0.5 %, and 0.005 % for $^{143}Nd/^{144}Nd$. During calculation of primary isotopic ratios and ε_{Nd}, ε_{Sr}, T_{Nd} (DM) values, modern parameters of model standard reservoirs CHUR ($^{143}Nd/^{144}Nd$ = 0.512638, $^{147}Sm/^{144}Nd$ = 0.1967) and depleted mantle (DM) ($^{143}Nd/^{144}Nd$ = 0.51315, $^{147}Sm/^{144}$ Nd = 0.2137) were applied [2].

Table 3. Isotope structure of Sm-Nd in the gabbro of the University massif

Exemplar	Concentration, mkg/g		Isotope relations		T_{DM}, million years	$\varepsilon_{Nd}(T)$
	Sm	Nd	$^{147}Sm/^{144}Nd$	$^{143}Nd/^{144}Nd$		
			C36/147			
WR	3.42	15.27	0.1353	0.512808±9	498	+7.3
Pl	1.531	9.46	0.0978	0.512709±16		
Ol	4.49	13.23	0.2050	0.513051±10		
Px	4.18	15.02	0.1682	0.512922±8		
			C41/84.0			
WR	1.769	7.46	0.1433	0.512907±12	492	+8.7
Pl	0.588	3.44	0.1033	0.512797±9		
Px	2.43	7.99	0.1841	0.513041±25		
Ol	3.95	11.020	0.2165	0.513160±12		

Note. C36/147 – melanocratic gabbro, C41/84.0 – leucocratic gabbro.

Isochrones were built by D. York's method [14] using Isoplot/Ex software [6]. More precise Sm-Nd isochrones were obtained by mineral phases and whole rock compositions of two samples of subalkaline gabbroic (*Fig. 7*).

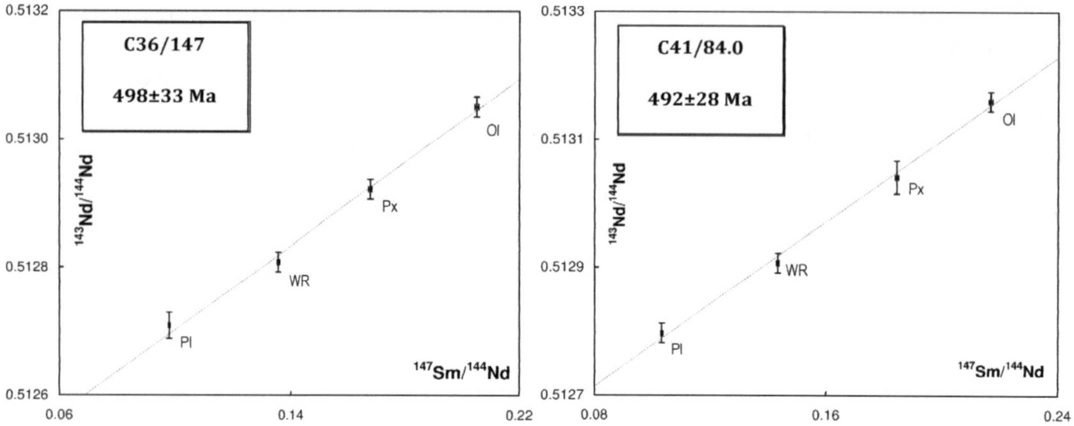

Figure 7. Sm-Nd mineral isochrones on subalkaline gabbroic of the University massif. Measurements were taken by a reference technique [8].

One of the biggest surprises during this research was the actual age of studied gabbroids, which is 498-492 ±23-28 Ma. Such expression of alkaline magmatism in this region took place exclusively in western part of Mariinsk Taiga and was recorded in alkaline-gabbroid association of Upper-Petropavlovsk Massif [11]. Dating alkaline rocks by Sm-Nd isotope method nowadays is quite complicated because variations in $^{147}Sm/^{144}Nd$ values do not exceed 0.09-0.11 range. However, we can conclude that their formation occurred in Devonian (380 – 430 Ma), and this age corresponds to the main stage of foydolite expression in this part of the region. An important element in studying University massif is its intermediate structural location between western and eastern sectors of areas of high-alkalinity magmatism expression on the Northern slope of Kuznetsk Alatay, which were formed during long period of time (507 -265 Ma) according to recent geochronological research [12]. We assume a case of several stages pulses of high-alkalinity magmatism within this geological object. More detailed studying will promote design of geodynamic model of high-aluminous alkaline magmas formation under settings of mantle plume activity in folded structures of ancient continent margins.

6. Conclusion

The results of conducted research let us make following conclusions, which reflect necessity of further studying of alkaline-gabbroic University massif. First, this object is of big interest for justifying locations of nepheline ore deposits in this region. Specifics of its geological composition and location at the border between two associations of carbonate and volcanogenic series, which have very different dislocation levels and petrographic material compositions, allow us to understand study composition of upper part of intrusive front of alkaline-gabbroic magmatism. Second, broad petrographic and petrochemical rock-forming variety provides for reconstruction of genesis process of high-aluminous minerals in magmatic complexes of this region. Third, detail geochemical testing promotes development of a more accurate geodynamic model of high-alkalinity magmatism expression in folded structures of Northern and Central Asia.

This work was conducted as a government task of the Ministry of Education and Science of the Russian Federation Project Number 5.2352.2017/4.6

References

[1] *Correlation of magmatic and metamorphic complexes of the western part of the Altai-Sayansk folded region* / S.P. Shokalsky, G.A. Babin, A.G. Vladimirov, etc. Novosibirsk: Siberian Branch of the Russian Academy of Science publishing house, GEO branch, 2000. **187** p.

[2] Faure G. *Fundamentals of isotope geology*. – M.: World, 1989. – **590** p.

[3] Gertner I. F., Vrublevsky V. V., Voitenko D. N. et al. Plume-related alkaline basic magmatism of the Kuznetsk Alanau: The Goryachegorsk complex // Magmatism and metallogeny of the

Altai and adjacent large igneous provinces with an introducing essay on the Altaids. *IAGOD Guidbook Series 16*. London, UK: CERCAMS/NHM, 2007. P. **141-153**.

[4] Gertner I. F., Vrublevsky V. V., Tishin P.A., etc. Temporary boundaries, sources of magmas and the formational status Paleozoic the feldshpatoidnykh of intrusions of the Northeast of Kuznetsk Alatau // *Geodynamic evolution of a litosphere of the Central Asian relative frame belt (from the ocean to the continent)*: Meeting materials. Issue 11. – Irkutsk: Institute of crust of the Siberian Branch of the Russian Academy of Science, 2013. Page **71-73**.

[5] Langmuir, C.H., Hanson, G.N., 1980. An evaluation of major element heterogeneity in the mantle sources of basalt // *Philos.Trans.* R. Soc. A 297. – London – 1978 – P. **383-407**.

[6] Ludwig K.R. *User's manual for Isoplot/Ex, Version 2.10*. A geochronological toolkit for Microsoft Excel. Berkeley Geochronology Center Special Publication, 1999, v. 1, **46** p.

[7] Osipov P. V., Makarenko N.A., Korchagin S.A., etc. New alkaline gabbroid the ore bearing massif in Kuznetsk Alatau//*Geology and geophysics*. – 1988. - No. 11. – Page **79-82**.

[8] Serov P.A., Ekimova N.A., Bayanova T. B., Mitrofanov F. P. Sulphidic minerals – new geochronometers at Sm-Nd dating of a rudogenez of the intrusions of Baltic Shield stratified the mafit-ultramafitovykh // *The Litosphere*. 2014. No. 4. Page **11-21**.

[9] Sun S., McDonough W.F. Chemical and isotopic systematics of oceanic basalts: implications for mantle composition and processes // *Magmatism in the ocean basins* / Eds. Saunders A.D.& Norry M.J. Geol. Soc. Special Publ. № 42. – 1989. – P. **313-345**.

[10] Voytenko D.N. Mikrostruktur of urtit of Kia-Shaltyrsky Pluton (Kuznetsk Alatau)//Geologists of the 21st century: *Thesis The All-Russian scientific conference – Saratov*: SO EAGO publishing house, 2003. Page **28-31**.

[11] Vrublevskii V. V., Gertner I. F., Gutierrez-Alonso G. et al. Isotope (U-Pb, Sm-Nd, Rb-Sr) geochronology of alkaline basic plutons of the Kuznetsk Alatau. *Russian geology and Geophysics (Geologiya I Geofizika)* 55. 2014. P. **1264-1277**.

[12] VrublevskiiV. V., Grinev O. M., Izokh A. E., Travin A. V. Geochemistry, isotope triad (Nd-Sr-O), and ^{40}Ar-^{39}Ar age of Paleozoic alkaline mafic intrusions of the Kuznetsk Alatau (by the example of the Beloya Gora pluton) // *Russian Geology and Geophysics*. 2016. Vol. 57, № 3. P. **464-472**.

[13] Yesin S. V., Korchagin S.A., Yesina O.A., Gertner I. F. Nepheline ore rocks of sites University 1 and 2 (Kuznetsk Alatau)//*Alkaline and subalkaline breeds of Kuznetsk Alatau*: The collection of articles / Under the editorship of M.P. Kortusov. – Tomsk: Thom publishing house university, 1987. Page **74-82**.

[14] York D. Least squares fitting of straight line // *Can. J. Phys*. V. 44. – 1966. – P. **1079-1086**.

IOP Conf. Series: Earth and Environmental Science **110** (2017) 012017 doi:10.1088/1755-1315/110/1/012017

New data on geology, petrochemistry and geochemistry of rocks of the Nizhnetarlashkynsky ultramafite massif (Western Sangilen, Eastern Tuva)

Ch K Oidup [1], F P Lesnov [2], Ye K Druzhkova [1]

[1] Tuvinian Institute for Exploration of Natural Resources of SB RAS, Internatsionalnaya, 117 a, Kyzyl, 667007, Russia
[2] Institute of Geology and Mineralogy of SB RAS, Ac. Koptyuga ave, 3, Novosibirsk, 63009, Russia

E-mail: oydup_ch@mail.ru, lesnovfp@list.ru

Abstract. Mafic-ultramafic massifs of Tuva territories are located both within linear belts tracing deep faults and as "disordered" areas associated with some belts sufferedthe later fold-block dislocations. Sometimes fragments of mafic-ultramafic bodies are present in granitoidal intrusive asxenogenic blocks of various sizes. Areal location of ultramafic massifs is particularly observed in the east of Tuva within the Khamsarynsky structural-formation zone as well as in the structures of the Tuvinian-Mongolian middle massif. Geochemical features of the ultramafic massi frock variety studies are continued for obtaining convincing conclusions on their formational affiliation. Ultramafic massifs of Tuva have isolated single data in terms of the geochemical characteristics which requires a purposeful work in this direction. The present paper summarizes for the first time petro-geochemical characteristics of dunites of the Nizhnetarlashkynsky ultramafic massif and fragments of individual picritic dikes among them.

Areal location of ultramafic massifs is not uncommon;their location isparticularly known in the east of Tuva within the Khamsyrinsky structural-formational zone as well as in structures of the Tuvan-Mongolian middle massif among the highly-metamorphic deposits of the Proterozoic. One of such mafite-ultramafite bodies is the Nizhnetarlashkynskyultramafite massif located within the Western Sangilen as well as some new data on geology, petrochemistry and geochemistry presented below.

The Nizhnetarlashkynsky ultramafite massif is outcropped both sides of the mid-channel valley of the Tarlashkynriver (50°26′58,28″N; 95°9′57,16″ E). Antalusite-sillimanite and cordierite-bearing crystalline schists and gneisses of the Mugursky suite (Proterozoic) outcropped within the massif. This ultramafic massif contactsthe Pravotarlashkynsky gabbroid complexon the right bank of the river, and it contacts the Bashkhymugursky granitoid massif in the northeast and east. There are many small blocks of marbles and marble limestones (the Morensky suite) as well as bodies of leucogranites among the ultramafic rocks.

The Nizhnetarlashkynsky massif is mainly composed of dunites where olivine grains and multiple grains are usually found in the talc-carbonate-serpentine matrix. Dunites have the following quantitative mineral composition (%): olivine (55–85), serpentine represented by chrysotile (15–40), carbonate (<1), talc (<1), chrome-spinel (up to 5), magnetite (1). Olivine content is about 90%in the composition of small lens-like separations. Olivine grains usually have an elongated one-axis shape. Serpentinites are present in the massif including dunites which form a narrow submeridionally oriented strip on the left-bank of the (southern) part of the massif. They have the following composition (%): serpentine — 60 %, olivine relics — up to 35 %, clay mineral on olivine — 5 %, chrome-spinel (2), carbonate (1), talc and tremolite (singular occurrence case).

Dunites of the described massif have the following average chemical composition according to the 9 analyses results carried out using X-ray fluorescence method (wt%): SiO_2 (37,18), TiO_2 (0,05),

IOP Conf. Series: Earth and Environmental Science **110** (2017) 012017 doi:10.1088/1755-1315/110/1/012017

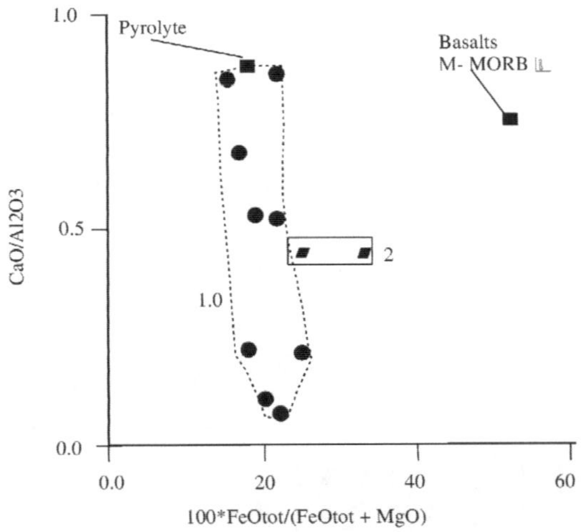

Figure 1. Diagram 100 * FeOtot/(FeOtot+MgO) – CaO/Al_2O_3 for dunites (1) and picrites (2) of the Nizhnetarlashkynky massif.

Al_2O_3 (1,12), FeO_{tot} (10,10), MnO (0,15), MgO (40,08), CaO (0,36), Na_2O (0,10), K_2O (0,02), P_2O_5 (0,01), BaO (0,01), SO_3 (0,07), V_2O_5 (0,01), Cr_2O_3 (3,01), NiO (0,25), Ignition loss (6,14) Sum (99,66). Dunites are characterized by stable relationships between the contents of the most components especially between MgO and FeO_{tot}. Dunites are comparable with pyrolyte according to the first named parameters (iron content) but generallythey differ from pyrolyte by lower values of the second named parameters according to the diagram coordinates $100*FeO_{tot}/(FeO_{tot}+MgO)$ – CaO/Al_2O_3 (Figure 1).

The total REE content in dunites of the Nizhnetarlashkinsky massif range within 0,27–1,60 g/t which is much higher than the average total content of these impurities in dunites from the Samail (Oman) massif (Table 1).

Table 1. Geochemical composition of dunites from the Nizhnetarlashkanskiy ultramafic massif, ppm

Elements	Numbers of samples									Samail massif
	14-1	14-2	14-4	14-6	14-7	14-8	14-9	14-10	14-11	
La	0,017	0,020	0,013	0,011	0,023	0,027	<0,01	0,012	0,017	0,003
Ce	0,041	0,047	0,031	0,024	0,050	0,053	0,012	0,029	0,046	0,006
Pr	<0,01	<0,01	0,022	<0,01	<0,01	<0,01	<0,01	<0,01	<0,01	0,002
Nd	0,026	0,037	0,031	0,011	0,053	0,036	<0,01	0,020	0,022	0,006
Sm	0,006	<0,005	0,007	<0,005	0,013	0,009	<0,005	<0,005	<0,005	0,004
Eu	0,003	0,003	0,005	<0,003	0,005	0,004	<0,003	0,003	0,004	0,001
Gd	<0,01	<0,01	<0,01	<0,01	0,021	0,015	0,016	<0,01	<0,01	0,005
Tb	<0,005	<0,005	<0,005	<0,005	<0,005	<0,005	<0,005	<0,005	<0,005	0,001
Dy	0,007	0,006	0,009	0,006	0,012	0,009	<0,005	<0,005	0,006	0,010
Ho	<0,005	<0,005	<0,005	<0,005	<0,005	<0,005	<0,005	<0,005	<0,005	0,003
Er	0,011	0,028	0,042	0,005	0,028	0,008	0,007	0,007	0,012	0,013
Tm	0,007	0,013	0,002	0,003	0,009	0,002	<0,002	0,002	0,008	0,003
Yb	0,006	0,006	0,009	0,005	0,012	0,009	0,004	0,003	0,003	0,026
Lu	<0,002	<0,002	0,002	<0,002	0,002	<0,002	0,003	<0,002	<0,002	0,006
Sum	0,156	0,342	0,193	0,105	0,248	0,194	0,231	0,118	0,208	0,08
(La/Yb)n	1,0674	1,3289	0,7650	0,2788	0,7267	0,8776	0,8041	0,2297	0,2297	0,10
(Eu/Eu*)n	<0,1	<0,1	0,12	<0,1	<0,1	<0,1	<0,1	<0,1	0,15	н.д.
Rb	<0,1	<0,1	0,12	<0,1	<0,1	<0,1	<0,1	<0,1	0,15	
Sr	0,74	0,64	0,72	0,32	1,98	0,79	0,45	0,58	0,51	
Y	<0,1	<0,1	<0,1	<0,1	<0,1	<0,1	<0,1	<0,1	<0,1	
Zr	1,63	1,15	0,82	0,76	1,17	0,59	1,01	0,60	0,75	
Nb	<0,05	0,065	<0,05	<0,05	0,069	<0,05	<0,05	<0,05	<0,05	
Cs	0,007	0,005	0,007	0,005	0,010	0,011	0,010	0,010	0,005	
Ba	0,58	0,53	0,56	0,46	0,62	0,95	0,45	0,67	0,39	
Hf	0,013	0,010	0,006	<0,005	0,011	<0,005	<0,005	<0,005	0,005	
Ta	<0,002	<0,002	<0,002	<0,002	<0,002	0,002	<0,002	<0,002	<0,002	
Th	0,014	0,016	0,002	0,004	0,013	0,006	0,002	0,002	0,005	
U	0,005	0,010	0,002	0,003	0,004	0,025	0,006	0,017	0,005	

Note. Analyzes were performed in the Analytical Center of IGM SB RAS by the ICP-MSc method by acid decomposition in an autoclave of the MARS-5 system (performed by I.V. Nikolaeva). For comparison, the average content of REE (9 assays) is given in dunites from the Samail (Oman) massif [1].

Chondrite-normalized REE patterns of all the dunite samples from the Nizhnetarlashkynsky massif have a negative slope and parameter values (La/Yb)n>1 which show their enrichment with light elements. Parameter values (Eu/Eu*)n for all these dunite samples are <1 which indicates on europium spent (Figure 2).

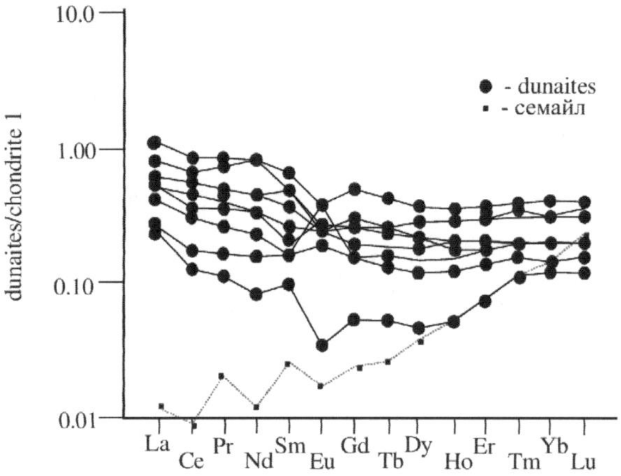

Figure 2. Chondrite-normalized REE patterns of dunites from the Nizhnetarlashkynsky massif (according to Table1) Spectra for the average content (6 analyses) for dunites of the Samail (Oman) [1] massif is considered ascomparison.

Abnormally high levels of light REE in the studied dunites which are not characteristic of restitogenic ultramafites are causally related to the presence of structurelessimpurity of these elements added by epigenetic fluids and concentrated in intergranular and intragranularrock microcracks. Decreased values of Rb, Ba, K, Nb, La, Ce, Pb and ratios Nb/Ta,Zr/Hf, U/Th, Ce/Yb of the initial mantle indicate the belonging of dunites (olivinites) of this massif to the depleted mantle.

It will be remarked that a small rock fragment cash indunite massdetermined by us as a picrate was found among the dunites of the investigated massif at paramerers (50°26′58,40″ N; 95°10′4,02″E). Xenoliths of spinel lherzolites were found as part of the kamtonite dikes of the Agardagsky complex within the neighboring Pravotarlashkynsky gabbroid complex among the granitoids of the Bashkhymugursky massifs and inclosing antalusite-sillimanite cordierite-containing shales and gneisses of the Mugursky suite [2]. Such formations were found for the first timeamong dunites. This macroscopically black fine-grained rock has a massive texture and poikilitic structure and has the following quantitative-mineral composition (%): clinopyroxene (35), olivine (30 %), serpentine (20), green spinel (10–15), ore mineral (1), carbonate (<1). There are inclusions of olivine and green spinel sized within 0,01–0,1 mm in clinopyroxene oikocrystalssized within 0,7–1,5 mm. In some sample parts fine-grained rock structure is changed to coarse-grained light rock with the same mineral composition but containing singular occurrence of green spinel. It was found that the picrites from the Nizhnetarlashkynsky massif are close to them for TiO_2, FeO, MgO, CaO, Na_2O and K_2O content but more depleted than SiO_2 and Al_2O_3(Table 2)

Table 2. Chemical composition of picrites from the Nizhnetarlakynsky massif, wt. %

Componenes	Spl.14-8	Spl.14-1	The composition of picrites(literary data)
SiO_2	**34,25**	**35,53**	**39,0–43,5**
TiO_2	0,34	1,57	0,3–1,5
Al_2O_3	11,57	11,36	3,5–8,5
Fe_2O_3	11,58	15,53	3,0–6,5
FeO	н.д.	Н.д.	3,5–13,5
MnO	0,16	0,22	0,1–0,3
MgO	30,51	28,29	20,0–32,0
CaO	5,16	5,05	2,5–7,5
Na_2O	0,15	0,21	0,2-0,5
K_2O	0,05	0,16	0,1–0,5
P_2O_5	0,01	0,10	н.д.
BaO	0,01	0,01	н.д.
SO3	0,17	0,08	н.д.
V_2O_5	0,017	0,077	н.д.
Cr_2O_3	0,210	0,200	н.д.
NiO	0,142	0,096	н.д.
LOI	6,13	1,66	н.д.
Sum	100,44	100,15	н.д.

Note. Analyzes were performed in the Analytical Center of IGM SB RAS by the RFA method.

in comparison with picrate chemical composition according to the Mountain Encyclopedia [Internet]. Picrites differ from the surrounding dunites by higher parameter value100*$FeO_{tot}/(FeO_{tot}+MgO)$ but much lower compared to N-MORB basalts. The graphs of the normalized N-MORB basalts of the petrochemical parameters of the studied picrites are like negatively inclined lines complicated by positive and negative anomalies of moderate intensity (Figure 3).

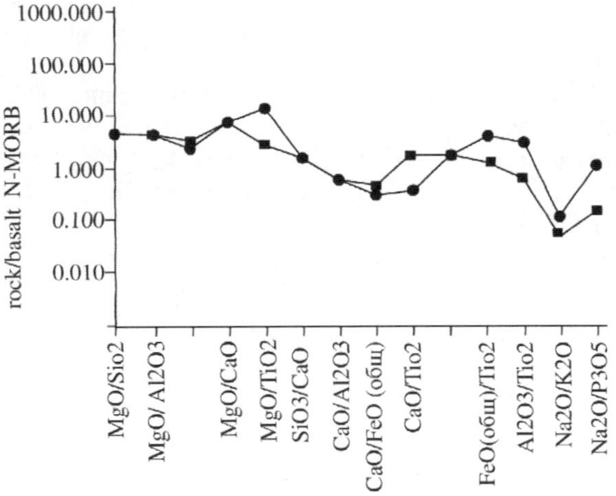

Figure 3. Graphs of normalized on N-MORB basalts of petrochemical parameters of picrites of the Nizhnetarlashkynsky massif (according to Table 2)

The two analyzed picrite samples differ according to the total REE content, in both cases sufficiently high REE is observed (Table 3).

Table 3. Content of rare-earth elements in picrites from the Nizhnetarlashkanskiy massif, g/t

Elements	Spl. 14-8	Spl. 14-1
La	0,42	1,63
Ce	0,89	4,50
Pr	0,14	0,74
Nd	0,90	5,00
Sm	0,34	1,96
Eu	0,18	0,30
Gd	0,59	2,50
Tb	0,10	0,54
Dy	0,62	3,80
Ho	0,13	0,91
Er	0,50	2,80
Tm	0,081	0,47
Yb	0,70	3,10
Lu	0,10	0,57
Sum	5,69	28,82
(La/Yb)n	0,41	0,36
(Eu/Eu*)n	1,22	0,41

Note. Analyzes were performed in the Analytical Center of IGM SB RAS by the ICP-MSc method with acid decomposition in an autoclave of the MARS-5 system (performed by I.V. Nikolaeva).

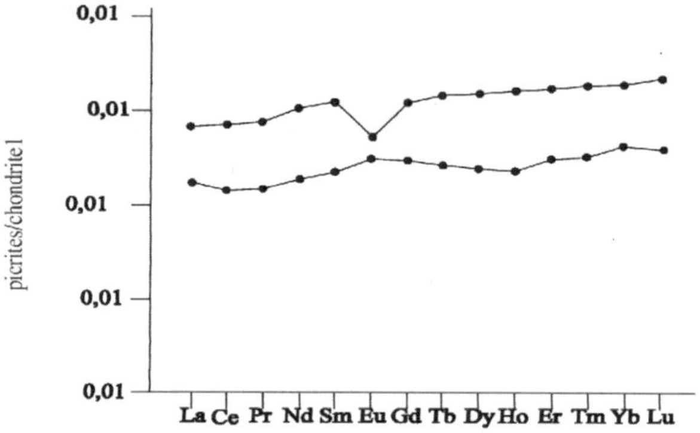

Figure 4. Chondrite-normalized REE patterns of picrites of the Nizhnetarlashkynsky massif (according to Table 3)

Positively inclined distribution spectra of REE content normalized in chondriteshave a similar configuration but they differ in position on the diagram as well as in the character of the Eu anomalies due to differences in the level of their accumulation (Figure 4). The resulting data can be assumed that the investigated picrites form one of the tectonically disintegrated dikes cutting the dunites of the Nizhnetarlashkynsky massif.

Nemtsovich has distinguished the Agardagsky dike complex within the Western Sangilen represented by lamprophyres and essexite diabases[3]. These dykes were further studied by [4] etc. and after by [2]. Dykes break through heterochronous metamorphic and magmatic formations and often contain deep xenoliths. It can be assumed that this xenolith is also a substance of the lithospheric mantleaccording to the geological position of the Nizhnetarlashkynsky ultramafic massif.

References

[1] Godard M Jousselin D Bodinier J-L 2000 Relationships between geochemistry and structure beneath a paleospreiading centre: a study of mantle section in the Oman ophiolite *Earth Planet. Sci. Lett.* **180** 133–148 http://www.gm.univ-montp2.fr/spip/IMG/pdf/Godard_EPSL_00.pdf.

[2] Gibsher A A 2009 Sostav i stroenieordovikskojlithospheric mantle Western Sangilena(Central Asian fold belt) from the data on the study of mantle xenoliths from the camptonite dikes of the Agardaghskyalkaline-basaltoid complex: Avtoref. dis.... doct.geol.-min. nauk. Novosibirsk 16 p. http://earthpapers.net/preview/306388/a?#?page=18.

[3] Nemtsovich V M 1967 Agardagsky Complex of Alkaline Basaltoids in the South-East of Tuva *Doklady AN USSR* **176-2** 81–85.https://elibrary.ru/contents.asp?issueid=1768949.

[4] Blyuman B A 2011 Dike complexes of alkaline basalts of Sangilen (South-Eastern Tuva) *Doklady AN USSR* **247-3** 672–674.https://elibrary.ru/contents.asp?issueid=1394766.

IOP Conf. Series: Earth and Environmental Science **110** (2017) 012018 doi:10.1088/1755-1315/110/1/012018

New data on the substantial composition of Kalba rare metal deposits

T. A. Oitseva[1], B. A. Dyachkov[1,5], A. G. Vladimirov[2,3,4], O. N. Kuzmina[1], O. V. Ageeva[6]

[1]D. Serikbaev East Kazakhstan State Technical University, Ust-Kamenogorsk, Republic of Kazakhstan
[2]V.S. Sobolev Institute of Geology and Mineralogy SB RAS, Novosibirsk, Russia
[3]Novosibirsk State University, Novosibirsk, Russia
[4] Tomsk State University, Tomsk, Russia
[5]Altai Geological and Ecological Institute, Ust-Kamenogorsk, Republic of Kazakhstan
[6] "GRK Topaz" LLP, Ust-Kamenogorsk, Republic of Kazakhstan

E-mail: tatiana.oitseva@gmail.com

Abstract. Geotectonic position, features of the geological structure and rare metal specialization of the Kalba-Narym granitoid belt formed in the Hercynian cycle in the post-collision (orogenic) geodynamic situation are considered. A geological-genetic model for the formation of the leading type of rare-metal pegmatite deposits (Ta, Nb, Be, Li, etc.) is presented. They are spatially and genetically related mainly to the granitoids of the 1st phase of the Kalba complex, P1 (Bakennoye, Jubilee, Belaya Gora, etc.). The rhythmically pulsating orientation of the process of pegmatite formation with the introduction of ore-bearing fluids (H2O, F, B, Cl, Ta, Nb, Be, etc.) is emphasized from the intracamera focus of a semi-closed magmatic system. The preferred location of ore pegmatite veins in granitoids of moderate basicity occupying an intermediate position in the petrochemical composition between normal granites and granodiorites geochemically specialized in Li, Rb, Cs, Sn, Nb, Ta. The leading ore-controlling role of the latitudinal deep faults of the ancient site in the distribution of rare-metal ore fields and deposits (Ognevsk-Bakennoye, Asubulak, Belogorsk, etc.) is determined. There is a zonal structure of pegmatite veins, a gradual development of mineral complexes from the graphic and oligoclase-microcline (non-ore) to microcline-albite and color albite-spodumene (ore). The mineralization of pegmatite veins is determined by the degree of intensity of the manifestation in them of metasomatic processes (microclinization, alibitization, greisenization, spodumenization, tourmalinization, etc.) and the identification of the main ore minerals (tantalite-columbite, cassiterite, spodumene and beryl). The diversity of the material composition of rare-metal pegmatites containing many unique minerals (cleavelandite, lepidolite, ambligonite, color tourmaline, spodumene, pollucite, etc.) is reflected, which brings them closer to the pegmatite deposits of foreign countries (Koktogai, Bernik Lake, etc.).
New results of the investigation of the material composition of ore-bearing granites, pegmatites and typomorphic minerals using electron microscopy reflecting the distribution of rare-earth, rare-metal, chalcophile and other elements in them are presented. Indicators of rare metal ore formation are rock-forming minerals of granites (quartz, microcline, biotite, muscovite), ore and associated minerals (cleavelandite, lepidolite, cassiterite, etc.). The most informative minerals include mica (muscovite, giltbertite, lepidolite), colored tourmalines and beryls of different composition and color. Identified typomorphic minerals and geochemical elements-

indicators of rare metal pegmatite formation are considered as a leading search criterion in assessing the prospects of the territory of East Kazakhstan.

1. Introduction
At present, Kazakhstan faces the problem of recreating its own mineral and raw materials base of rare metals and, first of all, tantalum, niobium, beryllium, lithium, rare earthsmetals, which are widely used in the sphere of high technologies. Kalba-Narym tectonic zone is the main rare-metal-bearing structure in the territory of the Great Altai. Many deposits and ore occurrences of rare metals of various genetic types are concentrated there [1, 3]. Among them, there are deposits of rare metal pegmatites relating to the leading geological-industrial type, which in the past years were developed by Belogorsk ore mining and processing enterprise, but now they are suspended. The research works carried out in recent years show that the region's perspectives for rare metals have not been exhausted yet and it is possible to identify new rare metal occurrences not only on the surface and but also hidden at depth and buried under the cover of loose deposits. The actual task is to resume prognostic-metallogenic and geological explorations for rare metals in order to assess new promising areas and sites, as well as a more detailed study of known ore-bearing structures and objects. According to these facts, regional and local forecasting and search criteria have been developed, among which the most important are geological-petrological and mineralogical-geochemical factors and prerequisites [3, 7].

2. Geological structure
According to the new geotectonic concepts, Kalba-Narym zone is characterized as anallogenic block of the Earth crust (terrane), a fragment of the continental plate drifting in the paleo-Asiatic ocean and joined to the Great Altai in the stage of the Hercynian collision of C_1. In the early stage of Hercynide (D_1-C_1), this was a large downfold made by carbon-carbonate-terrigenous formations (D_2-C_1) (Kystav-Kurchum, Takyr, Burabay and Dalankariy series). In the middle collision stage (C_1-C_3) molasse formation (Taubin series C_2), hypabyssal minor-intrusions and gabbro-diabase dikes (Karabirjuk complex C_{2-3}) and granodiorite-plagiogranites (Kunush complex C_{2-3}) were limitedly appeared. The late post-collision (orogenic) stage was marked by vertical arched and blocky uplifts and a flash of intense granitoid magmatism of Permian time with the formation of a large Kalba-Narym granitoid belt of rare metal specialization (Kalbin P_1 and Monastyr P_2 complexes). In the final stage of the Hercynian cycle, the north - eastern belts of dikes of gabbro-diabase-granite-porphyritic formation were localized (Mirolyubov complex P_2-T_1) [1].

According to the geological and geophysical data, Kalba-Narym zone is characterized by a sialic type of the Earth crust section with an increased depth of the meta- granitic layer up to 12 km and a reduced depth of the metabasalt up to 14-18 km with a metadiorite layer thickness of 12-14 km. In geophysical fields, it is characterized by negative magnetic anomalies and a regional gravitational minimum, which is explained by the immersion of the surface of M (42-48 km) and K (18-26 km).

Kalba-Narym granitoid belt (KNB) is the main rare metal structure of the Altai accretion-collisional system and is considered in the structure of the Great Altai [2, 3]. It is located between West Kalbin and Irtysh tectonic zones and is limited by deep faults - Kalba-Narym (in the north-east) and Terektin (in the south-west). The considering ore-bearing structure is linearly extended in the north-west direction by more than 500 km at a width of 20-50 km. The structural-metallogenic model of KNB reflects the relationship of the ore-magmatic systems with the deep zones of the Earth crust and the upper mantle, and, consequently, the granite belt was formed as a result of the prolonged deep evolution of the lithosphere substance. The centers of magma formation originated, judging by the composition of the granite smelting, in the metagranite layer or on its boundary with the metadiorite. Zones of transit heat-mass flows penetrated from the lower parts of the Earth crust and the upper mantle along the system of deep faults [3]. We also know the data of V.L. Khomichev that granites are the products of intra-chamber differentiation of the original "basic magma" (2010).

The ore-controlling role of latitudinal deep faults of ancient deposits and long-term activation is emphasized in the distribution of rare-metal pegmatite fields and deposits, especially at the nodes of their intersection with north - western, north - eastern or meridional disjuncts (Gremyachin-Kiin, Asubulak, Belogor, Mirolubov and others). Thus, Asubulak ore field is controlled by a latitudinal fault along which two ore-bearing strips of sublatitudinal strike can be traced: (1) Ungur (northern), including the ore objects of Carmen-Kuus, Akkesen, Ungursai, Plachgora and 2) Krasnokordon (southern), uniting the Yubileynoye industrial deposit of ore occurrences Krasny Cordon, Skalnoye and Budo in 1.5 km increments (Figure 1).

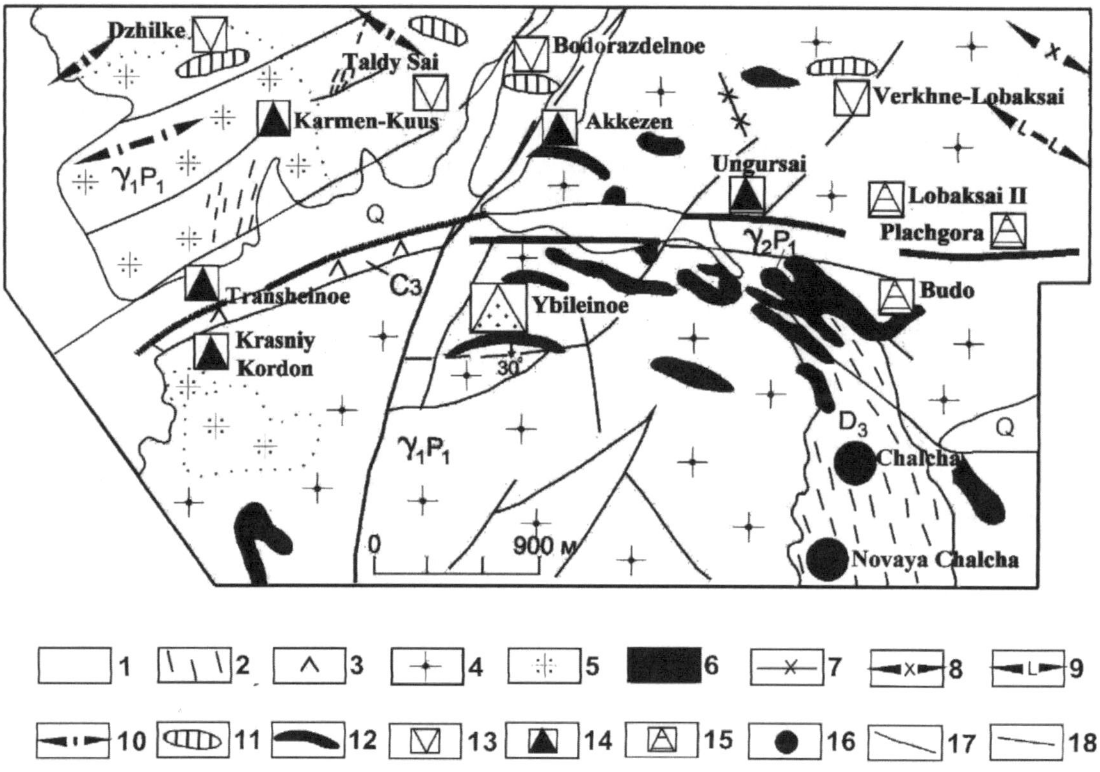

1 - quaternary sediments; 2 - shales of the takyr suite; 3 - diorites, granodiorites of the kunushsky complex; 4 - medium-coarse-grained porphyry-shaped biotite granites of the I phase, 5 - their contaminated differences and 6 - fine-medium-grained muscovite granites of the II phase of the kalbinsk complex; 7 - veined aplite-like granites, 8 - aplite-pegmatites, 9 - oligoclase-microcline pegmatites, barren, 10 - albitized pegmatites poorly ore-bearing, 11 - block microcline pegmatites, 12 - ore pegmatite veins; 13 - deposits of block columbite-beryl; 14-15 - deposits of rare metal pegmatites (14 - microcline-albite with petalite and 15 - albite with a colored mineral complex, spodumene-albite); 16 - quartz-veined tin; 17 - fragments of latitudinal deep fault; 18 - small disruptive disorders

Figure 1. Scheme of the geological structure of the Asubulak ore field

The model of formation of rare metal pegmatite deposits is determined by their genetic connection with the granites of Kalba complex P_1 and the spatial distribution of bodies mainly in medium-coarse-grained biotite granites of the I phase and their exocontacts [3, 6]. The processes of pegmatite

IOP Conf. Series: Earth and Environmental Science **110** (2017) 012018 doi:10.1088/1755-1315/110/1/012018

formation probably occurred in open or half-closed magmatic system with the rhythmically pulsating supply of ore-bearing distillates (H_2O, F, B, O, Ta, Nb, Be and others) from the intro- chambered center of granite massifs. The pulsating inflow of pegmatite-forming fluids determined the multi-rhythmic zonation of pegmatite veins, the phasic development of mineral complexes from graphical and oligoclase - microcline (barren) to microcline-albite and spodumene - containing (non-ferrous ore) with complex ores (Ta, Nb, Be, Li, Cs, Sn W).

3. Geological and petrological features

In the evolutionary range, granitoids of Kalba complex refer to the normal branch of differentiation and occupy an intermediate position between granodiorites and granites (Figure 2).

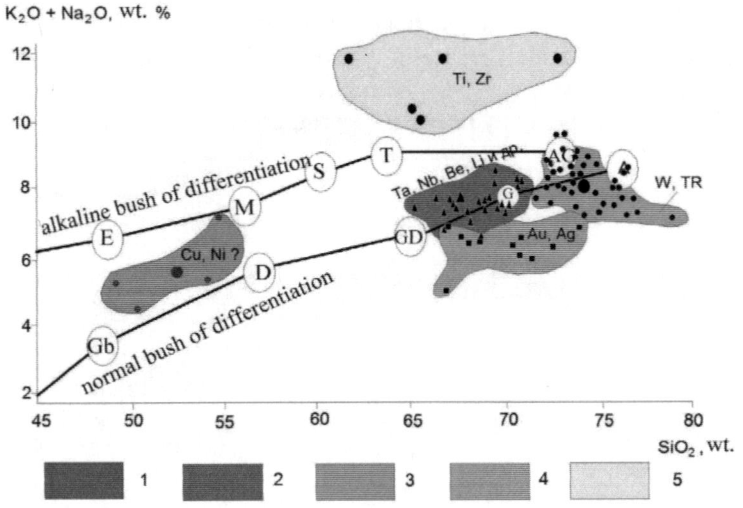

Intrusive complexes:
1 – karabiruk gabbro-diorite (C_{2-3}), 2 - kunush granodiorite - plagiogranite (C_3), 3 – kalba granite (P_1). 4 - monastirsky leucocytes (P_2), 5 - byran monzonite - syenite - alkaline - hydrocarbon (P_2).
Typical Rocks by R. Deli:
A - alyaskit,
G - granite,
GD - granodiorite,
D - granodiorite,
Gb - gabbro,
M - monzonite,
C - syenite,
T - trachite,
AG – alkaline granite, E - esseksit

Figure 2. Diagram of alkaline and flint acidity of Kalba granitoids

The first phase is characterized by an unbalanced composition of rocks (from normal granites to contaminated granites and granodiorites), accounting for about 70% of Kalba complex. It is represented mainly by medium-coarse-grained porphyritic biotite granites containing (%): plagioclase (No. 18-37) – 35-41, potassium feldspar – 15-20, quartz – 28-32, biotite up to 5-10, accessories - 0.5. Petrochemicals refer to the normal range of the sodium-potassium series ((Na2O: K2O = 0.78), low-plumazite (Ka = 0.69), moderate basicity (r – 0.34 kcal), and very high aluminous. Accessory specialization - garnet-ilmenite-apatite, accessory minerals - zircon, monazite, sphene, tourmaline and others; minerals - indicators of mineralization - apatite, tourmaline, cassiterite, spodumene.

Geochemically, granites are enriched with Li, Rb, Cs, Sn, Nb, the content of Ta in them varies from 1.2 to 4.8 g/t. The increased content of rare elements (Ta, Nb, Be, Li, Sn) is found in rock-forming minerals-quartz, feldspars, blotites and especially in muscovites, as well as in accessories - apatite, ilmenite, zircon and other minerals. The enrichment of granite melts with mineralizers - (extractors) of rare elements - B, F, C_1 has been revealed. In general, the geochemical type of granites is tin-tantalum (with lithium and beryllium) [3].

According to the results of mass spectrometry, the amount of rare-earth elements in granites of the first phase is 80.74 g/t. The enrichment of these granites with Nb, Sn and rare alkalis, especially Li (the ratio Cs: Rb: Li = 1: 1.6: 10), is established (Table 1).

Table 1. The content of rare elements in granites of the I phase of the Kalba complex (g / t)

Sample №	Ta	Nb	Be	Li	Rb	Cs	Sn	W	Mo
0-21 (1)	1,49	34,40	3,94	**294,50**	166,6	18,92	10,35	0,80	0,83
0-21 (2)	1,75	14,84	2,29	**289,60**	150,1	21,98	11,91	0,46	0,43
0-21 (3)	1,20	11,63	1,66	**256,00**	145,1	19,24	20,55	0,63	0,75
0-21 (4)	0,89	12,74	3,31	**107,50**	164,4	40,87	18,74	0,66	0,53
0-21 (5)	0,95	12,76	1,91	**295,00**	158,2	18,16	11,53	1,04	1,30
Average value	1,26	17,27	2,62	**248,52**	156,88	23,83	14,62	0,72	0,77

The calculated coefficients reflect a relatively uniform distribution of Rb and Li and a sharp increase in Rb/Cs ratio in later granitoids due to the increase in their potassium alkalinity (Figure 3). There is an increased tininess in regard to Ta and W, and the distribution of niobium and tantalum is uneven at a close ratio in granites of the I phase (1: 1) in all types of granitoids.

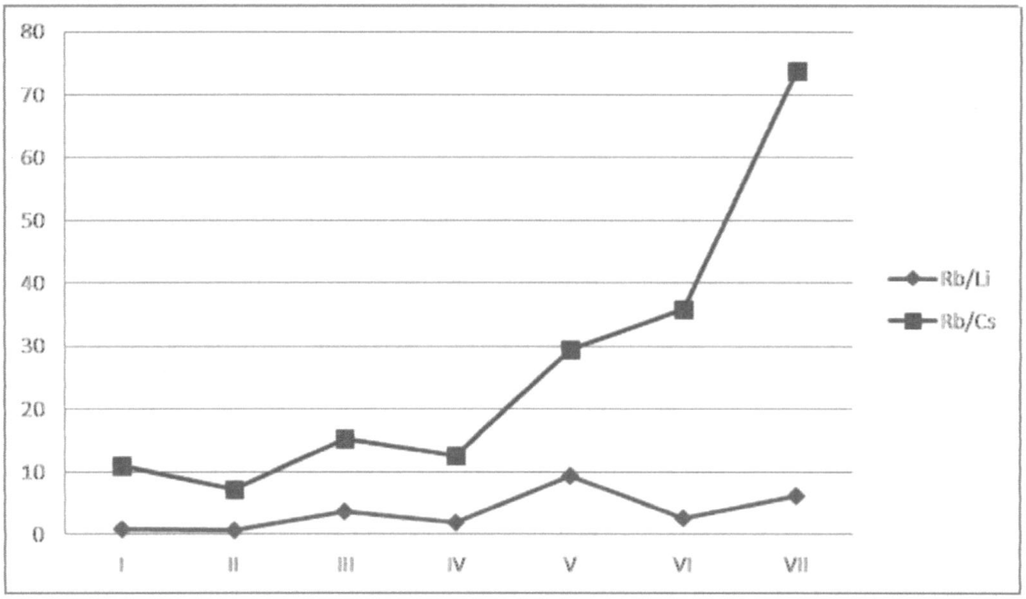

I-VII - intrusive complexes: I - kunush granodiorite - plagiogranite, P_1; II - IV kalba P_1 (II – porphyritic biotite granites of the first phase, III - biotite and IV - muskovite granites of the second phase); V - monastirsky leucocratic P_2, VI - keregetas-espinsky alkaline granite P_2; VII - mirolyubov dyke P_2-T_1

Figure 3. Diargam of distribution coefficients of the granitoid diagrams of granitoid Kalba and Zharma – Saur region

Studies show that the original granite melts, according to the prevailing content of rare earths in the light group and enrichment with lithophilic rare elements, had a crustal origin. By this fact, they differ from the granitoids of the East Sayan zone of mantle-crustal origin enriched with heavy rare earths [2]. At the same time, we determined that the ore-bearing granites of Kalba, in addition to rare elements,

IOP Conf. Series: Earth and Environmental Science **110** (2017) 012018 doi:10.1088/1755-1315/110/1/012018

have high concentration of Zn, Pb, Cu and significant values of As, Sb, Ag, Au, Pt, Bi and other elements reflecting mixed sources of the ore matter, associated, probably, with fluid flows from different level areas of magmatism. That was reflected in the formation in pegmatites, except lepidolite, spodumene, tantalite-columbite, beryl and cassiterite, associated minerals (pyrite, galena, arsenopyrite, antimonite, gold and others).

4. Features of mineral composition

Pegmatites are divided into three types: 1) oligoclase - microcline (barren), 2) microcline -block with beryl and columbite (slightly ore-bearing) and 3) albite with several mineral complexes (ore). In the ore veins, taking into account the works of A.I. Ginzburg, E.P. Pushko, O.D. Gavrilenko and our data, the following mineral associations or complexes are distinguished: microcline, albite - microcline, albite, quartz-albite-muscovite, spodumene - cleavelandite -quartz, lechidolite-pollucite- cleavelandite (non-ferrous).

Mineral and chemical composition of ores are very diverse. The main useful components are Ta, Sn, Li, Rb, Cs. The main vein and ore minerals are albite, microcline, quartz, muscovite, spodumene, beryl, tantalite-columbite, pollucite and cassiterite. Accessory minerals are apatite, tourmaline, garnet, gilbertite, morion, oncosin, cookeite, eucryptite, arsenopyrite, pyrite, calcite, microlite, mangan tantalite, montebrasite, amblygonite, petalite, molybdenite and others. Minerals that are rare in occurrence - sphene, epidote, biotite, chlorite, graphite, pyroxene, hornblende, struverite, ixiolite, bertrandite, ampangabeite, cyrtolite, pyrochlore, wolframite, scheelite, monazite, titano-ferrite, etc.

A great contribution to the study of the mineralogical composition of Kalba pegmatites was made by A.I. Ginzburg, V.D. Nikitin, S.G. Shavlo, N.A. Solodov, V.I. Kuznetsov, Yu.A. Sadovskiy, V.A. Filippov, and others. Pegmatites are characterized by a wide variety of minerals and contain more than 80 minerals in terms of their material composition and structural features. The main rock-forming minerals (%): albite, quartz, microcline and muscovite; accessory minerals are apatite, tourmaline, gilbertite, fluorite, calcite, and others. A unique complex of minerals of several generations (albite, cleavelandite, lepidolite, pollucite, spodumene, petalite, ambligonite, polychrome tourmaline, beryl, tantalite-columbite, cassiterite, and others) was found in a number of ore fields and deposits. The main ore minerals are tantalite-columbite, cassiterite, spodumene and beryl [1, 3].

New researches of the material composition of pegmatites were carried out by authors using electron microscopy in order to study typomorphic minerals and geochemical elements - indicators of rare metal pegmatite formation. On a number of typical deposits (Bakennoye, Yubileynoye, Medvedka, Tochka, and others) specimen and samples were taken from the inclosing granites, metasomatic formations and pegmatite veins. Analytical studies were carried out mainly in EKSTU "IRGETAS" laboratory on the ICP-MS mass spectrometer with inductively coupled plasma, which determines 73 elements with high sensitivity, and on the scanning electron microscope JSM-6390 LV. X-ray structural spectrometer SRV-1M was used to determine the composition of ore minerals. Isotope studies were carried out at the Institute of Geology and Mineralogy of SB RAS (Novosibirsk).

Mineralogical search criteria are determined by the degree of intensity of metasomatic processes in pegmatite veins (microclinization, albitization, greysening, and others) and identification of typomorphic minerals (microcline, albite, cleavelandite, muscovite, lepidolite, tourmaline, spodumene, and others). The conducted researches established that the leading minerals-indicators of rare metal ore formation are specular stones of different composition and colour, which are characterized by high concentration of Li, F and rare elements. Thus, for quartz- microcline – albitite pegmatites of Yubileynoye deposit, bladed clear muscovites of high fluorinity (F> 1-1.5%) enriched with Ta, Nb, Be, Sn and rare alkalis are typical (Table 2).

IOP Conf. Series: Earth and Environmental Science **110** (2017) 012018 doi:10.1088/1755-1315/110/1/012018

Table 2. Content of elements in muscovites from rare metal pegmatites of the Jubilee deposit (g / t)

№	Sample №	Ta	Nb	Be	Li	Rb	Cs	Sn	W	Mo
1	5953-6	27,78	199	56	6545	4869	894	502,2	4,35	0,06
2	4953	28,47	267	70,64	3996	4787	513,2	567,6	4,44	0,13
3	4980-6	23,98	218,8	38,28	5465	4849	639,4	557,8	6,59	0,86
4	4987-г	31,44	330,2	33,24	2190	4016	261,5	574,7	8,91	0,84
5	4994-г	39	214,7	29,28	9478	5837	1023	491,2	4,51	0,62
6	2234-1	121,14	147,37	23,14	4564	4114	547,6	323,4	21,69	6,98
7	2235-3	63,47	156,81	20,73	3299	3226	448,6	348,2	15,62	8,53
8	4215-a	153,1	185,1	22,14	413,9	1144,9	218,8	230,6	18,34	6,58
9	2212-1	121,6	301,1	28,17	2844	3872	400	403,9	10,27	6,07
10	2212-3	1795	406,1	32,6	2408	4064	600	721,5	7,91	4,41
11	4973-г	67,39	692	35,89	1010	1555	99,75	216,44	25,92 :	7,8

Note. The results of mass spectrometric analysis (ICP-MS). Analyst S.N. Polezhaev

Flaky muscovites of green colour, with higher Ta contents (up to 233 g / t) and Nb (up to 380 g / t), are characteristic of quartz-albite-muscovite (greysen) complex. For quartz- cleavelandite -spodumene (non-ferrous) complex, coarse-flaky lepidolites reflecting rich complex ores (Ta, Nb, Li, Cs, Sn) are enriched with Li (up to 16240) and Rb (up to 1718 g / t) (Table 3, Figure 4).

Table 3. Content of rare elements in mica (g / t).

№	Characteris tic of rock	Ta	Nb	Be	Li	Rb	Cs	Sn	W	Mo
1	Lepidolite	10.34	50.94	4.74	16240	1718	263.0	81.46	8.38	12.52
2	Gilberte	18.52	24.1	6.4	3835	5210	3328	148.5	1.93	0.55
3	Muskovite	224.76	283.47	35.46	3837.54	3848.54	513.26	448.87	11.69	3.90
	Average	84.54	119.5	15.5	7970.8	3592.2	1368	226.3	7.33	5.66

Note. The coefficients are calculated from the results of ICP-MS (g / t). Analyst S. N. Polezhaev

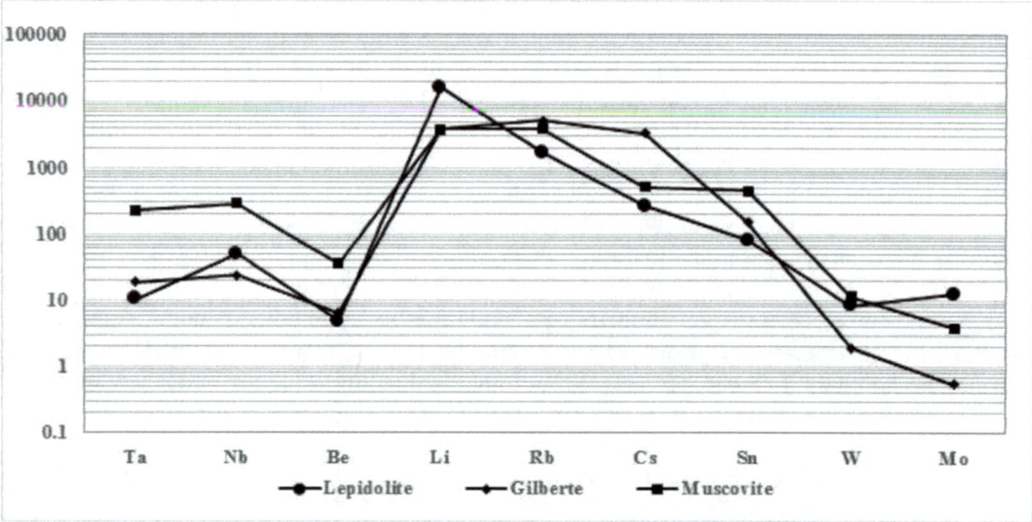

a - thincipper inclusion of tantalite and b - cassiterite, Medvedka deposit; c – microinclusion of tetrahedrite and d - fluorapatite, Bakennoe deposit. Analyst A. B. Rusakova

Figure 4. Microinclusion of minerals in muscovite of pegmatite deposits

The cryptocrystalline muscovite (zilbertite) differences of Carmen-Kuus ore occurrences are characterized by anomalous values of Cs (3328) with Li (3835) and Rb (5210 g/t). Microinclusions of tantalite, cassiterite, fluorapatite and other minerals were fixed in muscovites on the SEM image (Figure 4).

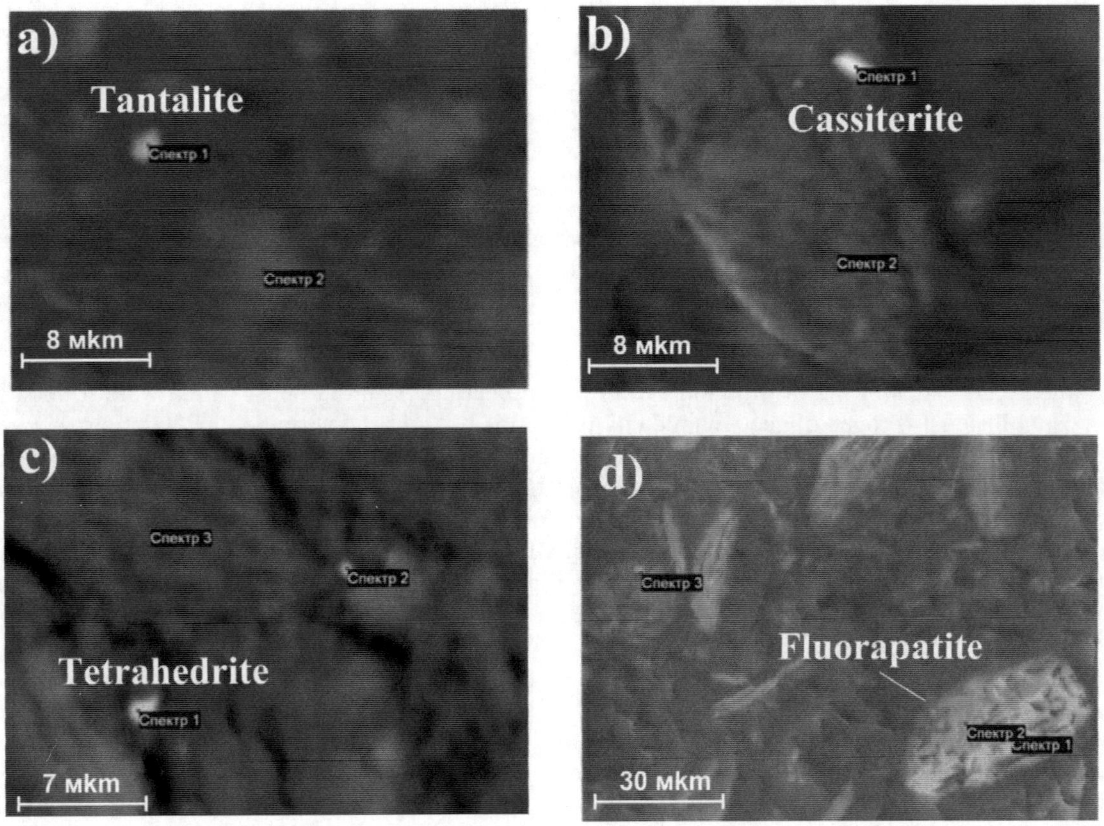

Figure 5. Diagram of distribution rare element in mica (g / t)

Informative minerals also include tourmalines of different composition and colour, among which the schorl with low values of rare elements reflect simple pegmatites, and polychrome tourmalines with anomalous contents of rare alkalies indicate ore mineral complexes (Table 4).

Table 4. Content of rare elements in tourmaline (g / t).

Minerals	Ta	Nb	Be	Li	Rb	Cs	Sn	W	Mo
Tourmaline (polychrome)	37.2	97.1	2.66	2986	605.6	135400	19.26	1.86	1.73
Rubellite (pink)	43.7	128.9	5	226.2	14.9	16.1	61.57	1.26	2.48
Sherl (Black)	7.6	22.6	2	135.4	6.7	22.1	4.87	0.82	1.93

Note. The results of mass spectrometric analysis. Analyst S. N. Polezhaev

In order to compare significant differences in beryl composition from different deposits are given (Table 5). Beryls from block microcline pegmatites are poor in Ta and Nb. Aquamarines of the

Delbegei massif are enriched with Ta and Nb, they contain more Sn with a low content of rare alkalis. Emeralds are characterized by low contents of rare elements and an anomalous value of Cr (636.4 g / t). The contents of Ta, Nb, Mo are higher for blue beryl, and white beryl with anomalous values of Li, Cs, Ta, Nb reflects rare metal pegmatites of Koktogay deposit (China).

Table 5. Content of rare elements in beryl (g / t)

	Ta	Nb	Be	Li	Rb	Cs	Sn	W	Mo	Cr
1	1.73	3.79	35377	294.77	60.35	320.2	14.76	2.52	1.34	25.44
2	1232	8080	15380	108.2	28.11	57.77	25.43	6.79	0.28	122.20
3	169.2	109.4	15780	50.04	17.5	48.89	45.86	4.77	0.25	117.80
4	0.4	9.5	30210	42.66	7.7	35.09	3.19	10.32	2.30	636.40
5	54.69	27.2	10390	240.8	29.5	177.7	2.26	1.56	495.3	44.32
6	284	96.95	12060	6782	205.2	15220	2.31	1.9	33.63	68.78

Note: 1 - beryl greenish-gray from block microcline pegmatites of Kalba, 2 - aquamarine from greisens of Delbegetey massif, 3 - aquamarine from quartz vein of Delbegetey massif, 4 - emerald, 5 - beryl blue, ore field of Koktogay, 6 - beryl white, ore Kotogay field (China).

5. Conclusion

The main regularities of the distribution of rare metal-bearing granitoid belts, which were formed in the post-collisional (orogenic) geodynamic environment (P_1-T_1), are determined in tectonic structures with the sialitic type of the Earth crust section. In Kalba-Narym belt, the leading deposits of rare-metal pegmatites are genetically related to granites of the 1st phase of Kalba complex (P_1). The ore-controlling role of latitudinal deep faults in the location of pegmatite ore fields and deposits is emphasized. The geological-genetic model of pegmatite formation reflects the gradual development of mineral complexes from the graphite and oligoclase- microcline (barren) to spodumene-containing one (non-ferrous) with rich complex ores (Ta, Nb, Be, Li, Cs, Sn). Mineralogical factors for the estimation of rare metal pegmatites are determined by the intensity degree of the metasomatic processes occurrences (microclinization, albitization, greysening, spodumenization, silicification) in them, the indicators of which can be typomorphic minerals (cleavelandite, muscovite, lepidolite, spodumene, colour tourmalines, pollucite, and others) similar for industrial pegmatite deposits of foreign countries (Koktogay, Bernik Lake, King Mountain, and others) [4, 5, 8, 9].

The leading role in ore formation is attached to albitization process, which is the initiator of all subsequent metasomatic transformations of pegmatite veins. Albitization is also associated with the formation of rare-earth mineralization in alkaline granites of Upper Espe deposit and rare metal mineralization in albitites at Vasylkovskoye gold-bearing deposit. The processes of albitization are also widely shown in other rare metal objects (Karasu, Alakha, Novo-Akhmirov, and others) [7]. The identification of micro- and fine-dispersed inclusions of rare and rare-earth minerals and geochemical elements (Au, Ag, Pt, Ir, U) in ores is of scientific importance, these results are recommended to take into account in the practice of geological explorations.

6. Acknowledgment

The article was prepared with the financial support of JSC "NATD" of the Republic of Kazakhstan (under grant No. 403).

References

[1] Sherba G N 1998 *Greater Altai (geology and metallogeny)* book 1-2 (Almaty: Gylym) p 304
[2] Vladimirov A G and Kosarev A M and et al 2016 *Correlation and geodynamic interpretation of igneous events and mineralization of the Southern Urals and Altai (Middle-Late Paleozoic)* (Geology and Geophysics) p 51
[3] Dyachkov B A 2012 *Genetic types of rare metal deposits of the Kalba-Narym belt* (Ust-Kamenogorsk: EKSTU) p 129

[4] Zagorsky V E, Vladimirov A G, Makagon V M, Kuznetsova L G and et al 2014 Large fields of spodumene pegmatites in the conditions of rifting and post-collisional shear-expansion deformations of the continental lithosphere *Geology and Geophysics* **2** 237-251.

[5] Zagorsky V E, Makagon V M, Shmakin B M and et al Rare metal pegmatite *Granite pegmatites* **2** 285

[6] Kotler P D, Chomikh S V, Vladimirov A g and et al New data on age and geodynamic interpretation of granitoids of the Kalba-Narym batholith (East Kazakhstan) *Reports of the Academy of Sciences* **462** 1-6

[7] Dyachkov B A, Mataibayeva I E, Zimanovskaya N A and et al *Efficient use of resources and environmental protection - key issues of mining and metallurgical complex development* Materials of IX International conference (Ust-Kamenogorsk: EKSTU) part 3 pp 42-49

[8] Morgan G B, London D 1987 *Alteration of amfibolitic wallrocks around the Tanco rare-element pegmatite, Bernic Lake, Manitova* American Mineral volume 2 pp 1097-1121.

[9] Zou Tianren Cao Huizhi, Wu Boging 1989 *Pegmatites of Altai province Xinjiang* Acta Geol. Sinica volume 2 p 45

IOP Conf. Series: Earth and Environmental Science **110** (2017) 012019 doi:10.1088/1755-1315/110/1/012019

Biostratigraphy and foraminiferal Assemblages of the West Siberian Albian

V M Podobina

National research Tomsk State University, Tomsk, Russia

Email: Podobina@ggf.tsu.ru

Abstract. New data of the Biostratigraphy and Albian foraminiferal assemblages from the sections of newly drilled wells in the Yamal peninsula have been obtained. This territory is assigned by the author to the Northern palaeobiogeographical district, the southern border of which goes along the latitudinal flow of the Ob River. The most diverse Albian foraminiferal assemblages have been discovered in the section of borehole 50 of the Malyginskaya area as well as in the sections of borehole 124 of the West-Tambeiskaya, boreholes, 201, 205 of the North-Tambeiskaya areas. Somewhat similar Albian foraminiferal assemblages were earlier studied by the author in 10 sections of the boreholes of the Samatlor area in the latitudinal Ob River Region. The Middle Albian assemblage is the most stable and diverse in its specific respect. It was discovered not only in the identified sections in the Yamal wells but also southwards in the earlier explored areas. The Albian foraminiferal species of the Yamal peninsula are illustrated in 5 photos and 1 drawing.
Key words. Biostratigraphy, Foraminifera, Albian, Western Siberia

1. Introduction

The Albian foraminiferal assemblages have been discovered recently in the Khanty-Mansiiskian Horizon in the Samatlor, Polarnaya, East-Lodochnaya and Sauthern Russkaya areas of Western Siberia [1]. Data on these assemblages were earlier obtained with regard to Transuralia [2]-[4]. The most detailed research with the description of many Albian species of Transuralia is presented by Z.I. Bulatova [2].

According to palaeobiogeographical division into districts within Western Siberia, these areas with the exception of Transuralia are related to the Northern district. The author regards Transuralia as the western district [5]. The Albian assemblages of the Northern district are studied in detail in 10 sections of the wells of the Samatlor area located in the middle part of the latitudinal flow of the Ob River. Only agglutinated quartz-siliceous foraminifera of a rather diverse specific composition are found in all these sections. V.A. Marinov also investigated the Transuralian Albian foraminifera and reflected his research in the regional stratigraphic scheme, 2005 [6]. Based on the results of the Albian foraminiferal finds in the Northern district, it was possible to compare them with the Transuralian as well as those from the Canadian Province [1] [7]-[10]. Within the Samatlor area, the author has established two foraminiferal zones: *Ammobaculites fragmentarius, Gaudryinopsis filiformis* (the Middle Albian) and *Ammotium braunsteini, Verneuilinoides borealis assanoviensis* (the Upper Albian). In countrast to Transuralia, the early Albian foraminifera have not been discovered yet in the sections of the Samatlor area but in the other areas of the Northern district there are the Lower Albian Foraminiferal assemblages.

In the latest regional stratigraphic scheme [6] with regard to the Albian of Western Siberia the species *Verneuilinoides borealis* Tappan *assanoviensis* (Zaspelova) is suggested by A.V. Marinov as one of index-species for the layers of the Middle and Upper Albian along with various species of the genus *Ammosiphonia* (*A. jamaica* – the Upper Albian, *A. beresoviensis* – the Middle Albian). In our view, these species as well as the genus in Transuralia and in other sections of the Northern district are not traced.

Based on the generic evidence including the chemical composition of the wall of a shell, the genus *Ammotium* characteristic of the Albian deposits of the West-Siberian and Canadian Provinces is present. The index-species *Verneuilinoides borealis* Tappan *assanoviensis* (Zaspelova) is mainly assigned to the upper layers of the Khanty-Mansiiskian Horizon and it can be an index-species for the Upper Albian just for the same zone. Z.I. Bulatova [2] was the first to distinguish the Upper Albian zone with the indicated index-species which coincides with the authors opinion [1]. To the species *V. borealis* Tappan *assanoviensis* (Zaspelova) the author added the second index-species *Ammotium braunsteini* (Cushman et Applin) which is no fewer than an index-species characteristic of the Upper Albian [1] [10]. As it is pointed out, the early Albian foraminiferal assemblage is not discovered in 10 sections of the Samatlor area. Here the assemblages of the Middle and Upper Albian are well-known.
The foraminiferal species: *Hyperammina pulverea* Bulatova, *Reophax troyeri* Tappan, *Haplophragmoides reconditus* Bulatova, *Recurvoides leushiensis* Bulatova, *Ammobaculites fragmentarius* Cushman, *Ammomarginulina cragini* Loeblich et Tappan, *Spiroplectammina cognata* Podobina, *Gaudryinopsis filiformis* (Berthelin), *G.* cf. *oblongus* (Zaspelova), *Pseudoverneuilina albica* Podobina are present in the Middle Albian deposits of the Samatlor sections in the foraminiferal zone Ammobaculites fragmentarius, Gaudryinopsis filiformis. The number of the samples of the species identified is not similar. The shells of the genera: *Haplophragmoides, Recurvoides, Ammobaculites, Pseudoverneuilina* and *Gaudryinopsis* are dominant. The preservation of the forms is also various, but, however, the shells preserved from the destruction make it possible to establish a major systematic composition of the Middle Albian index-species as well as *Recurvoides leushiensis* Bulatova and a species *Pseudoverneuilina albica* Podobina of the genus *Pseudoverneuilina* Podobina, 2013, are the most characteristic ones. To some extent, this species is similar with the genus *Verneuilina Orbigny*, 1840, but they are different in the quartz-siliceous wall composition (non-calcareous), a rapidly enlarging lower shell in the form of a pyramid with distinct three side angles and closely adjacent lower chambers.
The representatives of the genera: Haplophragmoides, *Ammomarginulina*, *Ammotium*, *Verneuilinoides* and *Gaudryinopsis* are the numerous ones in the Samatlor sections in the late Albian assemblage with *Ammotium braunsteini*, *Verneuilinoides borealis assanoviensis*. Among them, *Ammomarginulina obscura* (Loeblich), *Spiroplectammina sibirica* Podobina, *Gaudryinopsis oblongus* (Zaspelova) are the most distinctive ones with the exception of index-species. The washed out sediment is almost completely made up of coarse and middle-grained foraminiferal shells. Shell preservation is various, however, the specific composition characteristic of the Upper Albian in the sections of the Samatlor area is also revealed despite many destroyed and deformed forms. Pseudomorphoses of calcareous foraminifera which make it difficult to determine their systematic position are traced in individual samples. In Transuralia Albian shells of agglutinated and calcareous-secretion foraminifera are well preserved and the previous authors distinguished systematically relatively diverse assemblages of these organisms. Probably, the Transuralia basin formed by the boreal Albian transgression was much more deeper with a normal hydrological regime resulting in a diversity and good foraminiferal preservation.
More impoverished, mainly, middle Albian foraminiferal assemblage with *Ammobaculites fragmentarius*, *Gaudryinopsis filiformis* is discovered in the other areas of the Northern district – Polyarnaya, borehole 1 and East-Lodochnaya, borehole 1, South-Russkaya, 4 boreholes. Characteristic Middle Albian species which are similar to those of the Samatlor are defined except index-species. Such a similar specific composition and occurrence of course foraminiferal shells are indicative of shallow conditions of dwelling – the abundant Boreal transgression in the Middle Albian to the latitudinal flow of the Ob River.
Many species of two Albian assemblages (Middle and Upper) of the Samatlor and the other areas of the Northern district are vicariants or common species of those of the Canadian Province (Western Canada, Northern Alaska) which along with the West-Siberian Province are related to the Arctic palaeobiogeographical realm of just the same circumpolar belt.

The data obtained on the foraminiferal assemblages from the sections of the Yamal peninsula wells allow us to judge the systematic composition and the number of individuals of each species. The Middle Albian foraminiferal assemblage is distinguished more closely with regard to its specific composition in the sections of the wells. Early, and especially the Late Albian assemblage don't include diverse species the shells of which are preserved to a variable degree. Although calcereus-secretion well-preserved shells were investigated in the middle Albian assemblages, they were individual or were not found at all. Obviously, the Middle Albian transgression in the Yamal peninsula as well as in the territory of all the Northern district was vast and the basin assigned to this territory was much more deeper with a good hydrological regime for foraminiferal dwellings.

2. Material and Methods

Several wells have been drilled in the Yamal peninsula and foraminifera, ostracod nuclei have been obtained from the kern in four of them by means of microfaunistic analysis. Agglutinated quartz-siliceous and isolated calcareous-secretion foraminifera shells are found in the sections of borehole 50 of the Malyginskaya area, borehole 124 of the West-Tambeiskaya and boreholes 201, 205 North-Tambeiskaya areas from the deposits of Khanty-Mansiiskian Horizon. When studying foraminifera, it became clear that most discovered samples included the Middle Albian foraminiferal assemblages. The finds of the Early and Late Albian assemblages are isolated and of a limited occurrence in their specific relation.

Numerous kern samples (nearly 70) were investigated from the identified sections of the wells using the microfaunistic method. Thus, foraminifera and ostracod nuclei were in five kern samples from the depths interval 1820,0-1725,0 m of borehole 50 of the Malyginskaya area. Content rocks consist of dark-grey argillites with some interlayers of grey aleurolites of the Khanty-Mansiiskian Horizon. Basic agglutinated quartz-siliceous forms of the Early Albian are selected in these samples. Foraminifera of more diversified species composition and of a better preservation related to the Middle Albian are selected from the washed out powders in the above-mentioned seven samples of kern rocks of just the same lithological composition from the depths internal of 1725,0-1670,0 m. Along with the foraminifera numerous and diverse ostracod nuclei are also present.

In the section of borehole 124 of the West-Tambeiskaya area in one sample at a depth of 1850,6 m. quartz-siliceous, calcareous-secretion foraminifera and ostracod nuclei are isolated. A content rock consists of dark-grey argillite with lens-like interlayers of grey aleurolite. In contrast to agglutinated quartz-siliceous forms, calcareous-secretion ones are well-preserved. The age of the rocks of the sample concerned in defined as the Early Albian.

The two samples of this section (borehole 124) selected from the interval at the depths of 1675,8-1673,4 m. include diverse in the specific relation agglutinated and calcareous-secretion foraminifera of the Middle and Late Albian.

Foraminifera and ostracod nuclei are established in the sample from a depth of 1675,8 (borehole 124). The species of the agglutinated foraminifera with a quartz-siliceous wall are distinguished in the interval of the rocks 1675,8-1673,4 m. They are diverse in the specific respect and abundant in terms of the number of specimens of each species. Just in the same section in the above selected one sample (a depth of 1655,85 m) and poorly preserved agglutinated quartz-siliceous foraminifera which are few in number and a lot of ostracod nuclei (more than 50 samples per 100 g of a rock) are discovered. Among the foraminifera the genera *Ammotium*, *Verneuilinoides* characteristic of the Upper Albian of Western Siberia are present.

In the section of borehole 201 of the North-Tambeiskaya area 22 kern samples are selected from the interval of 1631,7-1545,0 m where foraminifera and numerous ostracod nuclei are isolated from the washed out sediment. In the foraminiferal assemblage (at a depth of 1634,0 m) foraminifera are few in number, and besides agglutinated quartz-siliceous forms assigned to the Lower Albian nuclei of calcareous-secretion shells of the order of Rotaliida are found. In 20 samples from the interval at a depth of 1630,7-1548,7 m foraminifera are of a diverse systematic composition are present. Well-preserved agglutinated foraminifera predominate among which the Middle Albian species including index-species are distinguished. Along with them, characteristic calcareous-secretion forms are selected.

3. Results of the research

The foraminiferal assemblage obtained from the kern samples of 4 well sections – 50 Malyginskaya area, 124 West-Tambeiskaya, 201, 205 North-Tambeiskaya areas include agglutinated quartz-siliceous and to a smaller extent calcareous-secretion foraminifera. The determination of the systematic composition of these foraminifera made it possible to establish three substages of the Albian layer in the Khanty-Mansiiskian Horizon. The Middle Albian substage was the most stable one which consisted of characteristic and abundant foraminiferal species. It is distinctly represented in its specific composition in all the four identified well sections. The Early and Late Albian foraminiferal assamblages stand out and they include especially the Late Albian few poorly preserved foraminifera.

3.1. The Albian biostratigraphy

Cretaceous system – K
Lower subsystem – K_1
Albian stage – K_1al
Lower substage – K_1al_1
Khanty-Mansiiskian Horizon

The Early Albian foraminiferal assemblage is established in the section of borehole 124 of West-Tambeiskaya area at a depth of 1850,0 m on the basis of the finds of quartz-siliceous and calcareous-secretion shells of foraminifera. Few ostracod nuclei are established along with foraminifera. Content rocks – dark-grey argillites have lens-like sublayers of grey aleurolites.

The species which higher in the Middle Albian became considerably diverse and got a great number of samples of each species are also specified. Here (1850,0 m), calcareous-secretion forms are of comparatively small sizes. But they, however, in contrast to agglutinated foraminifera are of a rather good preservation. The species: *Haplophragmoides* aff. *topagorukensis* Tappan, *Ammobaculites* cf. *fragmentarius* Cushman, *Gaudryinopsis* sp. *indet*., *Marginulina planuiscula* (Reuss), *Saracenaria solita* Bulatova, *Gavelinella stictata* (Tappan) are determined in the composition of the assemblage. Other poorly preserved shells assigned to the families of Haplophragmoididae and Ataxophragmiidae are also obtained.

It is difficult to judge the more exact age of the assemblage but the fact whether it is Albian or the Early Albian is hardly doubted.

The Early Albian foraminifera are discovered in the North-Tambeiskaya area of borehole 201 in dark-grey argillites, grey aleurolites and light-grey millstone grits selected from the interval depths of 1647,0-1631.0 m. Numerous ostracod nuclei are found along with them. Agglutinated coarse quartz-siliceous foraminiferal shells are of comparatively big sixes and the preservation of some samples is satisfactory.

Quartz-bearing pseudomorphoses of calcareous forms predominantly of the order of Rotaliida are also established.

The foraminiferal species of the following composition: *Saccammina micra* Bulatova, *Labrospira rotunda* Podobina, *Haplophragmoides* aff. *variabilis* Podobina, *Ammobaculites wenonahae* Tappan, *Ammoscalaria* aff. *cenomanica* Podobina, *Ammomarginulina obscura (Loeblich)*, *Flabellammina* aff. *acuminata* Podobina, *Spiroplectammina longula* Podobina, *Verneuilinoides* aff. *borealis* Tappan, *Gauryinopsis* aff. *teilleuri* Tappan, *Trochammina imiatensis* Tappan are determined in the foraminiferal assemblages at a depth of 1634,4 m (borehole 201). According to the species composition, the assemblage being studied is for the most part close to the Albian assemblages discovered to the interfluve of the Taz and Pur Rivers (the South-Russkaya area, borehole 55). Judging by the systematic foraminiferal composition and position in the section, it dates back to the Early Albian. (Fig. 1)

The Albian species known in the formation of Topagoruk of Northern Alaska [8].

Middle substage – K_1al_2
Khanty-Mansiiskian Horizon

The most diverse and numerous agglutinated quartz-siliceous and calcareous-secretion assemblages are discovered in the section of borehole 50 of the Malyginskaya area. The foraminifera are found in the depths intervals of 1680,05-1675,60 m. In the samples from this particular interval the author of the paper has astablished the Middle Albian assemblage with *Ammobaculites fragmentarius, Gaudryinopsis filiformis*. The species of *Haplophragmoides topagorukensis* Tappan prevail in the composition of the assemblage. The species: *Psammosphaera laevigata* White, *Labrospira* aff. *rotunda* Podobina, *Haplophragmoides topagorukensis* Tappan, *Recurvoides* aff. *leushiensis* Bulatova, *Ammobaculites fragmentarius* Cushman, *Pseudobolivina contorta* Bulatova, *Gaudryinopsis filiformis* (Berthelin), *Miliammina manitobensis* Wickenden, *Lenticulina topagorukensis* Tappan, *Saracenaria solita* Bulatova, *Gavelinella* aff. *strictata* (Tappan) are determined in the assemblage at a depth of 1675,6 m. In the assemblage along with the dominating agglutinated quartz-siliceous forms, calcareous-secretion shells of the genera *Lenticulina, Saracenaria, Gavelinella occur*. A similar diverse assemblage with well-preserved foraminiferal shells is first discovered among the analogous assemblages of the Northern palaeobiogeographical district. Many species of this assemblage are characteristic of the Albian assemblage of the Canadian Province [8] [9]. Fig. 2, 2a.

Agglutinated quartz-siliceous as well as calcareous-secretion foraminifera selected from the depths intervals of 1675,8-1673,4 m in the section of borehole 124 of the West-Tambeiskaya area are investigated. Separate ostracod nuclei are found in a sample at a depth of 1675,8 m.

The species: *Saccammina* aff. *sphaerica* (M. Sars), *Reophax* aff. *sherborniana* (Chapman), *Reophax* aff. *inordinatus* Young, *Haplophragmoides topagorukensis* Tappan, *Protobolivina contorta* Bulatova, *Gaudryinopsis filiformis* (Berthelin), *Lenticulina topagorukensis* Tappan, *Gavelinella* aff. *strictata* (Tappan) are determined in the middle Albian foraminiferal assemblage with *Ammobaculites fragmentarius*, *Gaudryinopsis filiformis* (Fig. 3). A similar Middle Albian assemblage is well known in the other sections of the Yamal peninsula wells. As it was pointed out earlier, the most abundant and diverse assemblage concerned is referred to as the section of Malyginskaya borehole 50 at a depth of 1675,0 m. According to its specific composition, the assemblage has much in common with that of the Canadian Province (general, vicarious species, geographical subspecies).

Numerous foraminifera and ostracod nuclei are discovered in the section of boreholes 201 and 205 of the North-Tambeiskaya area (the Yamal peninsula) in the intervals at the depths of 1630,7-1548,7 m (borehole 201) and 1881,6-1825,0 m (borehole 205) from the samples of dark-grey argillites. Calcareous-secretion shells are studied in addition to agglutinated quartz-siliceous ones. *Ammobaculites fragmentarius* and *Gaudryinopsis filiformis* are the index-species for the foraminiferal assemblages from these sections.

The species: *Psammosphaera laevigata* White, *Saccammina complanata* (Franke), *Hyperammina aptica* (Dampel et Mjatliuk), *Reophax inordinatus* Young, *Labrospira* aff. *collyra* (Nauss), *Haplophragmoides topagorukensis* Tappan, *Ammobaculites fragmentarius* Cushman, *Trochammina reinwateri* Cushman et Applin, *Gaudryinopsis nanushukensis* Tappan, *Gaudryinopsis filiformis* (Berthelina), *Uvigerinammina manitobensis* (Wickenden), *Miliammina awunensis* Tappan, *Saracenaria projectura* Stelk et Wall, *Pallaimorphina ruckerae* Tappan, *Eponides morani* Tappan are established in the composition of the assemblage along with *Ammobaculites fragmentarius, Gaudryinopsis filiformis* (the North-Tambeiskaya area, borehole 201) (Fig. 4). The representatives of the genus of *Haplophragmoides*, especially of the species *H. topagorukensis* Tappan are abundant. The shells are mainly well-preserved and they species constitute the Middle Albian assemblage known in the previous sections of the Yamal peninsula wells and some other sections of the North district of the West-Siberian Province as well as in the Western district – Transuralia. A similar foraminiferal assemblage is observed in the section of borehole 205 at a depth of 1836,45 m (the North-Tambeiskaya area). The composition of the assemblage includes the following species: *Haplophragmoides topagorukensis* Tappan, *Ammobaculites* aff. *fragmentarius* (Cushman), *Ammomarginulina obscura* (Loeblich), *Haplophragmium* aff. *ivlevi* Podobina, *Flabellammina* aff. *acuminata* Podobina, *Uvigerinammina manitobensis* (Wickenden), *Dentalina* aff. *basiplanata* Cushman, *Marginulina* aff. *curvatura* Cushman, M. *similis* Orb. *obliquenodes* Bandy, *Marginulina sphaerica* Podobina et Orlov, *Saracenaria solita* Bulatova, *Saracenaria projectura* Stelk et Wall, *Eponides morani* Tappan, *Rosalina* ? *interposita* Mjatliuk. The last 8 species are calcareous-secretion brownish shells belonging to the orders of Lagenida and Rotaliida (Fig. 5).

Somewhat different Middle Albian foraminiferal assemblage along with *Ammobaculites fragmentarius*, *Gaudryinopsis filiformis* is discovered just higher in the section at a depth of 1830,6 m. The species: *Reophax inordinatus* Young, *Ammobaculites fragmentarius* (Cushman), *Ammomarginulina obscura* (Loeblich), *Uvigerinammina manitobensis* (Wickenden), *Trochammina* aff. *wetteri* Stelk et Wall *tumida* Podobina, T. imiatensis Tappan, *Verneuilinoides* aff. *borealis* Tappan *assanoviensis* (Zaspelova), *Gaudryinopsis filiformis* (Berthelin), G. aff. *nanushukensis* Tappan, *Marginulina* aff. *similis* Orb. *obliquenodes* Bandy are determined in its composition. The shells in this assemblage are predominantly agglutinated coarse grained with a quartz-siliceous wall. Calcareous-secretion forms – the species of *Marginulina* aff. *similis* Orb. *obliquenodes* Bandy. But, however, both index-species which are characteristic of the Middle Albian occur [10].

All the investigated shells are well-preserved and known in the Transuralian deposits, as well as Topagoruk formation of Northern Alaska [8].

Upper substage – K_1al_3
Khanty-Mansiiskian Horizon

One sample selected from a depth of 1655,85 m (borehole 124 of the West-Tambeiskaya area) is situated in the upper-lying deposits of the Khanty-Mansiiskian Horizon. Slightly aboundant, well-preserved agglutinated quartz-siliceous foraminifera and a great number of ostracod nuclei are found in it.

Haplophragmoides cf. *variabilis* Podobina, *Ammotium* cf. *braunsteini* (Cushman et Applin), *Verneuilinoides* sp. *indet.* and poorly preserved taxa of the families of Haplophragmoididae and Ataxophragmiidae are determined along them. The assemblage in view of the representatives of the genera of *Ammotium* and *Verneuilinoides* is typical of the Upper Albian of Western Siberia.

4. Conclusion

The new foraminiferal finds from the drilled four sections of the Yamal peninsula wells have provided an opportunity to extend our knowledge about the occurrence of the Albian sea deposits in the Northern palaeobiogeographical district of Western Siberia. All the Albian assemblages of foraminifera are discovered in dark-grey argillites with the interlayers of grey aleurolites of the Khanty-Mansiiskian Horizon. Foraminifera mainly consist of agglutinated quartz-siliceous and to a smaller extant calcareous-secretion shells. The Early Albian foraminiferal assemblage was not completely diverse and it was discovered in two sections of borehole 124 of the West-Tambeiskaya and borehole 201 of the North-Tambeiskaya area.

Fairly diverse foraminiferal assemblages integrated in one Middle Albian assemblage with *Ammobaculites fragmentarius*, *Gaudryinopsis filiformis* are explored in all the four sections of the wells: 50 Malyginskaya area, 124 West-Tambeiskaya and 201, 205 of the North-Tambeiskaya areas. Many species of the Middle Albian assemblage (the Samatlor and the other areas) of the West Siberian Northern palaeobiogeographical district are also discovered in the assemblage from the Western one. At the same time, in comparison with the Middle Albian Transuralian assemblage [2], the Yamal, Samatlor assemblages are significantly impoverished in their specific composition. The Boreal transgression in Transuralia seemed to have formed a much more deeper and comparatively warmer basin in which the conditions for foraminiferal existence were comparatively good. In the Yamal shallow basin the conditions of dwelling were less favourable. Many foraminiferal shells, however, are well-preserved and representative samples of calcareous-secretion forms are also found. The Late Albian assemblage of foraminifera (borehole 124, the West-Tambeiskaya area) is established in the upper layers of the Khanty-Mansiiskian Horizon. The assemblage of interest includes the species of the genera of *Ammotium* and *Verneuilinoides*. But their poor preservation and separate finds do not allow us with confidence to assign their age to the Late Albian. Many species of the Albian assemblages, especially those from the Middle Albian of the West-Siberian Province have their analogs or similar species in the Albian deposits of the Canadian Province [8] [9]. Judging by the similarity in the foraminiferal composition, both of the provinces are related to the Arctic palaeobiogeographical realm of just the same circumpolar belt.

References

[1] Podobina V.M. 2013 Biostratigraphy of the Albian of the West Siberia Samatlor area (foraminiferal data) *Vestnik of Tomsk State University* [Bulletin of TSU] **374** pp 188-198 4 pls. (in Russian)
[2] Bulatova Z.I. 1976 Stratigraphy of the Aptian-Albian gas-bearing deposits of the West-Siberian plain in terms of foraminifera. Nedra [The Bowels of the earth] 152, 21 pls. (in Russian)
[3] Amon E.O. 2005 Agglutinated foraminiferal assemblages from the Khanty-Mansiisk suite (the Albian, Lower Cretaceous) in the Middle and Southern Transuralia *Lithosphere* **2** pp 97-134 (in Russian)
[4] Zakharov V.A., Marinov V.A., Agalakov S.E. 2000 The Albian layer of Western Siberia *Geology and geophysics* **41-6** pp 769-791. (in Russian)
[5] Podobina V.M. 1995 Paleozoogeographic regionalization of Northern Hemisphere Late cretaceous basin based on foraminifera *Proc. 4th Int. Workshop on Agglutinated Foraminifera. Spec. Publ.* **3** pp 39-47.
[6] Regional stratigraphic scheme of the Cretaceous deposits of Western Siberia (the Aptian – the Albian – the Cenomanian) Regional stratigraphic schemes of the Cretaceous deposits of Western Siberia. Adopted by VI Interdepartmental stratigraphic session, October 16, 2003. Affirmed by the Interdepartmental stratigraphic committee, April 8. 2005. Novosibirsk, 2005. (in Russian)
[7] Podobina V.M. 2000 Foraminifera and biostratigraphy of the Upper Cretaceous of Western Siberia. Tomsk. Scientific and Technological Literature Press. 388, 80 pls. (in Russian)
[8] Tappan H. 1962 Foraminifera from the Arctic slope of Alaska / Pt. 3, Cretaceous Foraminifera *U.S. Geol. Survey. Prof. Paper*. **236** pp 91-209, pls. 29-58.
[9] Wall J. 1967 Cretaceous Foraminifera of the Rocky Mountain Foothills, Alberta // Research Council Alberta. Bull. 20. 185 p., 15 pls.
[10] Podobina V.M. 2015 New data on middle and late Albian foraminifera and biostratigraphy of the northern palaeobiogeographical district of western Siberia *Geologos*. **21-1** pp 71-78, 4 pls.

Figure 1. Early Albian foraminiferal assemblage Western Siberia, North-Tambeiskaya well 201, depth. 1634,3 m; Khanty-Mansiiskian Horizon, Lower Albian

1 – *Saccammina micra* Bulatova; 2-6 – *Labrospira rotunda* Podobina; 7-13 – *Haplophragmoides* aff. *variabilis* Podobina; 14-20 – *Ammomarginulina obscura* (Loeblich); 21-23 – *Trochammina imiatensis* Tappan; 24 – *Flabellammina* aff. *acuminata* Podobina; 25 – *Spiroplectammina longula* Podobina; 26-

33 – *Verneulinoides* aff. *borealis* Tappan; 34 – *Gaudryinopsis* aff. teilleuri Tappan; 35-38 – kwarz-siliceus foraminiferal kernels of Rotaliida

Figure 2. Foraminiferal assemblage with *Ammobaculites fragmentarius, Gaudryinopsis filiformis*. Western Siberia, Malyginskaya well 50, depth 1675,0 m; Khanty-Mansiiskian Horizon, Middle Albian

1 – *Psammosphaera laevigata* White; 2-3 – *Labrospira* aff. *rotunda* Podobina; 4-22 – *Haplophragmoides topagorukensis* Tappan; 23-25 – *Recurvoides* aff. *leushiensis* Bulatova; 26-29 – *Ammobaculites wenonahae* Tappan; 30-31 – *Pseudobolivina contorta* Bulatova; 32-35 –

Gaudryinopsis filiformis (Berthelin); 36-40 – *Miliammina manitobensis* Wickenden; 41-43 – *Lenticulina topagorukensis* Tappan; 44-45 – *Saracenaria solita* Bulatova; 46-49 – *Gavelinella* aff. *stictata* (Tappan)

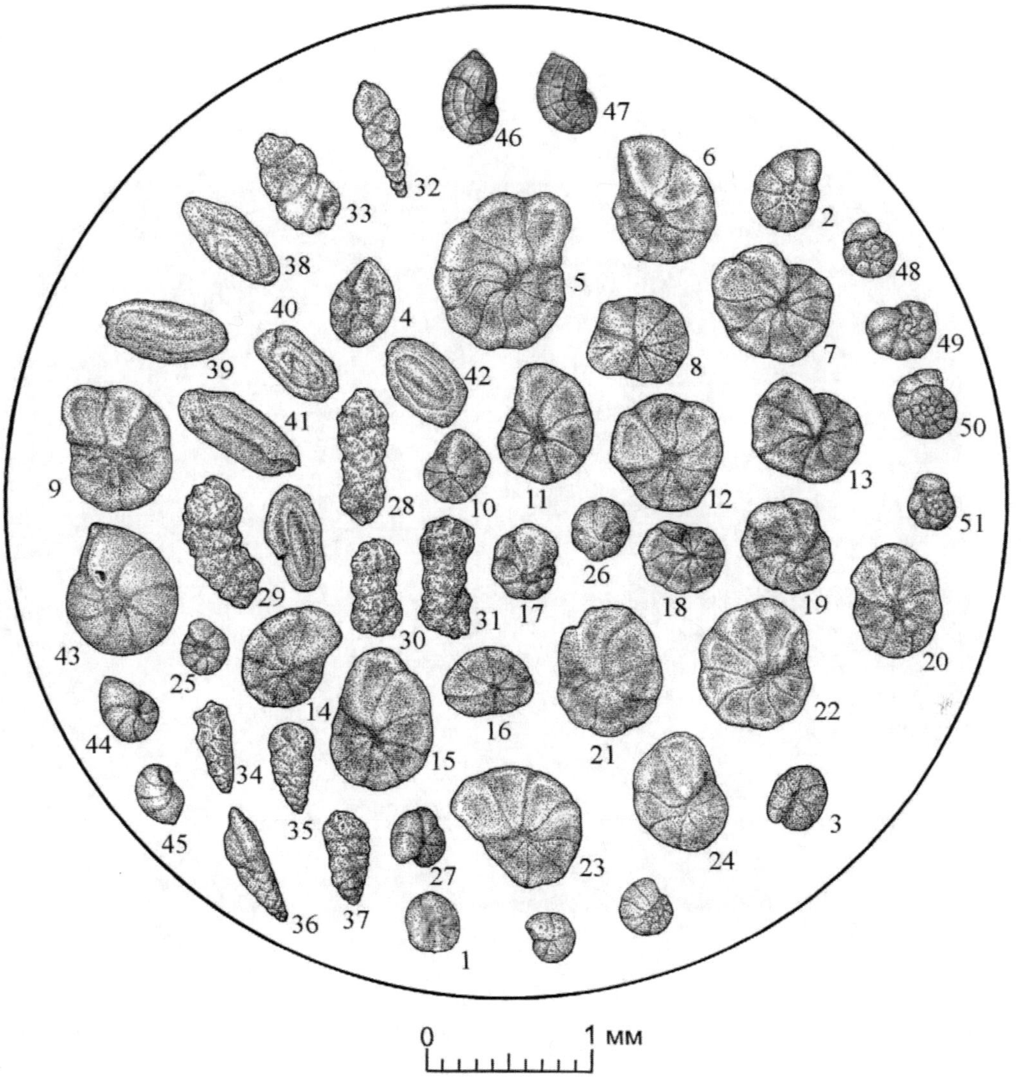

Figure 2a. Foraminiferal assemblage with *Ammobaculites fragmentarius, Gaudryinopsis filiformis*. Western Siberia, Malyginskaya well 50, depth 1675,0 m; Khanty-Mansiiskian Horizon, Middle Albian

1 – *Psammosphaera laevigata* White; 2-3 – *Labrospira* aff. *rotunda* Podobina; 4-24 – *Haplophragmoides topagorukensis* Tappan; 25-27 – *Recurvoides* aff. *leushiensis* Bulatova; 28-31 – *Ammobaculites fragmentarius* Cushman; 32-33 – *Pseudobolivina contorta* Bulatova; 34-37 – *Gaudryinopsis filiformis* (Berthelin); 38-42 – *Miliammina manitobensis* Wickenden; 43-44 –

Lenticulina topagorukensis Tappan; 45-47 – *Saracenaria solita* Bulatova; 48-51 – *Gavelinella* aff. *stictata* (Tappan)

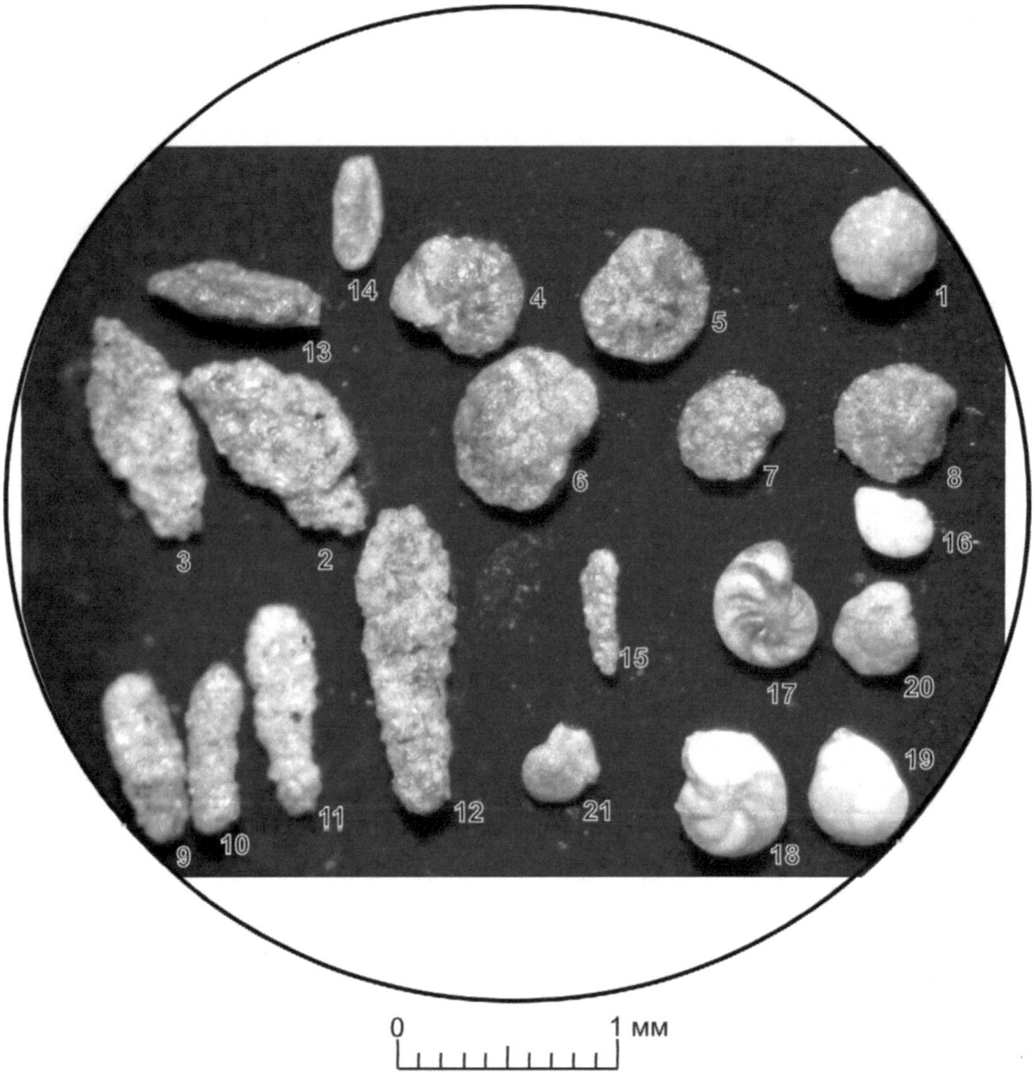

Figure 3. Foraminiferal assemblage with *Ammobaculites fragmentarius*, *Gaudryinopsis filiformis*. Western Siberia, West-Tambeiskaya well 124, depth 1673,40 m; Khanty-Mansiiskian Horizon, Middle Albian

1 – *Saccammina* aff. *sphaerica* (M. Sars); 2 – *Reophax* aff. *sherborniana* (Chapman); 3 – *Reophax* aff. *inordinatus* Young; 4-8 – *Haplophragmoides topagorukensis* Tappan; 9-12 – *Ammobaculites*

fragmentarius Cushman; 13-14 – *Miliammina manitobensis* Wickenden; 15 – *Gaudryinopsis filiformis* (Berthelin); 16-19 – *Lenticulina topagorukensis* Tappan; 20-21 – *Gavelinella* aff. *stictata* (Tappan)

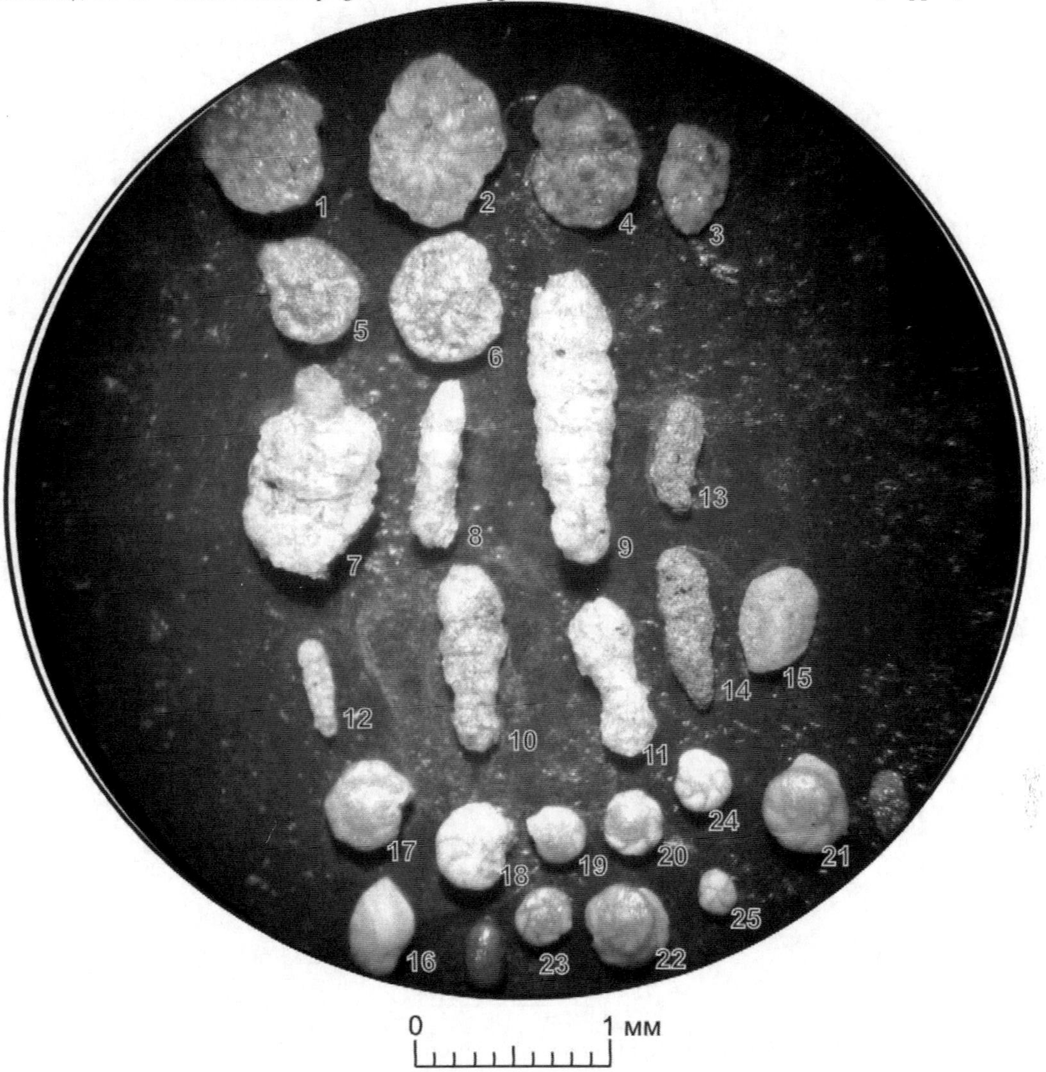

Figure 4. Foraminiferal assemblage with *Ammobaculites fragmentarius, Gaudryinopsis filiformis*.
Western Siberia, North-Tambeiskaya well 124, depth 1627,08 m;
Khanty-Mansiiskian Horizon, Middle Albian

1-3 – *Labrospira* aff. *collyra* (Nauss); 4-6 – *Haplophragmoides topagorukensis* Tappan; 7 – *Uvigerinammina manitobensis* (Wickenden); 8-11 – *Ammobaculites fragmentarius* Cushman; 12-14 – *Gaudryinopsis filiformis* (Berthelin); 15 – *Miliammina awunensis* Tappan; 16 – *Saracenaria projectura* Stelk et Wall; 17-23 – *Eponides morani* Tappan; 24-25 – *Pallaimorphina ruckerae* Tappan

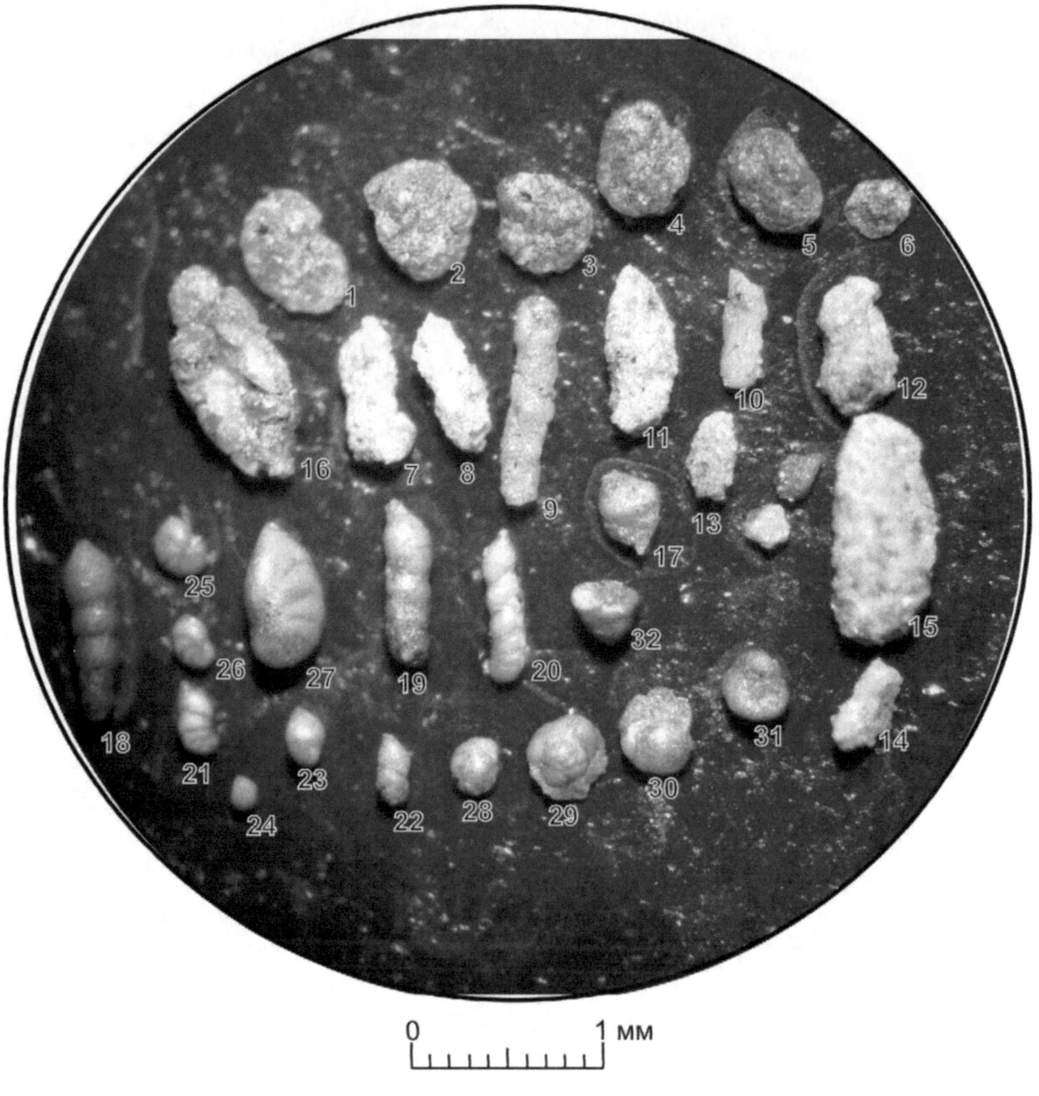

Figure 5. Foraminiferal assemblage with *Ammobaculites fragmentarius, Gaudryinopsis filiformis*.
Western Siberia, North-Tambeiskaya well 124, depth 1836,45 m;
Khanty-Mansiiskian Horizon, Middle Albian

1-6 – *Haplophragmoides topagorukensis* Tappan; 7-10 – *Ammobaculites fragmentarius* Cushman; 11-14 – *Ammomarginulina obscura* (Loeblich); 15 – *Haplophragmium* aff. *ivlevi* Podobina; 16-17 – *Uvigerinammina manitobensis* (Wickenden); 18-19 – *Dentalina* aff. *basiplanata* Cushman; 20 – *Marginulina* aff. *curvatura* Cushman; 21-22 – *M. rimilis* Orb. *obliquinoides* Bandy; 23-24 – *Marginulina sphaerica* Podobina et Orlov; 25-26 – *Saracenaria solita* Bulatova; 27 – *Saracenaria projectura* Stelck et Wall; 28-31 – *Eponides morani* Tappan; 32 – *Rosalina* ? *interposita* Mjatliuk

IOP Conf. Series: Earth and Environmental Science **110** (2017) 012020 doi:10.1088/1755-1315/110/1/012020

Petrology of Ortsog-Uul peridotite-gabbro massif in Western Mongolia

M. Shapovalova[1, 2], N. Tolstykh[1], R. Shelepaev[1, 2] M. Cherdantseva[1, 2]

[1]V.S. Sobolev Institute of Geology and Mineralogy SB RAS, 3 Koptug avenu, Novosibirsk 630090, Russia
[2]Novosibirsk State University, 1 Pirogova, Novosibirsk 630090, Russia

E-mail: shapovalovam@igm.nsc.ru

Abstract. The Ortsog-Uul mafic-ultramafic massif of Western Mongolia is located in a tectonic block with overturned bedding. The massif hosts two intrusions: a rhythmically-layered peridotite-gabbro association (Intrusion 1) and massive Bt-bearing amphibole-olivine gabbro (Intrusion 2). Intrusions 1 and 2 have different petrology features. Early Intrusion 1 (278±2.5Ma) is characterized by lower concentrations of alkalis, titanium and phosphorus than late Intrusion 2 (272±2Ma). The chondrite-normalized REE and primitive mantle-normalized rare elements patterns of Ortsog-Uul intrusions have similar curves of elements distribution. However, Intrusion 2 is characterized higher contents of REE and rare elements. High concentrations of incompatible elements are indicative of strong fractionation process. It has been suggested that Intrusions 1 and 2 derived from compositionally different parental melts. Model calculations (COMAGMAT-3.57) show that parental melts of two intrusions were close to high-Mg picrobasaltic magmas. The concentration of MgO in melt is 16.21 (Intrusion 1) and 16.17 (Intrusion 2). Isotopic data of Ortsog-Uul magmatic rocks exhibit different values of εNd (positive and negative) for Intrusion 1 and 2, respectively.

1. Introduction

Many mafic-ultramafic layered sulfide-bearing intrusions of the southeastern Siberian Region [1,2], the Vietnam [3], the China [4], and the Mongolia [5] is studied recently. A brief summary of the Ortsog-Uul massif based on previous geological reports in Russian by Izokh et al. [6]. The massif located on the northern slope of the Khangai uplands, on the left bank of the Tamiryn-Gol River (figure 1) in Western Mongolia. The Ortsog-Uul massif (S=5 km^2) as the adjacent Dulan-Uul mafic-ultramafic intrusion belongs to the Tamir Complex and it intrudes the metamorphic rocks, quartzites, gneisses, schists and amphibolites (figure 1). A more representative collection of samples was obtained by us in recent years to identify the genesis of the Ortsog-Uul massif. Our research also motivated by recent findings of Ni-Cu-PGE minerals included in the rocks. The data confirmed that Ortsog-Uul massif consists of two intrusions [7]. The identification of differences in the conditions of formation of these intrusions is the good prerequisite for understanding of the petrological processes.

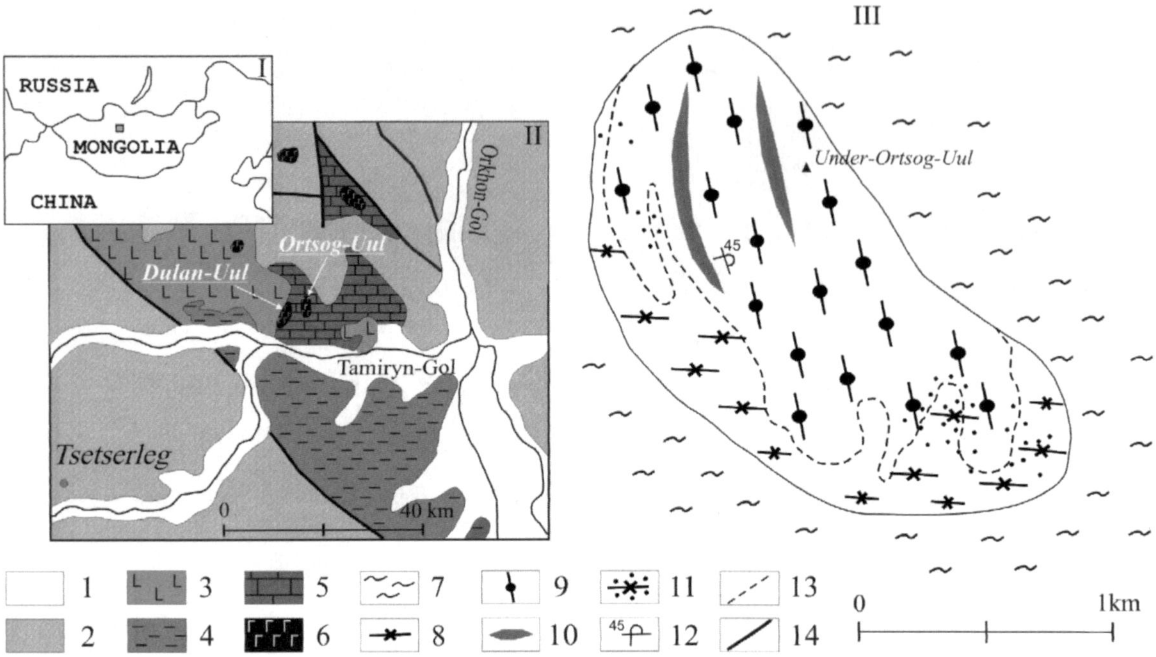

Figure 1. Simplified geological map of the Khangai uplands and the Ortsog-Uul massif: 1 – quaternary deposits; 2 – sedimentary terrigeneous rocks (PZ_3- MZ_2); 3 – basaltic rocks (PZ_3); 4 – green schists (R_3-$Є_1$); 5 – carbonaceous rocks (R_3- $Є_1$); 6 –Tamir Complex (PZ_1); 7 – host rocks (crystal slates and gneisses); 8 – Intrusion 2 gabbroids; 9-10 – Intrusion 1: 9 – olivine gabbro, gabbronorite and troctolite; 10 – melanotroctolite and plagioclase peridotite; 11 – sulphide mineralization; 12 – overturned layering; 13 – contact between Intrusions 1 and 2; 14 – fault.

2. Analytical methods

Following types of analysis were carried out in the Institute of Geology and Mineralogy in Novosibirsk, Russia. Major element oxides were determined by XRF using an ARL 9900. Trace elements and REE were estimated by ICP-MS using an Element-I Finnigan MAT. Electron microprobe analyses of all minerals were analyzed by Cameca Camebax Micro and Jeol JXA8100 electron microprobes (WDS). Argon isotope ratios of biotite were measured using a GV Instruments ARGUS. SHRIMP-II zircon dating was carried out in VSEGEI's Centre of Isotopic Research (St. Petersburg). Sm-Nd isotopic analyses of whole rock samples were determined using TIMS method in Geological Institute Kola Science Centre (Apatity).

3. Results

The Ortsog-Uul massif hosts two intrusions separated by a NW-SE distinct contact: a rhythmically-layered peridotite and gabbro (biotite-free) which is located north of the contact (Intrusion 1) and a massive biotite-bearing amphibole and olivine gabbro, south of the contact (Intrusion 2). The both intrusions are Permian in age. The Ar/Ar age of Intrusion 1 (Mg-hornblende from olivine mezogabbro) is 278.7 ± 2.5 Ma. A U-Pb zircon age of Intrusion 2 determined for olivine mezogabbro is 272 ±2 Ma (SHRIMP II) that is similar to Intrusion 1. However, the Ar/Ar age of Intrusion 2 (biotite from Bt-bearing olivine mezogabbro) is 257 ± 6.5 Ma.

The concentrations of SiO_2 and alkalis positively correlate suggesting progressive accumulation in residual melt during fractional crystallization. These are increase with decrease of MgO contents. The MgO versus Al_2O_3 plot (figure 2) shows the compositional variations of both intrusions due to olivine and plagioclase fractionation. The MgO ranges from 4 wt. % (plagioclase peridotite) to 33 wt. % (gabbronorite) in rocks of Intrusion 1 during fractional crystallization. The MgO range (17-25 wt. %)

is not extended in gabbro of Intrusion 2, while contents of alkalis, in particular, K_2O (up to 0.8 wt. %) are high. Samples of Intrusion 2 show a discrete vertical compositional field at 20 wt. % MgO in the plot of MgO versus (Na_2O+K_2O). The rocks of intrusion 2 are enriched in TiO_2 and P_2O_5 (up to 0.5 wt. % and 0.06 wt. % respectively) compared to Intrusion 1 resulting the presence of accessory titanomagnetite and apatite in the rocks of Intrusion 2.

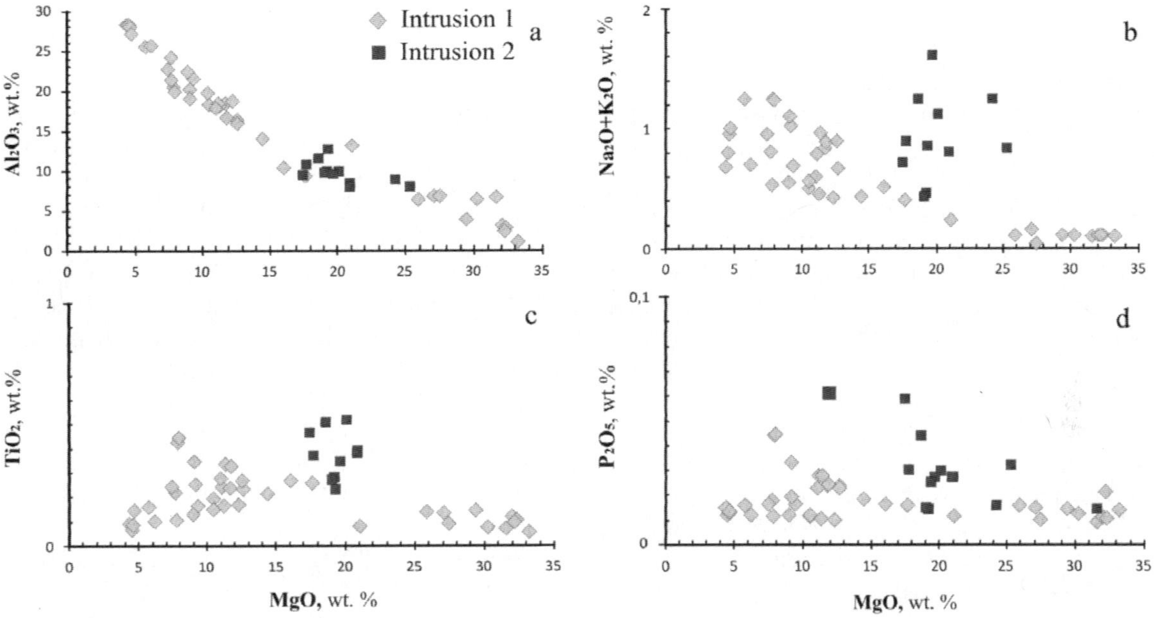

Figure 2. MgO versus Al_2O_3 (a), Na_2O+K_2O (b), TiO_2 (c), and P_2O_5 (d) plots of Intrusions 1 and 2.

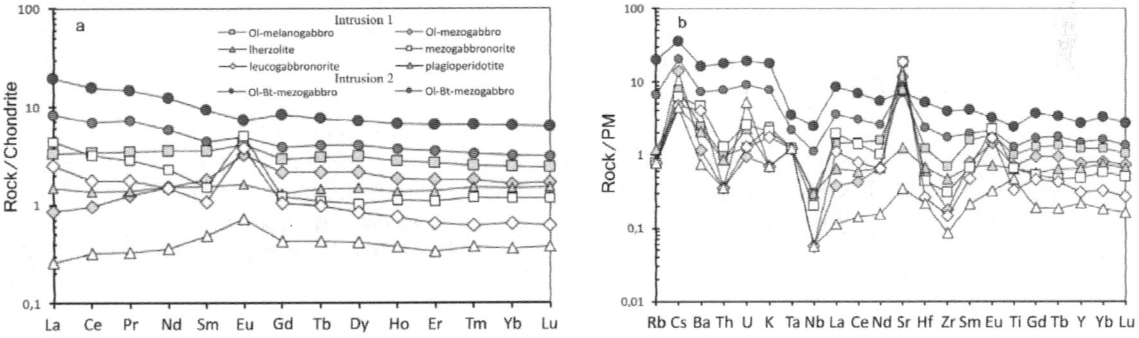

Figure 3. Chondrite-normalized [8] rare earth element (a) and primitive mantle-normalized [9] trace element (b) patterns of the Ortsog-Uul massif.

All Ortsog-Uul rocks have similar chondrite-normalized REE patterns [8] with slightly LREE enrichment ($(La/Yb)_N$ = 2.24–3.05), and weak positive Eu anomalies (figure 3 a). The Layered Series of Intrusion 1 displays a general increase of trace element contents from plagioclase peridotite to olivine melanogabbro. Most of the rocks of Intrusion 2 are characterized by higher concentrations of REE, HFSE and LILE normalized on the primitive mantle [9] than the Intrusion 1 as shown on the spidergrams. Gabbros from Intrusion 2 are REE-rich suggesting their derivation from a source which has the higher degree of fractionation compared to Intrusion 1 rocks. Both intrusions show positive anomalies of Sr and Eu (Eu/Eu* = 1.2 to 3.8), and negative anomalies of HFSE (figure 3 b). The εNd

IOP Conf. Series: Earth and Environmental Science **110** (2017) 012020 doi:10.1088/1755-1315/110/1/012020

is determined as positively (13.5; 270 Ma) and negatively (−4.3; 270 Ma) by isotopic analyzes for Intrusion 1 and 2, respectively.

Using the COMAGMAT-3.57 [10] software, we have found the compositions of the parental melts of the Ortsog-Uul intrusions, which might have crystallized at 2 kbar and the oxygen activity controlled by the QFM buffer. The initial melts estimated for both intrusions relate to high-Mg picrobasaltic magmas. The calculated concentration of MgO in melt is 16.21 (Intrusion 1) and 16.17 (Intrusion 2).

4. Discussion and conclusion

The age of the Ortsog-Uul massif is debatable. The age of Intrusion 2 determined by U-Pb method (SHRIMP II) is 272 ± 2 Ma whereas the Ar/Ar age of this intrusion is 257 ± 6.5 Ma. The both dating are disagree with the PZ_1 age shown on the geological map [11,12]. Nevertheless, the dating of Intrusion 2 by different methods has range 7-15 Ma. There are two explanations for this range. Firstly, closing time of isotope systems was possible various during slow cooling of massif or repeated warming up of the massif up to lower temperatures [13]. Secondly, zircon grains were probable xenogenic (trapping from the host rocks). In the latter case, the age of zircons will show the age of early crystallization of Intrusion 2 (not ancient, than 272 ± 2 Ma). Thus, the obtained data indicate the formation of the Ortsog-Uul massif in the Permian Period. In addition, the Nomgon massif (256±2.1 Ma) and a few basic rocks of the Selenga complex [5] are a similar age. The Nomgon massif and the Selenga complex are enriched in platinum group elements (PGE) [14,15]. This suggests that the Ortsog-Uul, Nomgon and Selenga massifs belong to the same PGE-enriched magmatic province. All aforementioned intrusions are located in common structure of the Central Asian Fold Belt (CAFB).

The each rhythm of Layered Series consists consistently of gabbro, troctolite, melanotroctolite and plagioclase peridotite from bottom to top. Such rhythmic sequence in the Ortsog-Uul massif disagrees with those of layered intrusions showing a normal-type stratification with mafic layers at the base and more leucocratic layers in the top in agreement with crystal fractionation process [16,17]. Consequently, it can be assumed that the Ortsog-Uul massif is included an overturned tectonic block. The overturned bedding of Ortsog-Uul rocks agrees with earlier geological data. The folded surface with elements of overturned bedding is shown at Early Proterozoic, at Triassic and Cretaceous systems of Early Mesozoic on the geological map L-48-I,II [11]. It confirms that studied region, including the Ortsog-Uul massif, have been involved in folding.

The MgO content smoothly decreases usually into the each cyclic unit of layered intrusions. The content of MgO increases abruptly during the transition to the next cyclic unit [18,19]. While it was revealed that the MgO contents gradually increases in each separate cyclic unit of Ortsog-Uul Layered Series up a section. This could be the case if plagioclase crystallizes earlier than olivine, but the euhedral crystals of plagioclase have not been observed in melanocratic rocks. Consequently, the overturned bedding of the tectonic block including Ortsog-Uul massif is the only reason for the reverse sequence.

Intrusions 1 and 2 have close MgO contents and different contents of incompatible elements. Firstly, rocks of Intrusions 1 and 2 form two separated trends (figure 2) on the MgO versus Na_2O+K_2O plots. The high contents of Na_2O+K_2O (up to 1.8 wt. %), TiO_2 (up to 0.6 wt. %), P_2O_5 is observed in rocks of the Intrusion 2 resulting the presence biotite, Ti-bearing clinopyroxene, titanomagnetite and apatite. Rocks of the Intrusion 2 are characterized by the high contents of REE, LILE and other rare earth element (except Sr, Eu) compared with Intrusion 1 (figure 3). It has been suggested that intrusions were formed from different portions of melt witch have not the equally processes of fractionation [20]. The isotopic data show positive εNd for Intrusion 1 and negative εNd for Intrusion 2. It confirms impossibility of sequential crystallization of both intrusions rocks from single melt [21,22]. Different contents of incompatible elements with the same concentration of MgO prove the various mantle sources of parental melts for these two intrusions. Based of isotopic data of Ortsog-Uul magmatic rocks have been determined that Intrusion 1 was formed from the depleted mantle source (εNd = 13.5; 270Ma) and Intrusion 2 was formed from the enriched mantle source (εNd= −4.3;

270Ma) [23]. The rocks of Ortsog-Uul massif are result fractionating of the picrobasaltic melts with the different content of alkalis and TiO_2 by COMAGMAT-3.57

From geological, mineralogical, geochemical and isotopic data, we can conclude that:

1) The Ortsog-Uul massif is located in the large Tamir Complex hosted within an overturned tectonic block;

2) There are two intrusions: an early [278.7 ± 2.5 Ma] layered Intrusion 1 and a late [272 ±2 (257?) Ma] massive Intrusion 2;

3) The wide difference of incompatible element contents shows that intrusions of Ortsog-Uul massif derived from compositionally different parental melts: Intrusion 1 from depleted source and Intrusion 2 from enriched source;

4) MgO contents in melts were estimated as 16.21 (Intrusion 1) and 16.17 (Intrusion 2); this proves that magma has the high-Mg picrobasaltic composition.

Acknowledgments
The research is financed by the Russian Foundation for Basic Research (grants No 16-35-00100, 16-05-00980, 16-05-00945)

References

[1] Kislov E.V. 1998 *The Yoko–Dovyren Layered Pluton*. (Ulan-Ude: Buryatia Scientific Center of the Siberian Branch of the RAS) p 268 [in Russian]

[2] Podlipsky M.Yu., Mekhonoshin A.S., Tolstykh N.D., Vishnevskiy A.V. and Polyakov G.V. 2015 Mineralogy and geochemistry of the Tartai massif, East Siberian metallogenic province *Geology of Ore Deposits* **57**(3) 195–220

[3] Svetlitskaya T.V., Tolstykh N.D., Izokh A.E. and Thi P.N. 2014 PGE geochemical constraints on the origin of the Ni-Cu-PGE sulfide mineralization in the Suoi Cun intrusion, Cao Bang province, Northeastern Vietnam. *Mineralogy and Petrology* **109**(5) 161-80

[4] Luo W., Zhanga Z., Santosha M., Houa T., Huanga H., Zhua J., Wanga X. and Fua X. 2014 Petrology and geochemistry of Permian mafic–ultramafic intrusions in the Emeishan large igneous province, SW China: Insight into the ore potential *Ore Geology Reviews* **56** 258–275

[5] Izokh A.E., Polyakov G.V., Gibsher A.S., Balykin P.A., Zhuravlev D.Z. and Parkhomenko V.A. 1998 The high-aluminous layered gabbroid of the Central Asian folded belt: geochemical features, Sm-Nd isotopic age and geodynamic conditions of formation *Geology and geophysics* **39**(11) 1565-77 [in Russian]

[6] Izokh A.E., Polyakov G.V., Krivenko A.P., Bognibov V.I. and Bayarbileg L. 1990 *Gabbroid formation of the Western Mongolia* (Novosibirsk: Science) p 269 [in Russian]

[7] Shapovalova M., Shelepaev R., Tolstykh N., Kalugin V. and Safonova I. 2015 Petrology of the Ortsog-Uul Gabbro-Petidotite PGE-Bearing complex, Western Mongolia *Conf. 13th SGA Biennial Meeting* **3** 983-85

[8] Boynton W.V. 1984 Geochemistry of the rare earth elements: meteorite studies. *Rare earth element geochemistry* ed P.Ed. Henderson (Amsterdam: Elsevier Science Published Company) pp 63-114

[9] McDonough W.F., Sun S.S., Ringwood A.E., Jagoutz E. and Hofmann A.W. 1992 Potassium, Rubidium and Cesium in the Earth and Moon and the evolution of the mantle of the Earth *Geochimica et Cosmochimica Acta* **56**(3) 1001-12

[10] Ariskin A.A., Frenkel M.Yr., Barmina G.S. and Nielsen R.L. 1993 Comagmat: a fortran program to model magma differentiation processes *Computers & Geosciences* **19**(8) 1155-70

[11] Khosbayar P., Byamba B., Binderya T., Gansukh Z. 1987 *Geological map: Ugiynuursky district of L-48-I, II scale: 1:200000*

[12] Izokh A.E. 1998 The layered ultramafic-mafic associations as indicators of geodynamic situations (on the example of the Central Asian folded belt) *Ph.D. thesis* (Novosibirsk) p 403 [in Russian]

[13] Dodson M.N. 1973 Closure temperature in cooling geochronological and petrological systems *Contributions to Mineralogy and Petrology* **40** 259-74

[14] Izokh A.E., Mayorova O.N. and Lavrentyev Yu.G. 1992 Minerals of platinum metals in the Nomgon troktolite-anortozite-gabbro intrusion (Mongolian) *Geology and geophysics* **1** 104-10 [in Russian]

[15] Sotnikov V.I., Ponomarchuk V.A., Berzin A.P. and Travin A.V. 1995 Geochronological boundaries of magmatism of Erdenet-Obo copper-molybdenum-porphyritic deposit (Mongolia) *Geology and geophysics* **6**(3) 78-89 [in Russian]

[16] Wager L.R. and Brown G.M. 1968 *Layered Igneous Rocks* ed W.H. Freeman & Co (San Francisco) p 587

[17] Sharkov Ye.V. 1980 *Petrology layered intrusion*s (Moscow: Science) p 183 [in Russian]

[18] McBirney A.R. 1996 The Skaergaard Intrusion *Developments in Petrology* **15** 147-80

[19] Polyakov G.V., Izokh A.Э, Bognibov V.I. Early 1984 Paleozoic formation of the layered peridotite-pyroxenite-gabbronorite massifs of Northwest Mongolia *Geology and geophysics* **1** 50-62 [in Russian]

[20] Bogatikov O.A., Bogdanova S.V., Borsuk A.M. etc. 1987 *Magmatic rocks* (Evolution of magmatism in the history of Earth vol 6) (Moscow: Science) p 438 [in Russian]

[21] Vrevsky A., Krimsky R. and Svetov S. 1996 Isotopic (Nd, O) and geochemical (REE) heterogeneity of the Archaean mantle, Baltic Shield *Geological Society Special Publication* **112** 43-54 [in Russian]

[22] Polat A., Kerrich R. and Wyman D. 1999 Geochemical diversity in oceanic komatiites and basalts from the late Archaean Wawa greenstone belts, Superior Province *Precambrian Research* **94** 139-73

[23] White W.M. and Hofmann A.W. 1982 Sr and Nd isotope geochemistry of oceanic basalts and mantle geochemistry *Nature* **296** 821-25

IOP Conf. Series: Earth and Environmental Science **110** (2017) 012021 doi:10.1088/1755-1315/110/1/012021

Mineralogical and geochemical feature of the disseminated ores of the southern part of the Noril'sk 1 deposit

N D Tolstykh[1], G I Shvedov[2], A A Polonyankin[3] and S A Zemlyansky[4]

[1] VS Sobolev Institute of Geology and Mineralogy, SB RAS, Novosibirsk, Russia
[2] LLC "Geological Technologies", Krasnoyarsk, Russia
[3] OOO "Russian platinum", Moscow, Russia
[4] Open Company "Novogeo", Krasnoyarsk, Russia

E-mail: tolst@igm.nsc.ru

Abstract. The new data on the composition of disseminated sulphide ore in the southern part of the Noril'sk 1 deposit are obtained. It was found that a sulfur-poor assemblage: as Fe-rich (cubanite, mackinawite) and Cu-rich (talnakhite, bornite, chalcocite, native copper) minerals due to the evolution of two different fractions of sulphide melt is characteristic of picritic gabbro-dolerite. It has been formed at relatively low sulfur activity ($\lg fS_2$ range from −12 to −11). And a high-sulfur assemblage (monoclinic pyrrhotite, pyrite and Ni-rich pentlandite) formed at higher sulfur activity ($\lg fS_2$ range from −10.5 to −10) due to the pyrrhotite-chalcopyrite fractionation of the sulphide melt is characteristic of olivine gabbro-dolerite. The ratios of Ni/Cu in the rocks decrease from picritic (Ni/Cu = 1.24 average) to olivine gabbro-dolerites (Ni/Cu = 0.69 average). But the Ni/Fe ratios in pentlandite are increasing down the cross-section due to increase of sulfur fugacity during the evolution of ore-forming system. The Pd-parageneses vary along the cross-section top down as follows: Sn → Pb →As →Bi(Te) in picritic gabbro-dolerite and Sn →Sb →As →Te(Bi) in olivine gabbro-dolerite. There is a clear correlation between the sulfide and PGM assemblage and the type of the host rocks.

1. Introduction

Platinum-copper-nickel deposits of the Noril'sk region are located in the northwestern part of the Siberian platform and are associated with flood-basalt of Perm-Triassic age [1-3]. The Noril'sk 1 deposit is located in the northern part of the same name intrusion, confined to the Noril'sk-Kharaelyakh fault in the Noril'sk Trough. Intrusion of Noril'sk I breaks the Carbon-Permian terrigenous deposits of the Tungusska Series and basalts of the Upper Permian-Lower Triassic age [2].

Numerous works are devoted to the ores of the Noril'sk 1 deposit [5,7-10]. Most of the works devoted to the massive ores or low-sulphide horizons, and only some of them are considering the disseminated ore of the Main Ore Horizon composed of picritic, taxitic, and lower olivine gabbro-dolerites [7,11,12]. The features of sulfide mineralization in each unit have been characterized previously [2], however the task of establishing a correlation between the types of the host rocks and the chemical and mineral composition of ores is still relevant [3].

The new data on the composition of disseminated sulphide ore in the southern part of the Noril'sk 1 deposit are obtained on the basis of the samples study of the PH-14 borehole (figure 1). This borehole crosses all the rocks, alternating in a certain stratigraphic sequence which have been described earlier [1,2]. We studied samples of the lower part of the intrusion Noril'sk 1, composed of picritic and lower olivine gabbro-dolerites containing the disseminated Cu-Ni sulphide with the minerals of platinum group elements (PGM). The zoning (changes of geochemical feature, mineral parageneses and compositions of sulfides) from picritic gabbro-dolerite (821.2-902.0 m) to olivine gabbro-dolerite (902.0-938.1 m) and to the contact with siltstones has been established. The more common contact rock of the Noril'sk 1 intrusion is taxitic gabbro-dolerite. But the lower olivine gabbro-dolerire instead of taxitic gabbro-dolerite occur in this cross-section of the PH-14 borehole that is similar to the Northern intrusive body (OM-14) of the Maslovsky deposit (figure 1) [3].

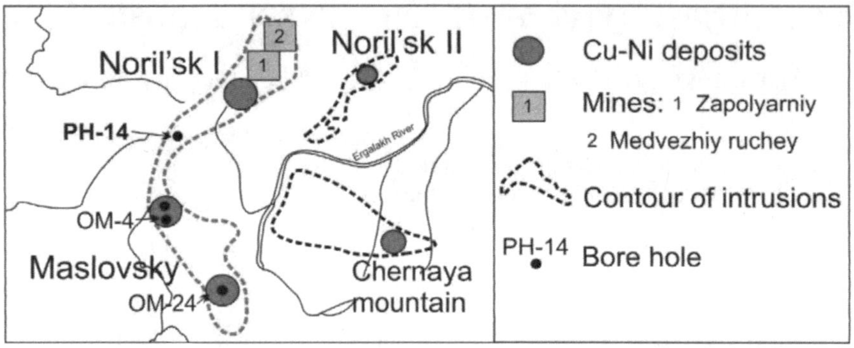

Figure 1. The scheme of location of ore-bearing intrusions and deposits of Noril'sk region

2. Result

2.1. Variations of ore elements

The absolute concentrations of Ni and Cu almost in the whole interval of the investigated cross-section do not exceed of 0.5 % in total (figure 2): the average content of Cu is 0.22 % in picritic gabbro-dolerite and 0.25 % in olivine gabbro-dolerite; the average content of Ni is 0.24 % and 0.17 %, respectively. But there is the single anomalies reaching 2-3% of Cu and Ni, which coincide with the presence of veins of sulphides. Content of Ni predominates over Cu (Ni/Cu is 1.24 average) in picritic gabbro-dolerite, but there are several horizons, where Ni/Cu <1, which coincide with the Pt-Pd anomalies. In olivine gabbro-dolerites, on the contrary, content of Cu prevails over Ni (Ni/Cu is 0.69 average). In general, Ni/Cu ratios vary considerably along section, gradually decreasing to contact with the host rocks (figure 2).

The concentrations of PGE vary in range: 0.11-3.80 % (Pt) and 0.27-7.36 % (Pd) in picritic gabbro-dolerite; 0.06-1.94 (Pt) and 0.15-5.80 (Pd) in olivine gabbro-dolerite; the total of Pd and Pt reach up to 10 ppm in massive ores (figure 2). The ratio of Pd/Pt as a whole increases downward along the cross-section: Pd/Pt = 2.50 (on average) in the upper part of picritic gabbro-dolerite (with CrSp), 2.79 – in the lower part of them, 3.08 – in olivine gabbro-dolerite, and 4.42 – in massive ores. It is consistent with the evolution of sulphide melt and fractionation of ore elements during its deposition to the bottom of the magmatic chamber [1,13].

2.2. Sulphide mineralization in picritic gabbro-dolerites

A considerable variety of ore minerals are characteristic of picritic gabbro-dolerite: pyrrhotite, chalcopyrite, pentlandite, cubanite, talnakhite, chalcocite, bornite, native copper, magnetite and ilmenite, which are spread unevenly. It is noticed that the amount of pyrrhotite, chalcopyrite and pentlandite decreases with the increase of cubanite (figure 3 a). But in the upper horizons (up to 840 m) the pyrrhotite is the most common. The prevalence of S-poor assemblage: as Fe-rich (cubanite, mackinawite) and Cu-rich (talnakhite, bornite, native copper) minerals are a typomorphic feature of the sulphide assemblage of picritic gabbro-dolerite. Chromspinels are common in the upper part of picritic gabbro-dolerites – "chromite horizon" of the thickness about 40 m (figure 2).

The drople-like, irregular and interstitial sulphide patches (aggregates) up to 5 mm are characteristic of picritic gabbro-dolerite (figure 4). Pyrrhotite is found in intergrowths with chalcopyrite, pentlandite and cubanite. Pentlandite occurs often as lamellas in the marginal parts of pyrrhotine grains (figure 4 d) due to decomposition of solid solution. Chalcopyrite is represented by small grains, often in association with bornite (figure 3c). Native copper occurs as the veins in sulphide aggregates (figure 4 a) and rims over sulphides and oxides. Chromspinels and magnetite occur as irregular impregnation in silicates, occasionally in association with sulphides (figure 4 f). Cubanite and chalcopyrite in the sulphide aggregates begin to dominate lower in the profile of the picritic gabbro-dolerite (figure 2), in which minor sphalerite, galena, native gold and PGM are included. The lamellae of cubanite are characteristic of chalcopyrite grains.

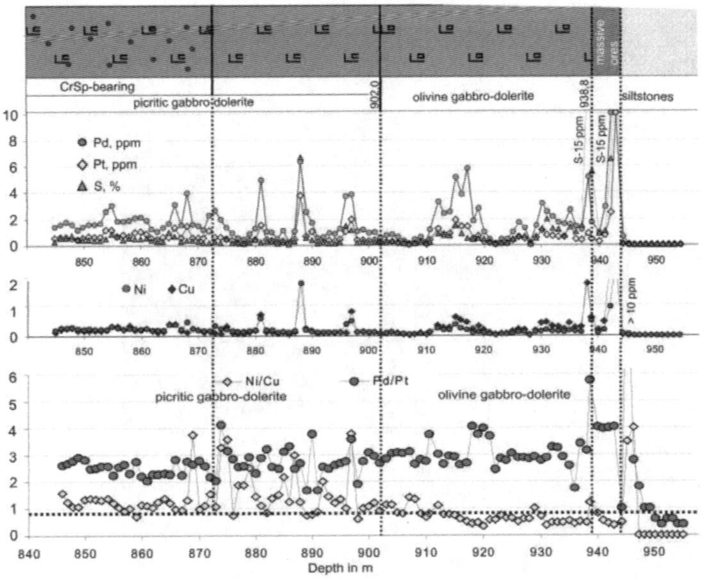

Figure 2. Variations in the content of ore elements, and ratio of Ni/Cu and Pd/Pt in picritic and olivine gabbro-dolerites along the cross-section of borehole PH-14

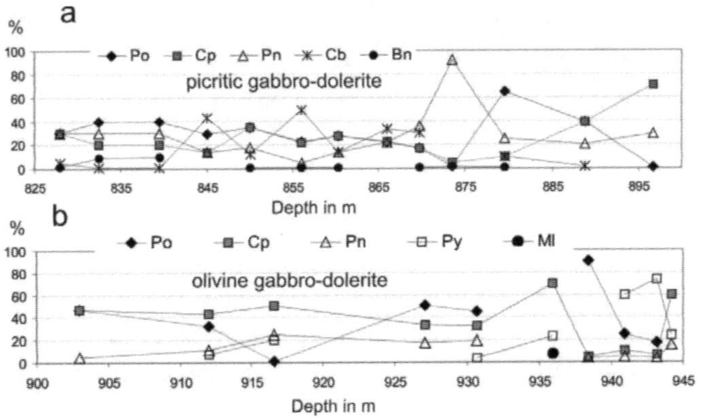

Figure 3. The variations in the amount of sulphides (%) along the cross-section of borehole PH-14: a – in picritic gabbro-dolerite, b – in olivine gabbro-dolerite. Po – pyrrhotite, Cp – chalcopyrite, Pn – pentlandite, Cb – cubanite, Bn – bornite, Py – pyrite, Ml – millerite

2.3. Sulphide mineralization in olivine gabbro-dolerite

Sulfides are represented by a high-sulfur association composed of chalcopyrite, pyrrhotite and pentanedite, as well as pyrite, which is occurs more often in the lower part of this interval (figure 3 b). Chromspinels and copper-rich minerals are absent in this interval. A positive correlation appears between amounts of chalcopyrite and pentlandite and a negative correlation of both these minerals with amount of pyrrhotite in olivine gabbro-dolerite (figure 3 b), which also distinguishes olivine gabbro-dolerites from picritic ones.

Chalcopyrite is the predominant sulphide in disseminated and vein-disseminated ores of olivine gabbro-dolerite (figure 3 b, figure 4 h). It together with other sulfides compose the xenomorphic interstitial aggregates of various sizes (figure 4 g-i). Pyrrhotite composes their internal parts and it is bordered by chalcopyrite. The pyrrhotite is almost completely replaced by pyrite-marcasite-magnetite aggregates and pentlandite – by violarite on some horizons of the lower part of olivine gabbro-

IOP Conf. Series: Earth and Environmental Science **110** (2017) 012021 doi:10.1088/1755-1315/110/1/012021

dolerites (figure 3 b). Millerite and unnamed phase of $(Fe,Ni)_2S_3$ occur in this assemblage (figure 5 a). Ilmenite and magnetite occur as uneven impregnation in silicates. Magnetite is also found as a secondary phase. Disseminated ores of olivine gabbro-dolerite change on the horizon of 839.8 m by massive sulphide ores, mainly pyrrhotite composition (85%), in which chalcopyrite and pentlandite are in a subordinate amount.

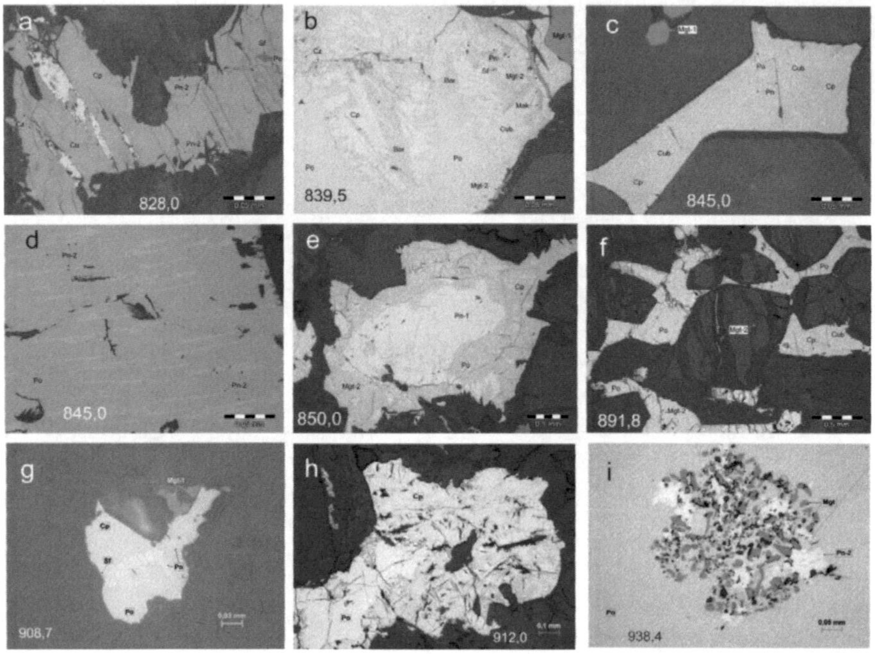

Figure 4. Photomicrographs in reflected light of the sulphide minerals of the picritic (a-f) and olivine (g-i) gabbro-dolerites.

2.4. Compositions of sulphides in picritic and olivine gabbro-dolerites

2.4.1. Fe(+Co) – S – Ni system

The compositions of sulfides in picritic and olivine gabbro-dolerites differ in sulfur and copper contents (figure 5). But in each unit they change with a depth. The pyrrhotite is close to troilite (FeS); pentlandite is Fe-rich (figure 5 a) and contains a minor Co (1.2-2.4 wt%); makinawite $[(Fe_{5.67}Ni_{2.84}Co_{0.21})S_{8.28}]$ is common at the upper horizons of *picritic gabbro-dolerite*. Below in the section, the compositions of pentlandite vary considerably from makinwite $(Fe_{5.37}Ni_{3.19}Co_{0.13})S_{8.31}$ to Fe-rich pentlandite $(Fe_{4.88}Ni_{3.53}Co_{0.12})S_{8.47}$. The Co content decreases to 0.60 wt.%. Below in the cross-section of picritic gabbro-dolerites the pyrrhotite is enriched in sulfur and pentlandite is enriched in nickel. And at the lowest horizon (896.7 m) pentlandite is represented by a nickel variety $(Ni_{4.54}Fe_{3.82}Co_{0.27})S_{8.37}$. The Fe content predominates over Ni and there is an Co as minor element (up to 6.2 wt.%) in the pentlandite on the upper horizons of *olivine gabbro-dolerite*. The compositions of the pentlandite are shifted to the nickel-rich chart area below in the section: $Ni_{4.76}Fe_{3.20}Co_{0.52}S_{8.32}$ (figure 5 a). In this case the pyrrhotite is most enriched in sulfur $Fe_{0.90}S_{1.10}$ (monoclinic species), and contains Ni < 1 wt.%; troilite is absent.

2.4.2. Fe(+Ni,Co) – S – Cu system

Compositions of Cu-minerals include chalcopyrite, cubanite, talnakhite, chalcocite and native copper in *picritic gabbro-dolerite* (figure 5 b). In this case, Fe predominates over Cu in chalcopyrite, and it occur as solid solutions with cubanite (figure 5 b). While the chalcopyrite becomes the less ferruginous, cubanite is not found, and pyrrhotite is the most enriched in sulfur (monoclinic species) in *olivine gabbro-dolerite* compared to the upper horizons.

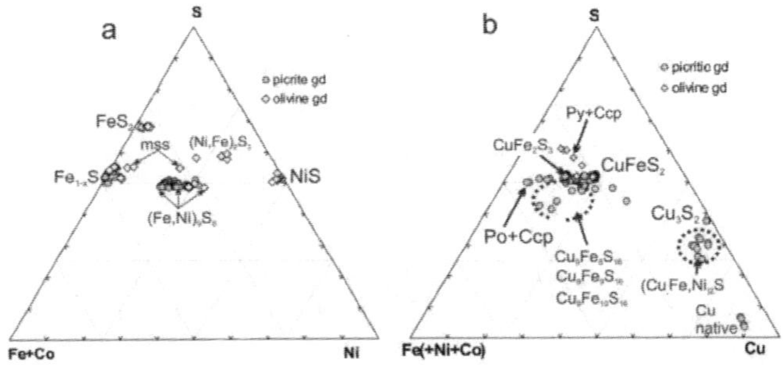

Figure 5. The compositions of Fe-Ni (a) and Cu-Fe sulfides in picritic and olivine gabbro-dolerites.

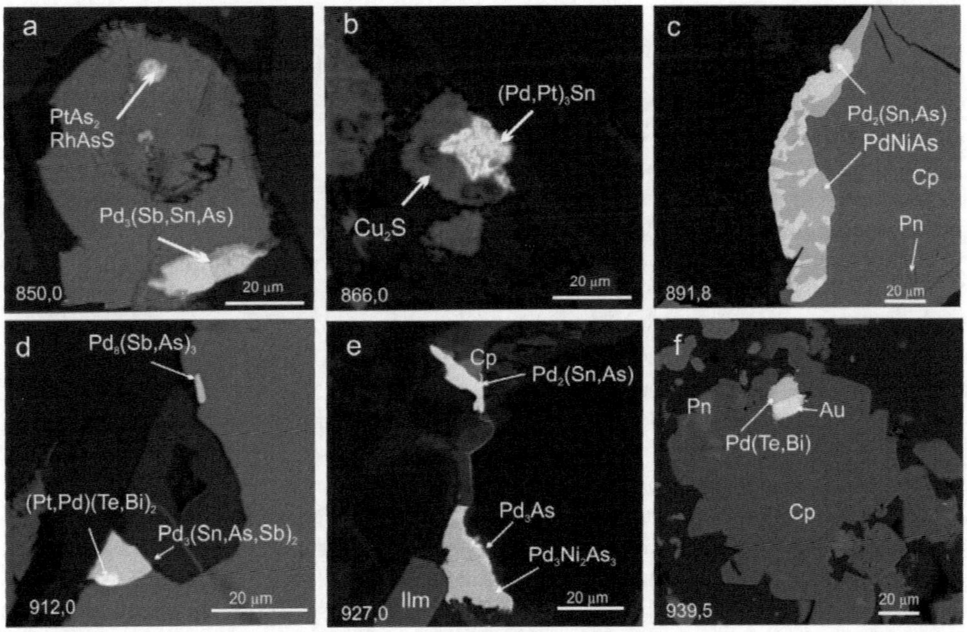

Figure 6. The compositions of Fe-Ni (a) and Cu-Fe sulfides in picritic (a-c) and olivine (d-f) gabbro-dolerites. Po – pyrrhotite, Cp – chalcopyrite, Pn – pentlandite, Ap – apatite, Au – gold.

PGM are the smallest single- and two-phase inclusions in all sulphides, silicates and on their contacts. The intergrowths PGM with gold (Au) observed often (figure 6). Sperrylite is found in all types and horizons of rocks (figure 7). The compounds of Pd with Sn and Pb [paolovite Pd_2Sn, atokite $(Pd,Pt)_3(Sn,As,Sb)$, unnamed $(Pd,Ag)_5Pb_3$] are typical for the upper horizons *of picritic gabbro-dolerite* (figure 7 a) which are changed by arsenides of PGE and As-containing PGM [hollingworthite RhAsS, majakite PdNiAs, As-paolovite $Pd_2(Sn,As)$] and further by Pd-Bi(Te) minerals [sobolevskite PdBi and michenerite PdBiTe] on the lower horizons (figure 7 b). The stannides and antimonides of PGE [paolovite $Pd_2(Sn,As)$, atokite $(Pd,Pt)_3Sn$, mertieite II Pd_8Sb_3] of the upper horizon *of olivine gabbro-dolerite* are changed by the Pd arsenides [menshikovite $Pd_3Ni_2As_3$, vincentite Pd_3As and unnamed Pd_5As_2], and finally, by Pd tellurides [kotulskite PdTe and Pd(Te,Bi)] on contact with massive ores (figure 6c). PGM are represented by PtFe in massive ores.

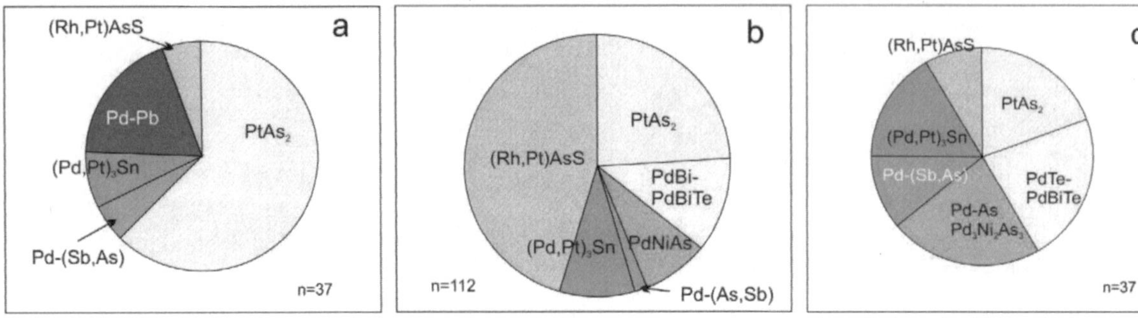

Figure 7. The ratio of PGM in different rocks: a – picritic gabbro-dolerite (chromite horizon), b – lower part of picritic gabbro-dolerite (horizon 891.8 m), c – olivine gabbro-dolerite; n – number of analyzes.

3. Discussion

Thus, the following mineralogical-geochemical regularities are observed in the disseminated ores of the southern part of the Noril'sk 1 deposit. The Ni/Cu ratios are decreasing in the rocks with depth: Ni/Cu is more than 1 in picritic gabbro-dolerite, whereas Ni/Cu is less than 1 in olivine gabbro-dolerite. The Ni/Fe ratio in the pentlandite increases with depth.

Iron-nickel ratios in pentlandite reflect the activity of sulfur ($\lg fS_2$) during its formation [14,15]. As the $\lg fS_2$ increases, the concentration of Ni in the pentlandite increases as well. The variation of k = Ni/(Ni+Fe) in pentanedite varies in the range 0.33 – 0.47 in picritic gabbro-dolerite that corresponds to the $\lg fS_2$ from −12 to −11. Whereas variation of k in pentlandites from olivine gabbro-dolerite varies from 0.49 to 0.57, therefore they crystallized under conditions of $\lg fS_2$ corresponding to an interval from −10.5 to −9. Thus, two different sulphide assemblage are characteristic of picritic and olivine gabbro-dolerites. In the first case, the "low-sulfur" minerals sequence is due to the evolution of the sulphide Cu-rich melt fraction separated from Fe-rich melt in the pre-crystallization time [5]. The following parageneses are formed during fractionation of mss and iss: $Po_{(h>tr)}+Cp+Pn_{(Fe>Ni)} \rightarrow Po_{(h>tr)}+Cp+Cb+Pn_{(Fe>Ni)} \rightarrow Cb\pm Cp+Pn_{(Fe>Ni)}\pm tr \rightarrow Th+Cb+Pn\pm Bn$ [1]. This agrees with the investigated parageneses in picritic gabbro-dolerite of PH-14 borehole. But disseminated ore hosed by picritic gabbro-dolerite has a Ni/Cu ratio >1. It means that the Cu-rich melt fraction was enriched in Ni also largely because the Ni tends to be an incompatible element in all plausible sulfide magma compositions [16]. The evolution of mineral parageneses in olivine gabbro-dolerite is: $Po_{(m>h)}+Cp+Pn_{(Fe<Ni)} \rightarrow Cp+Po_{(m+h)}+Pn_{(Fe<Ni)}$. It agrees with the experimental data on directional crystallization of melts [17] and is due to the pyrrhotite-chalcopyrite fractionation in conditions of increased sulfur fugacity.

A positive correlation between sulfur and copper for all types of rocks (figure 8 a) and a gradual increase in the concentrations of ore elements and sulfur towards the towards the bottom intrusion (see figure 2) is in agreement with the evolution of sulphide melt and fractionation of ore elements during its deposition [13]. So, there is a clear genetic relationship between the sulfide assemblages in disseminated ores and the type of the host gabbro-dolerites. A sulfur-poor and Cu-rich sulphide assemblage which has been formed at low sulfur activity is characteristic of picritic gabbro-dolerite. And a high-sulfur assemblage (monoclinic pyrrhotite, pyrite and Ni-rich pentlandite) formed at higher sulfur activity. Binary diagram of Cu versus Pd (figure 8 b) shows a difference in trends: a poorly defined linear for olivine gabbro-dolerite and indeterminate for picritic gabbro-dolerite.

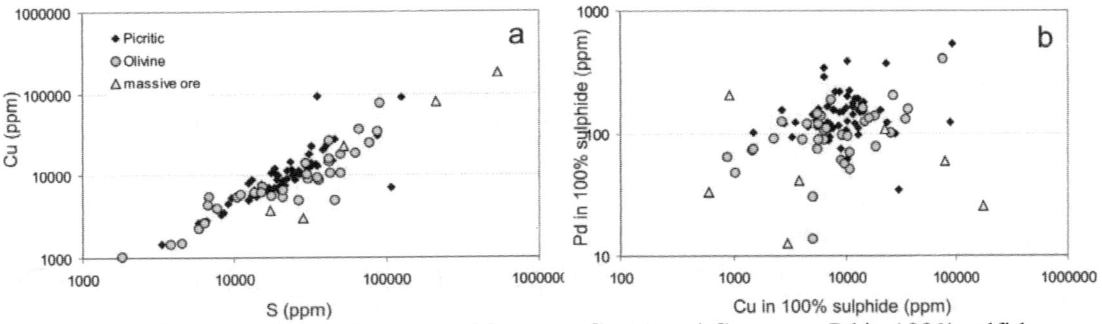

Figure 8. Binary diagrams of S versus Cu (a) and Cu versus Pd in 100% sulfides

4. Conclusions

Thus, on the basis of mineralogical and geochemical features it is established that:

– Ni/Cu ratios are decreasing in rocks from picritic to olivine gabbro-dolerites and toward the massive ores, that agreed with the increase of the chalcopyrite in this direction;

– Ni/Fe ratios in pentlandite are increasing from picritic to olivine gabbro-dolerites and to massive ores which indicates an increase in sulfur fugacity in mineral parageneses down the section of borehole;

– There is a clear correlation between the sulfide assemblages and the type of the host gabbro-dolerites: picritic gabbro-dolerites are characterized by a low-sulfur and Fe- and Cu-rich sulphide assemblage; olivine gabbro-dolerite is characterized by a high-sulfur assemblage of sulfides.

– The change of the compounds of Pd with semimetals is observed along the section: Sn → Pb → As → Bi(Te) in the picritic gabbro-dolerite and Sn → Sb → As → Te(Bi) in olivine gabbro-dolerite. The difference in evolutionary trends is due to the different schemes of fractionation of sulphide melt.

Acknowledgments

The work was supported by RFBR grants № 16-05-00884 and №16-05-00945. The authors are grateful to V.N. Knyazev and S.A. Silyanov for the performance of mineral analyzes.

References

[1] Naldrett A.J. 2003 Magmatic sulfide deposits of nickel-copper and platinum-metal ores (St. Petersburg: St. Petersburg State University) p 438 (in Russian)

[2] Likhachev A.P. 2006 Platinum-nickel-copper and platinum deposits (Moscow: Aslan) p 496 (in Russian)

[3] Krivolutskaya N.A. 2014 Evolution of trap magmatism and Pt-Cu-Ni ore formation in Noril'sk region (Moscow: The Partnership of Scientific Publications KMK) p 325 (https://books.google.ru/books?isbn=5873179204) (in Russian)

[4] Genkin A.D. 1968 Minerals of platinum metals and their associations in copper-nickel ores of the Noril'sk deposit (Moscow: Nauka) p 106 (in Russian)

[5] Distler V.V., Genkin A.D., Filimonova A.A., Khitrov V.G. and Laputina I.P. 1975 The zoning of copper-nickel ores of the Talnakh and Oktober Deposits *Geology of Ore Deposits* **2** 16–27 (in Russian)

[6] Distler V.V., Grokhovskaya T.L., Evstigneeva T.L., Sluzhenikin S.F., Filimonova A.A., Dyuzhikov O.A. and Laputina I.P. 1988 Petrology of magmatic sulphide ore-formation (Moscow: Nauka) p 327 (in Russian)

[7] Genkin A.D., Distler V.V., Gladyshev G.D., Filimonova A.A., Evstigneeva T.L., Kovalenker V.A., Laputina I.P., Smirnov A.V. and Grokhovskaya T.L. 1981 Cu-Ni sulphide ores of the Noril'sk deposits (Moscow: Nauka) p 234 (in Russian)

[8] Sluzhenikin S.F., Distler V.V., Dyuzhikov O.A., Kravtsov V.F., Kunilov V.E., Laputina I.P. and Turovtsev D.M. 1994 Low-sulfide platinum mineralization in the Noril'sk differentiated intrusive bodies *Geology of ore deposits* **36**(3) 195-217 (in Russian)

[9] Sluzhenikin S.F., Krivolutskaya N.A., Rad'ko V.A., Malitch K.N., Distler V.V. and Fedorenko V.A. 2014 Ultramafic-mafic intrusions, volcanic rocks and PGE-Cu-Ni sulfide deposits of the Noril'sk Province, Polar Siberia. Field trip guidebook *Conf. 12th International Platinum Symposium* ed O.N. Simonov (Yekaterinburg: IGG UB RAS) p. 80

[10] Spiridonov E.M. 2010 Ore-magmatic systems of the Noril'sk ore field *Russian Geology and geophysics* **51**(9) 1356–1378

[11] Komarova M.Z., Kozyrev S.M., Simonov O.N. and Lyul'ko V.A. 2002 The PGE mineralization of disseminated sulfide ores of the Noril'sk-Taimyr Region *The Geology, Geochemistry, Mineralogy and Mineral Beneficiation of Platinum-Group Elements* vol **54** ed L.J. Cabri (Quebec: Montreal) pp 547–567

[12] Kozyrev S.M., Komarova M.Z., Emelina L.N., Oleshkevich O.I., Yakovleva O.A., Lyalinov D.V. and Maximov V.I. 2002 The mineralogy and behaviour of PGM during processing of the Noril'sk-Talnakh PGE-Cu-Ni ores *The Geology, Geochemistry, Mineralogy and Mineral Beneficiation of Platinum-Group Elements* vol **54** ed L.J. Cabri (Quebec: Montreal) pp 757–791

[13] Duran C.J., Barnes S-J., Pleše P., Prašek M.Kudrna, Zientek M.L. and Pagé P. 2017 Fractional crystallization-induced variations in sulfides from the Noril'sk-Talnakh mining district (polar Siberia, Russia) *Ore Geology Reviews* DOI: 10.1016/j.oregeorev.2017.05.016.

[14] Vaughan D. and Craig J. 1981 *Mineral chemistry of metal sulfides* (Moscow: Mir) p 575 (in Russian)

[15] Peregoedova A.V. 1999 Physical and chemical behavior of Pt and Pd during the process of crystallization of Fe, Ni, Cu-containing sulfide melts and in subsequent subsolidus transformations. Author's abstract. diss. Novosibirsk (in Russian)

[16] Mungall J.E. 2005 Magmatic geochemistry of the platinum-group elements *Exploration for Platinum-Group Element Deposits* vol **35** ed J.E. Mungall (Mineralogical Association of Canada Short Course) pp 1–34

[17] Kosyakov F.I., Sinyakova E.F. and Distler V.V. 2012 Experimental simulation of phase relationships and zoning of magmatic nickel-copper sulfide ores, Russia *Geology of Ore Deposits* **54**(3) 179–208

Late-paleozoic granitoid complexes of the southwest Primorye: geochemistry, age and typification

A A Veldemar, G M Vovna

Far East Geological Institute Far Eastern Branch of Russian Academy of Sciences, 159, Prospekt 100-letiya, Vladivostok 690022, Russia

E-mail: veldemar@fegi.ru, gala1367@mail.ru

Abstract. The article presents the first data of geochemical studies of the Late Permian granitoids of the Gamov Complex located in the southwestern part of the Voznesenskiy terrane. The purpose of the study was to identify the main geochemical features of the Late Paleozoic granitoids of the southwestern Primorye, which in the future will allow us to draw conclusions about the petrogenesis of these granitoids. Elemental analysis of 20 samples was carried out, conducted statistical and mathematical processing of the data, have been constructed representative diagrams and graphs for this group of rocks. Elemental analysis was performed by atomic emission (ICP-AES) and inductively-coupled-plasma (ICP-MS) mass spectrometry, at the Analytical Center FEGI FEB RAS.

1. Introduction

The Late-Paleozoic Gamov complex of granitoids located in the south-west of Primorye is an object of investigation (figure 1). In the tectonic viewpoint, the Gamov complex is located in the southwestern part of the Voznesenskiy terrane. The complex is represented by a gabbro-diorite-granodiorite-plagiogranite-granite rock association that forms the Gamov batholith. Due to tectonic faulting and overlapping by later formations, it is exposed by scattered fragments of up to 120 km^2. The complex breaks, hornfels, sections granitizing the Reshetnikovskaya and Barabashsky formations of the upper Permian, and is overlapped by the Upper Triassic Talminskaya series [3].

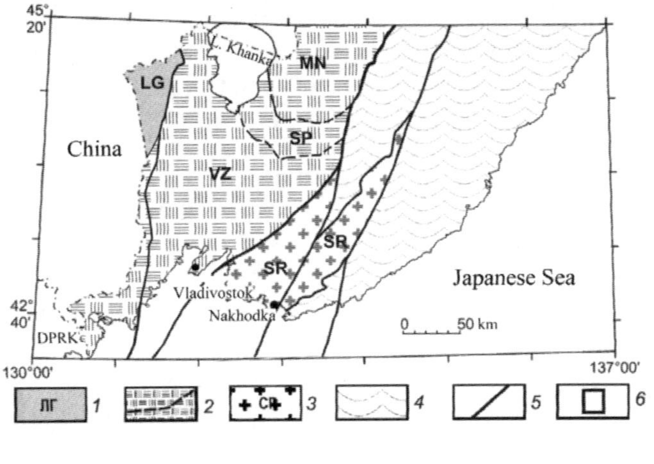

Figure 1. Scheme of terranes of the Southern Primorye region [2] with changes.
1 - Laoelin-Grodekovsky (LG) late Paleozoic terrane; 2 - terranes of the Bureya-Khanka Early Paleozoic orogenic belt: Matveevsko-Nakhimovsky (MN), Spassky (SP), Voznesensky (VZ); 3 - Sergeevsky (SR) terrain; 4 - terranes of the Late Mesozoic Sikhote-Alin orogenic belt; 5 - faults; 6 - area of research.

Three phases are distinguished in the structure of the Gamov granitoid batholith. The first is gabbro, gabbro-diorites and diorites. The second is biotite-hornblende-quartz granodiorites and plagiogranites. The third phase includes dikes and veins of aplite-like granites, leucogranites and pegmatites [3].

2. Materials and methods

Elemental analysis of 20 samples was carried out at the Analytical Center of the Far Eastern Geological Institute, FEB RAS. Major-element compositions (excepting SiO_2 and H_2O^-) were determined by atomic emission spectroscopy with inductively coupled plasma mass ICP-OES spectrometer iCAP 6500Duo (Thermo Scientific Corporation, USA). SiO_2 and H_2O^- contents were determined using a standard weighting (gravimetric) method. Trace element and REE abundances were measured by inductively coupled plasma mass ICP-MS spectrometry Agilent 7500C (Agilent Technologies, USA). The precision of the analyses was generally 2-5% for major oxides, and 5-10% for trace and REE elements. The correctness of the results is confirmed by the analysis of international and Russian standard samples of JG-3, JB-3, JA-2 (Japan) and SG-3 (GSO No. 3333-85, Russia).

3. Results and discussion

The contents of major components, rare and rare-earth elements in representative samples of the main species of rocks are given in the table (table 1). On the classification diagram ($Na_2O + K_2O$) - SiO_2, the points of the compositions of rocks of the Gamov massif are located in the fields of granodiorites, granites, leucogranites (figure 2). The rocks are mainly magnesian, peraluminous, potassium-sodium series. By the ratio of K_2O and SiO_2, the rocks refer to the calcic and calc-alkalic series.

According to the classification of Whalen [9] (figure 2E, F) and Maeda [6] (figure 2G) the figurative points of granitoids of the Gamov complex are located in the S and I types of granite.

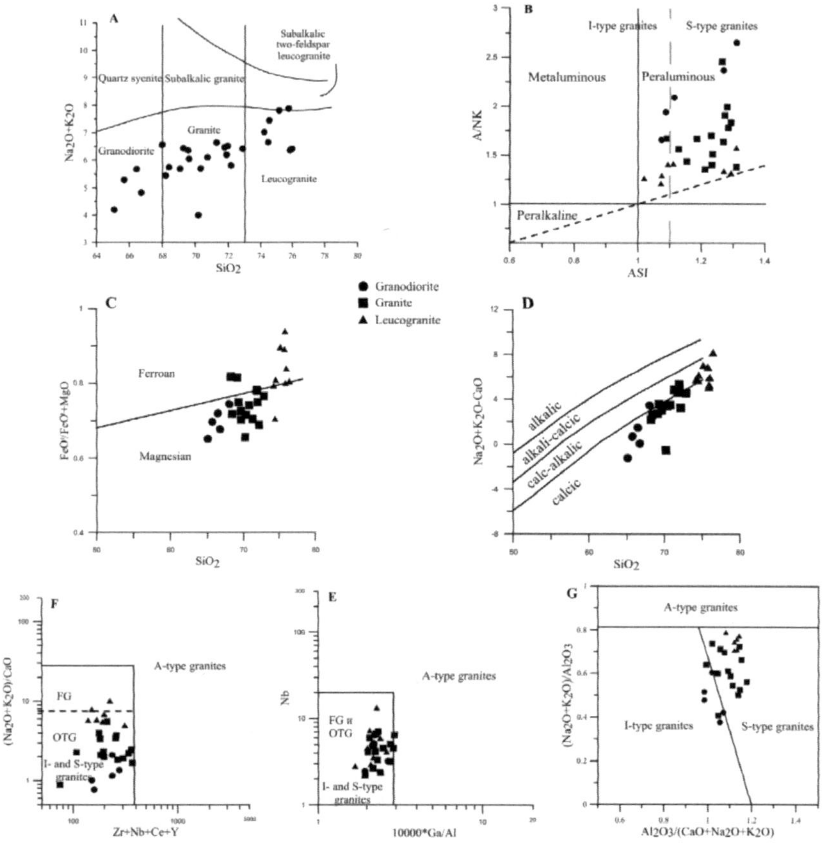

Figure 2. Geochemical discrimination diagrams for granitoids of the Gamov complex.
A - SiO_2 – ($Na2O+K_2O$) [1];
B - ASI – A/NK, где ASI = Al_2O_3/ ($Na_2O+K_2O+CaO-1.67P_2O_5$), A/NK = $Al_2O_3/(Na_2O+K_2O)$ [7];
C - SiO_2 – $FeO^t/(FeO^t+MgO)$ [5];
D - SiO_2 – Na_2O+K_2O+CaO [5];
E - 10000*Ga/Al – Nb [9];
F - Zr+Nb+Ce+Y – (Na_2O+K_2O)/CaO [9]: FG- fractionated felsic granites, OTG- unfractionated granites;
G - $Al_2O_3/(CaO+Na_2O+K_2O)$ – $(Na_2O+K_2O)/Al_2O_3$ [6].

Table 1. Chemical compositions of granitoids of the Gamov Complex

Element	А-54	А-55а	К-7-9	З-56а	К-7-7	Г-11-5	Т-52	Ал-18	Т-50	Л-1б	К-26а	К-7-8
	1	2	3	4	5	6	7	8	9	10	11	12
SiO_2	65,68	66,43	67,98	68,20	69,27	70,17	70,31	71,27	74,20	74,44	75,15	75,75
TiO_2	0,56	0,52	0,51	0,53	0,48	0,34	0,40	0,41	0,20	0,17	0,12	0,12
Al_2O_3	15,47	15,18	14,83	15,63	14,56	14,88	14,54	13,82	13,26	13,87	13,57	13,34
Fe_2O_3	5,03	4,90	4,60	5,10	4,18	2,94	3,45	3,31	1,92	1,92	1,68	1,43
MnO	0,09	0,08	0,07	0,07	0,06	0,04	0,05	0,05	0,04	0,02	0,02	0,02
MgO	1,97	1,72	1,43	1,02	1,26	1,39	1,23	1,25	0,46	0,73	0,18	0,16
CaO	4,59	4,19	3,10	3,25	2,88	4,50	2,30	1,78	1,26	0,96	0,76	0,99
Na_2O	2,97	2,98	3,29	3,44	3,12	3,08	3,55	4,32	3,23	5,89	3,63	3,56
K_2O	2,32	2,70	3,28	2,01	3,32	0,92	2,15	2,32	3,79	0,76	4,19	4,32
P_2O_5	0,08	0,08	0,08	0,11	0,09	0,04	0,09	0,09	0,04	0,02	0,04	0,02
LOI	0,90	0,86	0,42	0,51	0,57	1,43	1,45	1,22	1,25	0,97	0,46	0,3
Sum	99,79	99,63	99,85	100,11	99,84	99,81	99,66	99,89	99,65	99,78	99,79	100,14
Be	1,13	0,90	1,66	1,89	1,08	0,85	1,04	0,93	1,03	1,96	1,31	1,01
Sc	16,4	15,3	13,5	16,1	11,4	4,9	6,3	7,9	4,5	4,1	4,1	3,1
V	87,94	82,09	86,93	55,08	74,78	73,44	46,84	44,40	15,43	12,31	6,66	14,59
Cr	173,2	134,5	150,6	174,8	144,7	257,1	118,8	126,9	5,53	6,33	18,13	201,6
Co	11,22	10,78	9,78	6,14	8,36	9,47	6,60	6,02	2,15	2,28	1,21	1,67
Ni	6,70	4,72	7,26	8,92	7,09	9,79	6,06	3,67	2,49	2,12	2,42	5,64
Cu	10,13	45,22	26,82	42,54	15,16	25,53	4,86	8,32	1,57	25,53	6,73	9,37
Zn	49	50,5	59	87,8	47,7	37,2	49,2	24,5	28,9	32,1	18,8	25,9
Ga	17,56	17,70	17,51	19,34	15,96	15,31	22,58	14,84	16,93	15,55	16,31	14,65
Rb	74,97	90,37	191,44	73,74	164,89	30,17	95,22	49,11	105,02	19,81	117,49	138,34
Sr	192,3	195,3	169,9	205,3	165,8	378,4	437,7	346,7	209,4	334,6	126,9	98,3
Y	25,49	28,53	31,62	40,53	27,42	5,29	17,89	13,78	26,08	13,64	31,62	22,45
Zr	169,3	179,9	151,0	203,2	254,9	47,73	217,2	197,7	110,0	146,0	119,6	77,11
Nb	4,89	5,17	6,46	7,08	6,01	2,19	6,44	4,15	5,95	2,98	6,55	7,32
Mo	7,17	6,50	8,17	9,93	7,39	13,30	6,36	6,14	0,03	0,45	0,44	11,49
Cd	0,01	0,09	0,19	0,29	0,35	0,18	0,13	0,09	0,07	0,40	<DL	<DL
Sn	2,23	3,00	1,18	3,04	1,39	0,17	1,95	0,12	2,34	2,30	2,06	3,14
Cs	1,59	1,76	12,76	2,47	12,11	1,73	2,54	1,28	1,89	0,66	1,65	3,40
Ba	366,9	525,7	393,9	397,2	406,6	284,4	648,6	850,7	521,9	197,1	693,3	711,8
La	16,49	29,68	31,52	51,63	32,13	8,11	55,43	24,84	25,05	20,20	32,97	24,84
Ce	36,45	62,84	49,10	117,63	52,38	19,68	118,59	43,55	51,34	36,32	68,68	45,03
Pr	4,33	6,66	6,46	13,61	6,37	2,45	12,20	4,52	5,57	3,96	7,84	4,69
Nd	16,35	24,69	24,94	52,25	24,70	9,56	41,30	15,32	20,99	12,92	27,86	16,90
Sm	3,34	4,23	4,68	9,95	4,73	2,00	7,04	2,33	3,66	2,44	6,81	3,40
Eu	0,87	0,76	0,95	1,09	0,94	0,45	0,83	0,81	0,47	0,59	0,72	0,48
Gd	3,96	4,72	4,99	9,20	4,02	1,35	5,04	2,59	4,25	2,40	5,65	3,53
Tb	0,65	0,79	0,86	1,31	0,80	0,17	0,67	0,34	0,68	0,29	0,97	0,59
Dy	4,31	4,75	5,66	6,96	4,24	0,92	3,05	1,99	3,79	1,93	5,34	3,64
Ho	0,94	0,99	1,23	1,50	1,15	0,20	0,57	0,55	0,75	0,37	1,20	0,81
Er	2,69	2,83	3,83	4,39	2,67	0,44	1,94	1,45	2,45	1,42	3,16	2,21
Tm	0,36	0,44	0,51	0,65	0,40	0,08	0,23	0,18	0,39	0,14	0,50	0,32
Yb	2,44	3,02	3,44	4,29	3,25	0,52	1,64	1,41	2,72	1,09	3,38	2,00
Lu	0,38	0,42	0,47	0,54	0,34	0,11	0,25	0,25	0,40	0,17	0,47	0,31
Hf	4,75	5,11	4,86	6,17	7,65	2,35	5,43	5,16	3,99	3,44	4,05	2,78
Ta	0,48	0,49	0,56	0,72	0,57	0,26	0,45	0,33	0,61	0,05	0,72	0,77
W	5,88	5,27	6,31	7,60	6,49	9,05	4,21	4,85	0,34	0,21	0,52	9,69
Pb	9,41	9,39	19,40	11,54	17,60	9,93	12,45	3,09	16,19	8,38	16,02	23,06
Th	8,37	8,84	22,27	18,66	19,62	7,65	24,14	10,50	17,56	8,45	18,51	15,08
U	1,38	1,51	3,02	2,63	3,03	1,64	1,99	1,72	2,40	0,79	2,80	2,70

Note: The content of oxides in mass%, microelements in g / t.
<DL- value below detection limit
1-3 - granodiorites, 4-8 - granites, 9-12 - leucogranites

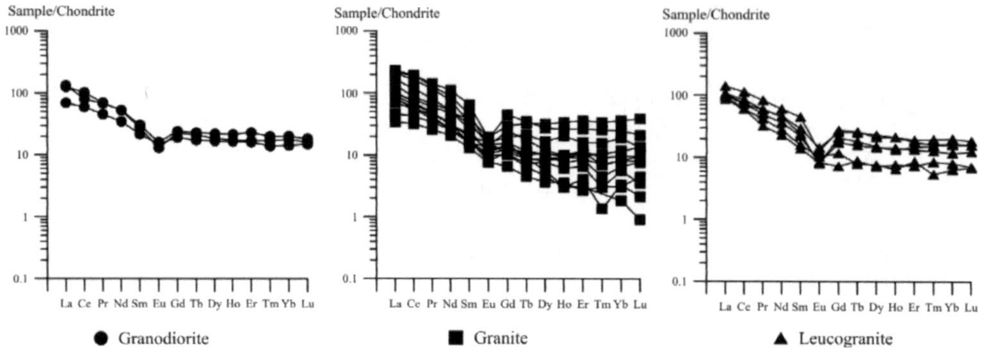

Figure 3. The REE distribution spectra in granitoids of the Gamov complex are normalized by the chondrite composition [8].

REE patterns of granites (figure 3) are characterized by an enrichment of light rare earth elements (LREE) $(La/Yb)_N$ = 2,44-32,92 and the presence of a negative Eu anomaly (Eu/Eu* = 0.35-0.84), which is typical for granites formed due to crustal sialic sources. Granitoids of the Gamov complex are significantly enriched with large-ion lithophiles (LILE) - Cs, Rb, Ba, K, and high field strength elements (HFSE) -Th and U (figure 4).

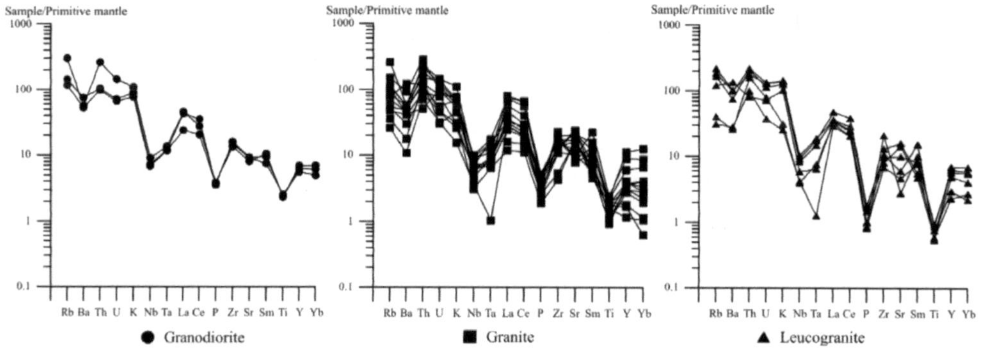

Figure 4. Trace element spidergrams for granitoids of the Gamov complex are normalized by the composition of the primitive mantle [8].

Radiological K/Ar dating in the petrotip (Gamov Peninsula), in biotite and amphibole, correspond to 251 and 252 Ma [3]. Definitions of Chinese geologists U/Pb method in contiguous territories give an age of 254 million years, which corresponds to the late Permian time [3]. The latest U-Pb isotope studies of zircons (SHRIMP-II method) [4] established that the age of the Gamov massif is 260 ± 2 million years, which confirms by the earlier obtained Late Permian age.

4. Conclusions

Late-Paleozoic granitoids of the Gamov complex form a series of granodiorite-granite-leucogranite. Refer to magnesian, peraluminous, calcic and calc-alkalic varieties. They have elevated concentrations of light rare earth elements and large-ion lithophile elements. Correspond to granites of I and S types.

References

[1] Bogatikov O A 1983 *Magmatic rocks* vol 1(Moscow: Science) p 768
[2] *Geodynamics, magmatism and metallogeny of the East of Russia* (Vladivostok: Dal'nauka) 2006 ed A I Hanchuk vol 1 p 572
[3] *State Geological Map of the Russian Federation. Scale 1: 1,000,000 (third generation). Leaf (L- (52), 53, (K-52, 53)-Lake Khanka.) Explanatory note* (St. Petersburg:

VSEGEI Cartographic Factory) 2011 ed S V Kovalenko p 684

[4] Kruk N N, Golozubov V V, Kasatkin S A 2015 *Russian Geology and Geophysics* Granitoids of the Gamov intrusion (southern Primorye), its peculiarities and indicative and geodynamic role **12** 1685

[5] Frost B R, Barnes C G, Collins W J, Arculus R J, Ellis D J, Frost C D 2001 *Journal of Petrology* A geochemical classification for granitic rocks **11** 2033

[6] Maeda J 1990 *Tectonophysics* Opening of the Kuril basin deduced from the magmatic history of Central Hokkaido, North Japan **3/4** 235

[7] Maniar P, Piccoli P 1989 *The geological Society of America Bulletin* Tectonic discrimination of granitoids **5** 635

[8] Sun S S, McDonough W F 1989 *Geol. Soc. Spec. Publ.* Chemical and Isotopic Systematics of Oceanic Basalts: Implications for Mantle Composition and Processes **42** 313

[9] Whalen J B, Currie K L , Chappell B W 1987 *Contributions to Mineralogy and Petrology* A-type granites: Geochemical characteristics, discrimination and petrogenesis **95** 407

Origin of carbonatites of the Matcha alkaline pluton from Turkestan-Alai ridge, Kyrgyz Southern Tien Shan

V V Vrublevskii

Department of Geology and Geography, Tomsk State University, Lenin Avenue 36, Tomsk, 634050, Russia

E-mail: vasvr@yandex.ru

Abstract. Postorogenic alkaline intrusions in the Turkestan-Alai segment of the Southern Tien Shan coexist with dikes and veins of carbonatites dated at ~220 Ma. They are primarily composed of calcite and dolomite (60–85 %), as well as sodic amphibole, phlogopite, clinopyroxene, microcline, albite, apatite, and magnetite, with accessory niobate, ilmenite, Nb-rutile, titanite, zircon, baddeleyite, monazite-(Ce), barite, and sulfides. The rocks share mineralogical and geochemical similarity with carbonatites that originated by liquid immiscibility at high temperatures above 500°C. Silicate and salt-carbonate melts are derived from sources with mainly negative bulk εNd(t) ~ from –11 to 0 and high initial $^{87}Sr/^{86}Sr$ ratios (~ 0.7061-0.7095) which may be due to mixing of PREMA and EM-type mantle material. Pb isotopic ratios in accessory pyrrhotite ($^{206}Pb/^{204}Pb$ = 18.38; $^{207}Pb/^{204}Pb$ = 15.64; $^{208}Pb/^{204}Pb$ = 38.41) exhibit an EM 2 trend. The intrusions bear signatures of significant crustal contamination as a result of magma genesis by syntexis and hybridism. Concordant isotope composition changes of $\delta^{13}C$ (–6.5 to –1.9 ‰), $\delta^{18}O$ (9.2-23 ‰), δD (–58 to –41 ‰), and $\delta^{34}S$ (12.6-12.8 ‰) in minerals and rocks indicate inputs of crustal material at the stage of melting and effect of hot fluids released during dehydration of metamorphosed oceanic basalts or sediments. The observed HFSE patterns of the oldest alkaline gabbro may be due to interaction of the primary mafic magma with IAB-type material. The isotope similarity of alkaline rocks with spatially proximal basalts of the Tarim large igneous province does not contradict the evolution of the Turkestan-Alai Triassic magmatism as the «last echo» of the Tarim mantle plume.

1. Introduction

Formation of carbonatites is usually associated with either fractional crystallization of CO_2-saturated alkali-silicate magmas, or intrusion and crystallization of alkali-dolomite magma from independent mantle sources [1–3]. Bear signature of crustal contamination of magmatic derivatives is recorded in volcano-plutonic complexes in Phanerozoic orogenic belts [4-7].

In Kyrgyz Southern Tien Shan, dikes and veins of carbonatites Ar–Ar dated at ~220 Ma complete forming postorogenic Late Triassic alkaline intrusions of the Turkestan-Alai Ridge. Plutons are composed by essexites, foidolites, nepheline and alkaline syenites [8–10]. The most abundant carbonatite veins are concentrated in the Delbek valley that cuts the North-Eastern contact of Matcha pluton (~ 30 km^2), they occur as up to 100–400 m long injection zones together with syentite offshoots among fenitized schists. Some veins are small (to n × 1 m) and often form branching systems and thick lenses. Carbonatites have massive or banded, schlieric-taxitic structures. Predominant carbonates (calcite and dolomite, 60–85 %) commonly occur as fine- to medium-grained aggregates, often with round porphyritic crystals that enclose particles of other typical phases (10–35 %): alkaline amphiboles, clinopyroxenes and feldspar, micas, apatite and magnetite. Accessory sulfides (pyrrhotite, pyrite, etc.), pyrochlore, monazite, titanite, zircon, baddeleyite, thorite, ilmenite, ilmenorutile, rutile, barite occur sporadically as scarce microcrysts or euhedral grains (~ 0.01–1 mm) in a silicate-carbonate matrix [11–12]. Observed signs of deformation with twinning bends in carbonates, along with kink-bands in mica flakes and banded structures as evidence of possible plastic flow. Geochemical characteristics suggest mantle-crustal matter source and liquid immiscibility origin of carbonatites.

2. Chemical composition of alkaline rocks and carbonatites

For silicate rocks of typical alkaline plutons (Matcha and Zardalek ones) from gabbro to foydolites and syenites, it is typical to see increase in concentrations of SiO_2 ~ from 40 to 66 wt. %, Al_2O_3 ~ from 12 to 29 wt. %, (Na_2O + K_2O) ~ from 2.3 to 19 wt. %. At the same time following concentrations decrease: MgO ~ from 11 to 0.3, CaO ~ from 20 to 2 and Fe_2O_3 ~ from 14 to 2 wt. % (table 1).

Table 1. Representative chemical composition of igneous alkaline rocks and carbonatites from Turkestan-Alai ridge[a].

Massif	Matcha			Zardalek						
Rock[b]	1[c]	2[c]	3[c]	4	5	6	7	8	9	10
SiO_2, wt. %	49.65	52.06	62.51	1.23	1.03	3.28	13.55	51.82	42.47	53.39
TiO_2	1.00	0.68	0.12	0.04	0.02	0.03	0.18	0.67	0.40	0.18
Al_2O_3	16.15	20.25	16.64	0.05	0.30	0.60	2.95	11.97	28.94	25.99
Fe_2O_3	10.99	7.70	5.10	1.97	1.77	1.53	1.01	8.51	3.91	1.82
MnO	–[d]	0.13	0.09	0.13	0.18	0.21	0.12	0.13	0.12	0.06
MgO	2.64	0.21	0.47	0.05	2.16	13.90	7.87	10.77	0.89	1.51
CaO	8.78	4.51	2.72	53.98	53.39	36.17	39.64	8.39	3.39	1.38
Na_2O	5.26	7.60	6.80	0.30	0.08	0.34	0.60	1.62	8.50	4.22
K_2O	4.86	4.92	6.22	0.20	0.08	0.32	1.85	1.37	7.51	9.23
P_2O_5	0.06	0.08	0.03	0.80	3.22	3.77	2.34	0.30	0.16	0.03
LOI	1.10	3.08	0.22	40.99	37.01	39.23	29.28	4.06	3.33	1.92
Total	100.49	101.22	100.92	99.74	99.24	99.38	99.39	99.61	99.62	99.73
Cr, ppm	–	9.5	9.8	2.7	7	1.5	3.7	596	5.2	3.5
Ni	–	1.6	2.6	13	25	18	19	63	3.1	2.4
V	–	9	21	6	2.2	10	18	164	40	13
Co	–	3.5	4.5	1.7	5	3.1	3.4	32	6	2.5
Sc	–	0.6	1.3	1.6	1.2	0.8	1.3	26	1.1	0.34
Pb	–	18	39	53	17	–	–	7.5	51	108
Cs	–	2	5.6	1.7	0.3	0.1	17	1.4	27	25
Rb	90	255	226	42	9.2	10	211	41	339	314
Ba	3100	715	854	1985	616	270	554	469	1190	1820
Sr	740	430	315	4056	2692	1426	2035	342	1011	1396
Y	24	14	7.5	182	40	8.5	32	13	15	3
Zr	120	124	249	25	1.8	1.8	6.6	84	110	64
Hf	2.7	3.8	6.7	0.3	0.1	0.12	0.28	2.2	1.6	0.92
Nb	24	51	44	12	9.4	11	14	9.6	12	20
Ta	1.5	2.3	2.1	0.8	0.9	0.93	1	3	0.83	1.2
Th	6.5	4	6.6	0.3	0.9	2.3	0.61	5.1	19	56
U	2.6	2.2	2.9	0.2	1.6	8.7	0.96	2.5	10	28
ΣREE	320.7	115.7	63.1	707.1	205	56.6	146.3	89.2	178.8	147.3

[a] The ICP-MS and XRF analyses were carried out at the National Research Tomsk State University (Tomsk), at the Institute of Geology and Mineralogy SB RAS (Novosibirsk) and the Institute of Mineralogy, Geochemistry and Crystal Chemistry of Rare Elements RAS (Moscow) on an Agilent 7500cx, Elan-6100 DRC, Finnigan MAT spectrometers and a Thermo Scientific ARL-9900XP instrument under standard operating conditions.

[b] 1, 8 = subalkaline and alkaline gabbro; 2, 10 = nepheline syenite; 3 = alkaline syenite; 9, 10 = foidolite; 4–7 = carbonatites: REE-enriched variety (4), REE-depleted variety (5, 6), melanocratic variety (7).

[c] after Nenahov, 2002 [9].

[d] (–) is not detected.

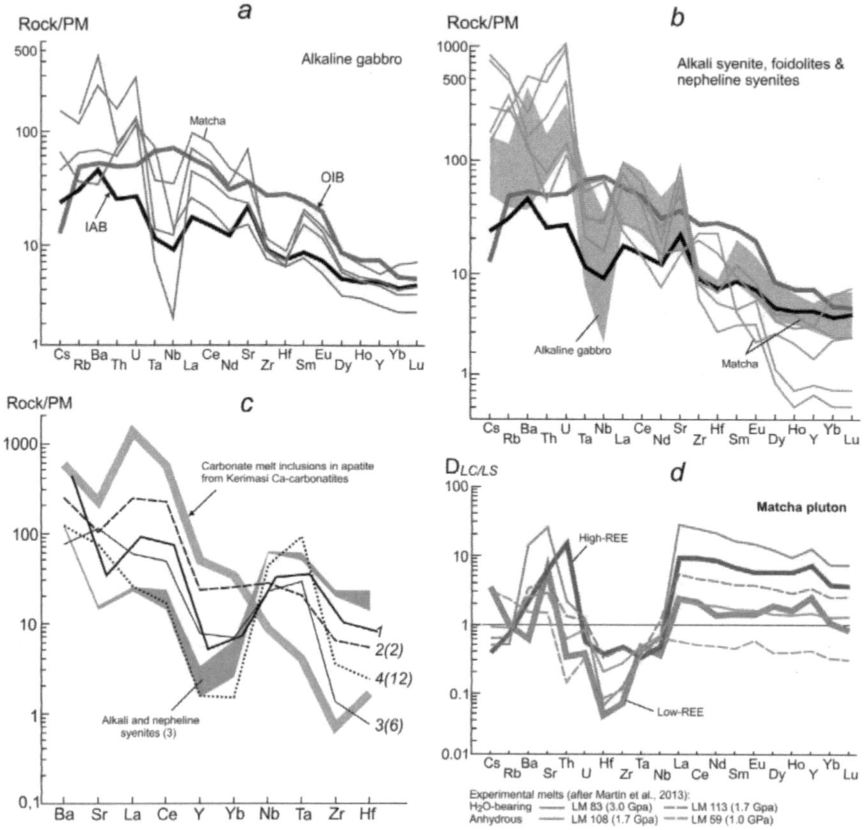

Figure 1. Primitive mantle-normalized [13] multi-element diagrams (*a–c*) of the Matcha alkaline rocks and carbonatites and experimental carbonate-silicate melt partition coefficients ($D_{LC/LS}$) of trace elements [15] and their ratios (*d*) for the Matcha carbonatites and nepheline syenite. In panels (a–b) shows OIB [13] and IAB [14] average compositions. In panel (*c*), trace elements are ordered according to their decreasing experimental partition coefficients between immiscible silicate-carbonate liquids: *1* = essexite; *2* = high-REE carbonatites; *3* = low-REE carbonatites (6); *4* = low-REE carbonatites (12); numerals in braces show number of analyses. In panel (*d*), lines «High-REE» and «Low-REE» refer to trace element ratios of high-REE carbonatites, average for two samples and low-REE carbonatites, average over six samples, with average nepheline syenite (two samples).

Nature of alkalinity (up to 2–9 wt. % $_2O$; 0.8–2.2 K_2O/Na_2O) suggests that the rocks are derivatives of shoshonite series of active continental margins (ACM). In more differentiated varieties, amount of siderophile elements (4–596 Cr, 2–63 Ni, 1.3–227 V, 3–38 Co, 0.3–26 Sc, ppm) decreases, enrichment in Cs, Rb, Ba, Th, U occurs at OIB level and higher, and less in Nb, Ta and LREE. Sr content reaches ~ 1400-1800 ppm, which could be caused by crust contamination of melts. Distribution of rare earth elements (average REE ~ 200 ppm; La/Yb_N 10-17) and other trace elements in gabbro is similar to island arc basalt (IAB) compositions and partially to oceanic island basalt ones (OIB) (figure 1a, b). In later foidolites and syenites, REE concentrations decrease on average down to 140 ppm at wide ranges of La/Yb_N ~ 2-150. It might be due to dilution of a melt by REE-depleted crustal material.

Carbonatites contain ~ 34–55 wt. % CaO, ~ 0.1–17 wt. % MgO, ~ 0.3–19 wt. % SiO_2, ~ 0.3–5.6 wt. % P_2O_5, ~ 0.8–3.5 wt. % Fe_2O_3, ~ 0.1–3 wt. % (Na_2O + K_2O). Content of typomorphic trace elements reaches up to ~ 1500–4000 ppm Sr, ~ 500–2400 ppm Ba, ~ 20–140 ppm Zr, ~ 10–70 ppm Nb, ~ 250–1030 ppm (REE + Y). Sr/Ba ratios, which are >1 (from 2 to 17), and Th/U ≈ 0.1 to 53 indicate relatively high-temperature settings for formation and possible heterogeneity of the rocks. There are three groups with REE contents: up to 700-1000, 150-270, 23-106 ppm. At the same time, distribution of La, Ce, Yb and Y, as well as Sr, Nb, Ta, Zr, and Hf in these rocks is complimentary in relation to multistage spectrum of alkaline rocks (figure 1c, d), which is noted in experimental systems with silicate-carbonate immiscibility and in some natural melts [15-18]. Comparing to syenites, carbonatites are enriched in REE and Sr, but depleted in Nb, Ta, Zr and Hf.

3. Isotope systematics of alkaline intrusions

The initial $^{87}Sr/^{86}Sr$ and $^{143}Nd/^{144}Nd$ ratios in alkaline rocks vary from 0.70601 to 0.70915 and 0.512035 to 0.512363, respectively (table 2). The carbonatite samples have similar Nd isotope ratios (0.511803–0.512143, $\varepsilon_{Nd}(t)$ from –10.8 to –4.1) but contain more radiogenic strontium ^{87}Sr (0.70705–0.70947, 40 to 74 $\varepsilon_{Sr}(t)$). This distribution may record the evolution of alkaline magmatism in the case of juvenile PREMA–type material mixed with that of both enriched lithospheric mantle (EM) and upper crust. Similar isotope parameters with +0.2 to−10.1 ε_{Nd} and 8–67 ε_{Sr} are observed in Mesozoic (about 120–220 Ma) alkaline rocks and carbonatites located within orogenic areas of Southern Mongolia, Gornyi Altai and Western Transbaikalia in the Central Asian and Qinling (Central China) belts (figure 2a).

Table 2. Sr–Nd isotopic composition for the alkaline rocks and carbonatites of the Turkestan–Alai[a].

Sample	Massif	Rock, mineral	$^{143}Nd/^{144}Nd$	± 2σ	$^{143}Nd/^{144}Nd(t)$	$\varepsilon_{Nd}(t)$	$^{87}Sr/^{86}Sr$	± 2σ	$^{87}Sr/^{86}Sr(t)$	$\varepsilon_{Sr}(t)$
1	*Z* [b]	Alkaline gabbro	0.512537	12	0.512363	0.16	0.70855	18	0.70833	57.99
3		Alkaline gabbro	0.512208	14	0.512035	–6.25	0.70706	17	0.70691	37.87
5		Foidolite	0.512277	8	0.512198	–3.06	0.70740	15	0.70669	34.73
9		Foidolite	0.512239	13	0.512183	–3.36	0.70824	16	0.70724	42.63
10		Nepheline syenite	0.512507	19	0.512340	–0.29	0.70705	18	0.70601	25.11
XXX	*M* [b]	Nepheline syenite	0.512318	8	0.512132	–4.36	0.71190	19	0.70915	69.71
1-8-2		Glimmerite	0.512148	6	0.511971	–7.50	0.70956	21	0.70872	63.60
1084		Carbonatite	0.512092	4	0.511932	–8.26	0.70779	11	0.70772	49.34
2-3		Carbonatite	0.512272	20	0.512108	–4.82	0.70788	16	0.70783	50.96
1085		Carbonatite	0.512332	14	0.512143	–4.14	0.70882	18	0.70876	64.10
12-5		Carbonatite	0.511966	14	0.511803	–10.8	0.71008	14	0.70947	74.31
1086		Carbonatite, *Ap*[c]					0.70749	9	0.707484	46.05
1080/2		Carbonatite, *Amp*[c]					0.70825	9	0.707045	39.82

[a] Sr–Nd isotopic composition was analyzed by the standard techniques on a Finnigan MAT 262 and an MI 1201–T mass spectrometers in the static mode at the Geological Institute of the Kola Science Center RAS (Apatity).
[b] *Z* – Zardalek intrusion, *M* – Matcha intrusion.
[c] *Ap* = apatite, *Amp* = amphibole.

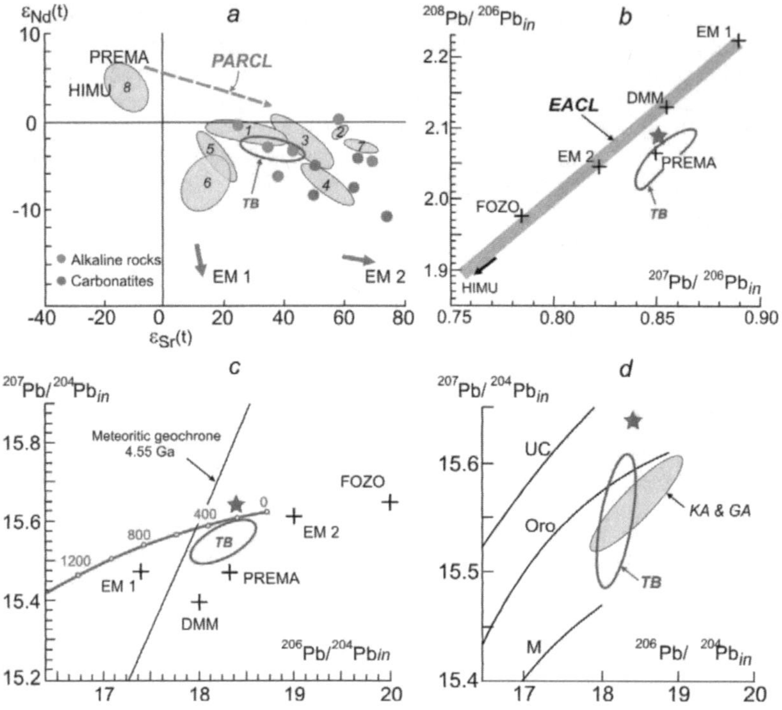

Figure 2. $\varepsilon_{Nd}(t)$ vs. $\varepsilon_{Sr}(t)$ plot (*a*) and Pb isotope composition (*b–d*) of pyrrhotite (blue star) for the Turkestan-Alai alkaline rocks and carbonatites, and some other carbonatite-alkaline complexes from the western Central Asian Orogenic Belt and Precambrian cratons. In panel (*a*), fields in greenish show predominant compositions of Mushugai-Khuduk and Bayan-Khushu (*1*), Lugin-Gol (*2*), Beltsin-Gol and Overmaraat-Gol (*3*) complexes in Mongolia; Darai-Pioz (*4*) complex in Tajikistan; Arshan, Khalyuta, Oshurkovo (*5*) complexes in Western Transbaikalia; Huayangchuan, Yuantou, Qinlongtou, Dashigou, and Shijiawan (*6*) complexes in Central China, lamproites in Gornyi Altai (*7*) and carbonatites from the Siberian, European and Northern American cratons (*8*) [19–23]. Contour *TB* is isotopic range of basalts from the Tarim large igneous province [24–26]; Western CAOB Paleozoic alkaline rocks and carbonatites line (*PARCL*) is according to (Doroshkevich et al., 2012; Vrublevskii et al., 2012; Vrublevskii, 2015) [5–7]; PREMA, HIMU, EM1 and EM2 mantle reservoirs are according to (Zindler and Hart, 1986; Stracke et al., 2005) [27, 28]. In panel (*b–d*), the elements of plumbotectonics, lead isotope evolution and mantle components of PREMA (Prevalent Mantle), FOZO (Focus Zone), HIMU (High-μ), DMM (Depleted MORB Mantle), EM1 (Enriched Mantle 1) and EM2 (Enriched Mantle) reservoirs are according to (Stacey and Kramers, 1975; Zartman and Doe, 1981; Zindler and Hart, 1986; Hart et al., 1992; Stracke et al., 2005; Armienti and Gasperini, 2007) [27–32]; East African Carbonatite Line (*EACL*) is after (Bell and Tilton; 2001) [33]; area *KA&GA* represents alkaline rocks and carbonatites of the Kuznetsk Alatau and Gornyi Altai in the western CAOB [34]. Abbreviations in panel (*d*) stand for UC = upper crust, Oro = Orogenic, M = Mantle.

Initial Pb isotope ratios in pyrrhotite from carbonatites ($^{206}Pb/^{204}Pb$ 18.38; $^{207}Pb/^{204}Pb$ 15.64; $^{208}Pb/^{204}Pb$ 38.41) indicate interaction with EM2 source (figure 2b, c). Enrichment of the mineral in radiogenic ^{207}Pb and ^{206}Pb also allows us to assume participation of upper crustal material in petrogenesis that is typical for orogenic areas (figure 2d). The isotopic composition of lead in pyrrhotite was measured by the standard techniques on a Thermo Scientific Neptune MC-ICP-MS mass spectrometer at the Institute of Geology of Mineral Deposits, Petrography, Mineralogy and Geochemistry RAS (Moscow).

Values $\delta^{13}C$ (from –6.5 to −1.9 ‰) and $\delta^{18}O$ (18–23 ‰) in carbonates correlate directly (table 3, figure 3a); such correlation is often noticed when mixing mantle-derived and sedimentary CO_2 [39, 40]. It looks like crustal contamination occurred at the melting stage, which caused simultaneous enrichment in ^{18}O silicates ($\delta^{18}O$ 11.9–16.5 ‰), apatite ($\delta^{18}O$ 17.2 ‰) and magnetite ($\delta^{18}O$ 11.9–13.5 ‰). Variations of δD in water-bearing silicates (from –41 to –58 ‰; table 3, figure 3b) coincide with range from –40 to –90 ‰ in amphiboles and micas from mantle rocks [38]. However, level of enrichment in ^{18}O corresponds better to metamorphic waters. Studied pyrrhotine is enriched in $\delta^{34}S$ = 12.6–12.8 ‰ comparing to meteorite standard $\delta^{34}S_{CDT}$ ~ 0 ‰, which is possible when mixing of juvenile mantle-derived and sedimentary sulfur. $\delta^{34}S$ values were determined at the isotope laboratory of the Geological Institute RAS, Moscow on a MI 1201–V mass spectrometer.

Table 3. Isotopic composition of O, C and H for the Turkestan–Alai alkaline rocks and carbonatites[a].

Massif	Rock	Sample	Material	$\delta^{13}C$, ‰	$\delta^{18}O$, ‰	δD, ‰	T, °C[d]
M [b]	Carbonatite	1086	*Cb*[c]	–3.9	20.7		
			Ap		17.2		403, 508[e]
	Carbonatite	1085	*Cb*	–4.4	18.0		
			Lpm		11.9	–58	
	Carbonatite	1084	*Cb*	–6.5	18.6		
	Carbonatite	12-5	*Cb*	–4.0	19.8		
	Carbonatite	12-1	*Cb*	–3.0	21.4		
	Carbonatite	1080	*Cb*	–2.7	22.1		
			Phl		16.4		
	Carbonatite	1080/1	*Cb*	–2.3	23.0		
			Mag		13.5		504
	Carbonatite	1080/2	*Cb*	–3.5	21.2		
			Phl		14.7	–57	
			Amp		16.5	–41	
	Carbonatite	1081	*Cb*	–1.9	21.8		
			Phl		16.1		
			Mag		11.9		488
Z [b]	Alkaline gabbro	1	*WR*		9.8		
	Alkaline gabbro	3	*WR*		9.2		
	Foidolite	5	*WR*		12.0		
	Foidolite	9	*WR*		12.4		
	Nepheline syenite	10	*WR*		9.4		

[a] The $\delta^{13}C_{V\text{-}PDB}$, $\delta^{18}O_{V\text{-}SMOW}$ and $\delta D_{V\text{-}SMOW}$ values were determined at the isotope laboratories of the Geological Institute RAS (Moscow) and the Geological Institute SB RAS (Ulan-Ude) on a MI 1201–V, GD–150, Delta V Advantage and a Finnigan MAT 253 mass spectrometers.

[b] *M* – Matcha intrusion, *Z* – Zardalek intrusion.

[c] *Cb* = carbonate, *Ap* = apatite, *Amp* = amphibole, *Phl* = phlogopite, *Lpm* = lepidomelane, *Mag* = magnetite, *Cc* = calcite, *WR* = whole rock.

[d] Isotopic temperatures were calculated from the relation: $1000 \ln\alpha$ (calcite–magnetite) = $5.74 \times 10^6/T^2$, after Clayton and Kieffer, 1991 [43], $1000 \ln\alpha$ (calcite–apatite) = $1.6 \times 10^6/T^2$, after Fortier, Luttge, 1995 [44].

[e] calculated as in Alberti et al., 1999 [45].

4. Discussion

4.1. Conditions of the petrogenesis

Important geochemical characteristics of studied rocks are their general depletion in REE and high values of $^{87}Sr/^{86}Sr(t)$ and $\delta^{18}O$. The patterns we have found out let us assume that evolution of parent alkali-mafic magma in the lithospheric mantle was concurrent with melting and assimilation of crustal rocks (syntexis), and that hybrid SiO_2-saturated melts formed with participation of REE-depleted substrate [9]. This could promote adding of extra LILE into magma. It looks like magma contamination by metasedimentary sediments resulted in additional carbonation and further liquid immiscibility of salted carbonate melt. Similar to model systems [15-17], during formation of Matcha pluton, conjugated silicate liquid with syenite composition retains predominantly Nb, Ta, Zr, Hf, but not REE (figure 1d). Based on experimental data, carbonate melt extract REE, Sr, Ba better when pressure and H_2O content increases. Thus, REE-depleted varieties of studied carbonatites could have been formed at a later stage of the magmatic process under conditions of decompression and degassing of the aqueous fluid.

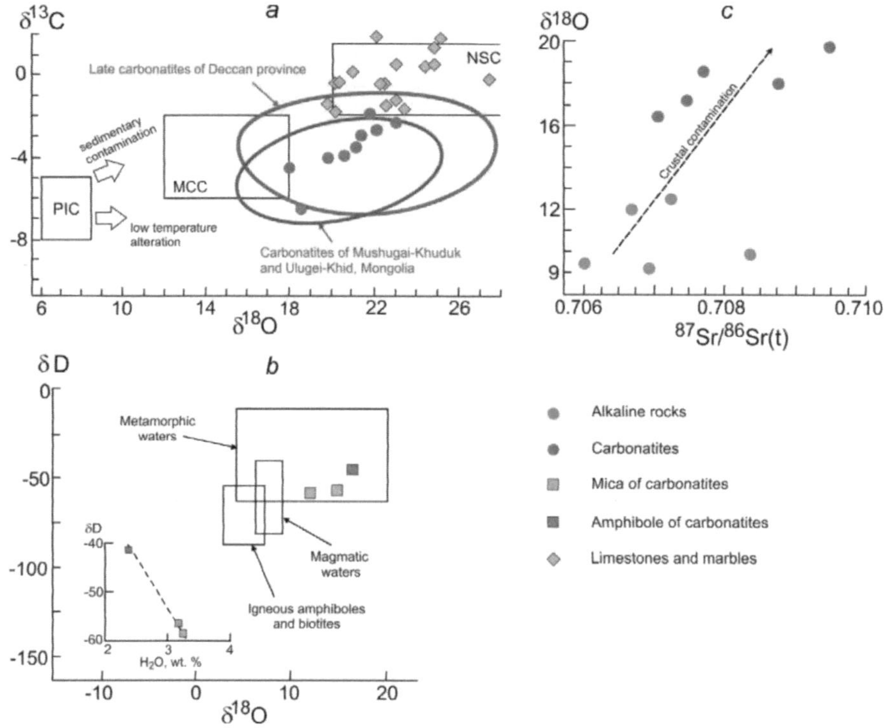

Figure 3. $\delta^{13}C$ vs. $\delta^{18}O$ (*a*), δD vs. $\delta^{18}O$ (*b*), and $^{87}Sr/^{86}Sr(t)$ vs. $\delta^{18}O$ (*c*) plots for the Matcha intrusive complex. PIC (primary igneous carbonatites) field and shifts of changes in the O-C isotopic compositions are according to (Conway and Taylor, 1969; Demény et al., 1998) [35, 36]; MCC = mantle-crustal carbonatites of the western CAOB [37]; NSC = normal sedimentary carbonates [38]. Shown for reference are: compositions of late carbonatites in the Deccan volcanic province in India and carbonatite dikes, veins, breccia and tuffs in Southern Mongolia [39, 40]; magmatic and metamorphic waters, igneous hornblende and biotite [38, 41, 42]. The data on carbonatite-alkaline magmatic complexes hosted by limestones and marbles in the western CAOB are from (Kuleshov, 1986; Doroshkevich et al., 2012; Vrublevskii et al., 2012; Vrublevskii, 2015) [5–7, 40].

Suggested magmatic nature of carbonatites from the Matcha massif is also shown as the earliest high-temperature paragenesis that includes predominant carbonates, Fe–Mg silicates, K–Na feldspars, apatite, and magnetite. In the absence of isotope reversal in the assemblage, the calculated temperatures of $\Delta^{18}O$ isotope fractionation between carbonate, magnetite, and apatite are assumed to be equilibrium and may indicate the end of crystallization at ~400–500°C. Extraction of accessory pyrochlore from the melt is indicated by presence of chemical zoning in grains with enrichment in UO_2 and lowering of Nb, Ca, Na, and F concentrations at a late magmatic stage of carbonatite genesis [12]. Therefore, VA number of vacancies at the A site ~ 0.1–0.2 apfu in crystal-chemical formula of the mineral does not exceed values typical for igneous carbonatites [46].

4.2. Mantle-crust sources and geodynamic setting of the alkaline magmatism
Compared to younger nephelinite-carbonatite volcanism of the East African Rift produced by mixing of magmas from the HIMU and EM1 mantle reservoirs [33], Triassic alkaline intrusions of Turkestan-Altai have less radiogenic isotope composition of Nd and are enriched in ^{87}Sr ($\varepsilon_{Nd}(t)$ ~ –11 to 0, $\varepsilon_{Sr}(t)$ ~ 25–74). Along with primary ratios $^{207}Pb/^{206}Pb$ = 0.851 and $^{208}Pb/^{206}Pb$ = 2.089 in pyrrhotite from carbonatites, this can indicate mixing of mantle plume components PREMA-type with material of enriched lithospheric mantle EM 2 (see figure 2).

Late Paleozoic-Early Mesozoic tectonic units in the adjacent Turkestan-Alai segment are mainly composed from accretionary-collisional clastic-carbonate and volcanic rocks. Therefore, intrusions in island arc complexes may be contaminated with material bearing such geochemical signatures. In fact, distribution of the most HFSE in studied gabbro inherits spectral profile of IAB and partially OIB (see figure 1), meanwhile varieties with the most potassium content correspond to shoshonite series from ACM. Taking this into account, relative enrichment of all rocks in Cs, Rb, Ba, K, Th, U, Sr and low REE content should be considered as the result of magma contamination by crustal components. Obviously, their participation is seen in enrichment of minerals in ^{207}Pb, ^{206}Pb, ^{34}S, ^{13}C, ^{18}O and direct correlation of values $\delta^{18}O$ c $\delta^{13}C$ и $^{87}Sr/^{86}Sr(t)$ (see figures 2, 3). We assume that metamorphic fluid source was buried brines that possibly became mobilized upon conductive heating of carbonate sediments by intruding magma [4].

Note that many basalts in the large igneous province of the northern Tarim craton margin adjacent to the Tien Shan have similar Nd–Sr–Pb isotope compositions (see figure 2). Its formation may result from the activity of the Early Permian (~290 Ma) plume [24]. The Turkestan-Alai postorogenic alkaline magmatism possibly acted in a setting of local extension and was the final stage ("last echo") of the Tarim plume.

5. Conclusion

(1) Postorogenic plutonism in the Turkestan-Alai segment of the Kyrgyz Southern Tien Shan produced alkaline gabbro and alkaline and nepheline syenites coexisting with high-temperature liquation carbonatites. (2) The alkaline rocks and carbonatites have a parental magma source, possibly, due to mixing of PREMA and EM 2-type material. Plume-lithospheric interactions led to syntexis and hybrid melting whereby magma became contaminated with crustal material. (3) The HFSE patterns in early alkaline gabbro bear signatures of interaction between the primary mafic magma and IAB-type material. Strong isotope similarity of alkaline rocks with basalts of the Tarim large igneous province suggests that they have similar plume source.

Acknowledgements
Author is grateful to V.S. Gurskii and A.V. Zhdan, as well as to colleagues from the National Research Tomsk State University (Russia) and from institutions of the Russian Academy of Sciences, who assisted in the research. This work was conducted with the support of Russian Ministry of Education and Science, project 5.2352.2017/PCh.

References

[1] Bell K, Kjarsgaard B A and Simonetti A 1998 Carbonatites – into the twenty-first century *Journal of Petrology* 39 1839–1845
[2] Gwalani L G, Moore K and Simonetti A 2010 Carbonatites, alkaline rocks and the mantle: a special issue dedicated to Keith Bell *Mineralogy and Petrology* 98 5–10

[3] Bell K and Simonetti A 2010 Source of parental melts to carbonatites–critical isotopic constraints *Mineralogy and Petrology* 98 77–89
[4] Pokrovskii B G, Andreeva E D, Vrublevskii V V and Grinev O M 1998 Contamination mechanisms of alkaline-gabbro intrusions in the southern periphery of the Siberian craton: evidence from strontium and oxygen isotopic compositions *Petrologiya* 6 237–251
[5] Doroshkevich A G, Ripp G S, Izbrodin I A and Savatenkov V M 2012 Alkaline magmatism of the Vitim province, West Transbaikalia, Russia: Age, mineralogical, geochemical and isotope (O, C, D, Sr and Nd) data *Lithos*, 152 157–172
[6] Vrublevskii V V, Krupchatnikov V I, Izokh A E and Gertner I F 2012 The alkaline and carbonatitic rocks of Gorny Altai (Edel'veis complex) as indicators of Early Paleozoic plume magmatism in the Central Asian Fold Belt *Russian Geology and Geophysics* 53 721–735
[7] Vrublevskii V V 2015 Sources and geodynamic setting of petrogenesis of the Middle Cambrian Upper Petropavlovka alkaline basic pluton (Kuznetsk Alatau, Siberia) *Russian Geology and Geophysics* 56 379–401
[8] Shinkarev N F 1978 *Origin of Igneous Rocks* (Leningrad: Nedra) p 304 (in Russian)
[9] Nenahov V M 2002 *Late Paleozoic collisional magmatism and metallogeny of the Turkestan-Alai. Metallogeny of Collisional Geodynamic Settings* (Book 2) ed N V Mezelovskii (Moscow: GEOS) pp. 125–220 (in Russian)
[10] Solomovich L I 2007 Postcollisional magmatism in the South Tien Shan Variscan Orogenic Belt, Kyrgyzstan: Evidence for high-temperature and high-pressure collision *Journal of Asian Earth Sciences* 30 142–153
[11] Mayorov I P and Gavrilin R D 1971 Carbonatites of the Late Palaeozoic Turkestan-Alai geosynclines *Sovetskaya Geologiya* 10 111–116
[12] Vrublevskii V V, Bukharova O V and Morova A A 2016 The Nb-Ti mineralization of calcite-dolomite carbonatites of the Matcha alkaline pluton, Alay Range, Southern Tien Shan. *Geosfernye Issledovaniya* 1 52–65
[13] Sun S. and McDonough W F 1989 Chemical and isotopic systematics of oceanic basalts: implications for mantle composition and processes *Geol. Soc. Spec. Publ.*, 42 313–345
[14] Kelemen P B, Hanghøj K and Greene A R 2003 One view of the geochemistry of subduction-related magmatic arcs, with an emphasis on primitive andesite and lower crust Treatise on *Geochemistry* Elsevier Ltd. 3 pp 593–659
[15] Martin L H J, Schmidt M W, Mattsson H B and Guenther D 2013 Element partitioning between immiscible carbonatite and silicate melts for dry and H2O-bearing systems at 1–3GPa *Journal of Petrology* 54 2301–2338
[16] Jones J H, Walker D, Pickett D A, Murell M T and Beattie P 1995 Experimental investigations of the partitioning of Nb, Mo, Ba, Ce, Pb, Ra, Th, Pa, and U between immiscible carbonate and silicate liquids *Geochimica et Cosmochimica Acta* 59 1307–1320
[17] Veksler I V, Petibon C, Jenner G A, Dorfman A M and Dingwell D B 1998 Trace element partitioning in immiscible silicate-carbonate liquid systems: an initial experimental study using a centrifuge autoclave *Journal of Petrology* 39 2095–2104
[18] Guzmics T, Zajacz Z, Mitchell R H, Szabó C and Wälle M 2015 The role of liquid-liquid immiscibility and crystal fractionation in the genesis of carbonatite magmas: insights from Kerimasi melt inclusions *Contributions to Mineralogy and Petrology* 169 article 17
[19] Vladykin N V, Morikiyo T, and Miyazaki T 2005 Sr and Nd isotopes geochemistry of alkaline and carbonatite complexes of Siberia and Mongolia and some geodynamic implications Sources of Deep Magmatism and Plumes Irkutsk 19–37
[20] Vrublevskii V V and Gertner I F 2005 Origin of carbonatite-bearing complexes from fold systems: isotopic evidence for the mantle-crust interaction Sources of Deep Magmatism and Plumes Irkutsk 38–58
[21] Nikiforov A V, Yarmolyuk V V, Kovalenko V I, Ivanov V G and Zhuravlev D Z 2002 Late Mesozoic carbonatites of western Transbaikalia: Isotopic–geochemical characteristics and

sources *Petrologiya* 10 146–164
[22] Xu C, Taylor R N, Kynicky J, Chakhmouradian A R, Song W and Wang L 2011 The origin of enriched mantle beneath North China block: evidence from young carbonatites *Lithos* 127 1–9
[23] Vrublevskii V V, Voitenko N N, Romanov A P, Polyakov G V, Izokh A E, Gertner I F and Krupchatnikov V I 2005 Magma sources of Triassic lamproites of Gornyi Altai and Taimyr: Sr and Nd isotope evidence for plume–lithosphere interaction *Doklady Earth Sciences* 405A 1365–1367
[24] Zhang Y, Liu J and Guo Z 2010 Permian basaltic rocks in the Tarim basin, NW China: implications for plume–lithosphere interaction *Gondwana Research* 18 596–610
[25] Li Y Q, Li Z L, Sun Y L, Santosh M, Langmuir C H, Chen H L, Yang S F, Chen Z X and Yu X 2012 Platinum-group elements and geochemical characteristics of the Permian continental flood basalts in the Tarim Basin, northwest China: Implications for the evolution of the Tarim Large Igneous Province *Chemical Geology* 328 278–289
[26] Wei X, Xu Y G, Feng Y X and Zhao J X 2014 Plume-lithosphere interaction in the generation of the Tarim large igneous province, NW China: geochronological and geochemical constraints *American Journal of Science* 314 314–356
[27] Zindler A and Hart S R 1986 Chemical geodynamics Ann. Rev. Earth Planet. Sci., 14 493–571
[28] Stracke A, Hofmann A W and Hart S R 2005 FOZO, HIMU, and rest of the mantle zoo *Geochemistry, geophysics, geosystems* 6 1–20.
[29] Stacey J C and Kramers J D 1975 Approximation of terrestrial lead isotope evolution by a two-stage model *Earth and Planetary Science Letters* 26 207–221
[30] Zartman R E and Doe B R 1981 Plumbotectonics – the model *Tectonophysics* 75 135–162
[31] Hart S R, Hauri E H, Oschmann L A and Whitehead J A 1992 Mantle plumes and entrainment: isotopic evidence Science 256 517–520
[32] Armienti P and Gasperini D 2007 Do we really need mantle components to define mantle composition? *Journal of Petrology* 48 693–709
[33] Bell K and Tilton G R 2001 Nd, Pb and Sr isotopic compositions of East African carbonatites: evidence for mantle mixing and plume inhomogeneity *Journal of Petrology* 42 1927–1945
[34] Vrublevskii V V, Krupchatnikov V I and Gertner I F 2017 The role of mantle in the development of Early Paleozoic oceanic islands volcanism (for the geochemical data of OIB from SE Altai Mountains) *Geosfernye Issledovaniya* 1 28–38
[35] Conway C H and Taylor H P 1969 18O/16O and 13C/12C ratios of coexisting minerals in the Oka and Magnet Cove carbonatite bodies *Journal of Geology* 77 618–626
[36] Demény A, Ahijado A, Casillas R and Vennemann T W 1998 Crustal contamination and fluid/rock interaction in the carbonatites of Fuerteventura (Canary Islands, Spain): a C, O, H isotope study *Lithos* 44 101–115
[37] Vrublevskii V V 2017 Typical δ13C–δ18O ratios in mantle-crustal carbonatites of the CAOB Abstr. 27 th Goldschmidt conference Paris
[38] Pokrovskii B G 2000 Crustal Contamination of Mantle Magmas: Evidence from Isotope Geochemistry (Moscow: Nauka) p 228. (in Russian)
[39] Ray J S, Ramesh R, Pande K, Trivedi J R, Shukla P N and Pate P.P. 2000 Isotope and rare earth element chemistry of carbonatite-alkaline complexes of Deccan volcanic province: implications to magmatic and alteration processes *Journal of Asian Earth Sciences* 18 177–194
[40] Kuleshov V N 1986 Isotopic composition and origin of deep carbonates (Moscow: Nauka) p 124 (in Russian)
[41] Sheppard S M F 1986 Characterization and isotopic variations in natural waters *Reviews in Mineralogy* 16 165–184
[42] Taylor H P Jr and Sheppard S M F 1986 Stable isotopes in high temperature geological processes *Reviews in Mineralogy* 16 227–269

[43] Clayton R N and Kieffer S W 1991 Oxygen isotopic thermometer calibrations *The Geochem. Soc. Spec. Publ.*, 3 3–10
[44] Fortier S M and Luttge A 1995 An experimental calibration of the temperature dependence of oxygen isotope fractionation between apatite and calcite at high temperatures (350–800oC) *Chemical Geology* 125 281–290
[45] Alberti A, Castorina F, Censi P, Comin-Chiaramonti P and Gomes C B 1999 Geochemical characteristics of cretaceous carbonatites from Angola *Journal of African Earth Sciences* 29 735–759
[46] Nasraoui M and Bilal E 2000 Pyrochlores from the Lueshe carbonatite complex (Democratic Republic of Congo): a geochemical record of different alteration stages *Journal of Asian Earth Sciences* 18 237–251

IOP Conf. Series: Earth and Environmental Science **110** (2017) 012024 doi:10.1088/1755-1315/110/1/012024

The problem of genesis and systematic of sedimentary units of hydrocarbon reservoirs

E N Zhilina [1], **O S Chernova**[2]

[1]National Research Tomsk State University, Tomsk, Russia
[2]National Research Tomsk Polytechnical University, Tomsk, Russia

E-mail: elena@ggf.tsu.ru

Abstract. The problem of identifying and ranking sedimentation, facies associations and their constituent parts - lithogenetic types of sedimentary rocks was considered. As a basis for paleo-sedimentary modelling, the author has developed a classification for terrigenous natural reservoirs,that for the first time links separate sedimentological units into a single hierarchical system. Hierarchy ranking levels are based on a compilation of global knowledge and experience in sediment geology, sedimentological study and systematization, and data from deep-well coresrepresentingJurassichydrocarbon-bearing formationsof the southeastern margin of the Western Siberian sedimentary basin.

At present, work in the field of reservoir sedimentology has increased noticeably in Russia and abroad. The modern studies of oil-bearing formationsfollow a long and well-established focus on genetic and historical-geological aspects, allowing for the determination of sedimentary conditions and the distribution patterns of sedimentary complexes capable of generating and retaining hydrocarbons. Therefore, the most important task of sedimentary oil and gas geology from an applied perspective is to examine the process of sediment generation in time and space in order to successfully predict the size and shape of the oil and gas reservoirs and which can be used in building the correct 3D geological models.

The basis for paleo-sedimentological modelling is that the morphology and filtration-capacity of productive horizons are to a large extent predetermined by the physical and geographic conditions of the ancient sedimentary environments, which are closely linked to the tectonics of the region, its geological age, and the direction and intensity of secondary processes.

The problem of identifying and ranking the sedimentary environments and their facies associations has been addressed over many years in numerous published works by many authors and groups of authors. The research hasidentifieda number of conditions that have a significant impact on the formation and placement of hydrocarbon deposits in the sedimentary space. Most authors have assigned the most important role to geomorphological, lithologic-stratigraphic and paleogeographic factors in determining all the basic characteristics of the final product of the sedimentation system. The paleo-sedimentology of the modelling in this area is based on sedimentological studies to identify and examine the various-ranked sedimentary elements of natural terrigenous systems emerging from certain natural processes.

There are several steps to successfully address the problem of developing a classification of various sedimentary units in the context of detailed elaboration of popular sedimentological models. Firstly, it is necessary, to have an identical understanding between researchers of the shared object of research, thesedimentology of natural reservoirs.Secondly, there must be consistency in the use ofspecial sedimentological nomenclature, the terms of which are often interpreted arbitrarily and thus have a mixed meaning.Thirdly, there is long overdue need to address the issues of classification ofsedimentation, facies and their component lithogenetic types, on a single methodical basis.

Abundant Russian and English geologic literature contains extensive, synthesized material on a variety of working classifications for the conditions of sedimentation, facies and types of rocks in different natural environments which were created according to the needs of authors. A singletaxonomy, which is a hierarchical structure linking separate sedimentological objects (sedimentological unit) into a strict hierarchical sequence, does not exist today. This is why there are e often significant differences and terminological difficulties in facies research into units belonging to different sedimentary levels.

The problem of defining and merging different sedimentary data into different groups arises in the early stages of the study-in the description and assigning of lithogenetic types. Subsequently, in the definition of facies associations, which in turn, compose large sedimentary cycles reflecting ancient landscape characteristics, the problem is only exacerbated.

In the analysis and prediction of the ancient conditions of sedimentation, the researcher faces the inevitable need to link multi-dimensional parts of the internal architecture of natural existing systems. The primary structures from the original sedimentation situation, recorded in the well core material are not fully visible and are significantly affected by millions of years of secondary changes of the sedimentary material, which have fundamentally changed primary character of the rocks. Therefore, the restoration of the ancient depositional environments using well core data is an inherently complex geological task. For its correct interpretation additional knowledge and understanding are required:

(1) Processes of sedimentation and the results of their formation *(what type of a deposit is formed as the result of certain sedimentary process?)*;

(2) Formation mechanisms of sedimentary processes;

3) Textural characteristics of sedimentary material;

(4) Basics ofpetrography (*composition of the units*) and geochemistry (*the conditions of the main diagenetic processes leading to the inevitable modifications of natural reservoirs at the micro and macro level*);

(5) Basic ofpaleontology and historical geology (*understanding of major groups of organisms and their living conditions).*

It is also necessary to understand the hierarchy of the main sedimentary units, to know their paragenesis in the transgressive and regressive development of the region, and to have an objective understanding of modern sedimentary processes and an understanding of the parameters controlling modern deposition environments.

When working with drill core material, it is important to understand the difference between modern and ancient paleofaciescomplexes: to know the typical facies associations, both in modern and ancient sediments; to have structural analysis skills (have experience in interpretation of textures and modern and lithified strata; and to know the identification criteria for each sedimentary environment.

In modern sedimentological and facies-cycle research on sedimentary horizons a unified conceptual approach is used for the interpretation of ancient deposition systems, based on studying and identification of genetic indications of their rocks specified on mechanisms and the conditions of their formation. It is obviously important to have the hierarchical scheme reflecting the full diversity of natural reservoirs.

Experts of petroleum engineering deal with products of ancient sedimentary systems, with the specific geological bodies representing existing systems, having certain orientation in the accommodation space, a defined stratigraphic age and certain morphological form. All existing systems (oil and gas reservoirs) have strict hierarchical association with each other. Knowing these associations (chains) and laws of their development it is possible to predict the occurrence of those or other facies associations and to recognize ancient deposition characteristics fromdrill core wells, i.e. to solve the reverse problem of geology. For palaeo-sedimentological modeling it is also extremely important to define the main sedimentary trends which are in close geomorphological and dynamic dependence.

Thus, natural reservoirs can be considered as fossil paleo-morphostructural forms with certain dimensions, capacity, and morphology of the inner structure, and the nature of contact with the host formations. All these parameters are controlled by the depositional environment and are directly connected with geomorphological and dynamic conditions in sedimentary basin

The hierarchical structure in question is the result of a synthesis of world experience with ancient sedimentary sequences, preliminary work in the systematization of the English and Russian sedimentological nomenclature, and lithological-facies and paleogeographic studies of Jurassic terrigenous sediments of south-eastern Western Siberia.

The classification of sedimentologicalobjectsis built on a hierarchical basis, including five levels based on parameters that mutually exclude each other in different classes.Each parameter can be objectively defined, identified, and uniquely valued. The criteria for the established hierarchical structure were: the deposition of the terrigenous material and its form of accumulation (geomorphological feature) and the position of the lithified sediment in the typical sedimentation profile.

The main sedimentary characteristics, "sedimentary environment", "macrofacies", "facies association", "facies", "lithogenetic type", "hierarchical rank" are considered by various authors, both in modern and ancient sedimentary environments. The concepts adopted in the classification are defined as follows.

*Sedimentary environment:*the ancient landscape, which includes the physio-chemical conditions of the sediment formation, which are characteristic of the removal area, and accommodation space.

Facies: the final (specific) product of a particular geological process operating in a particular region with geomorphological features, or a real geological body, with three-dimensional characteristics.

*Lithogenetic type:*lithological type of rock possessing a set of genetic features, caused by the processes of its formation (in English-speaking terminology - *lithofacies*).

Hierarchical rank: the level assigned to a specific sedimentological unit is considered when building the hierarchical relationships of sedimentary objects.

According to this approach, the sedimentation, which formed the main natural reservoirs of hydrocarbons during the Jurassic age, within the province of the south-eastern part of the Western Siberian Basin, are traditionally divided into five major groups, comprising most of the other groupings.1) ***continental environments*** *(alluvial fan, fluvial, lacustrine, poludal)*; 2) ***transitional environments*** *(deltaic, estuarine)*; 3) ***nearshore environments*** *(tidal and beaches*); 4) ***shallow-marine environment*** *(sandy bars, barriers and lagoons*); 5) ***marine environment*** *(shelf sandy ridges)*. The full scheme of sedimentological units is given in the article "The Systematics and Hierarchy of Natural Reservoirs as Foundation of Palaeo-sedimentological Modelling" [1].

In order to facilitate understanding and to eliminate contradictions, the terms are presented in both Russian and English. Aeolian, glacial and volcanogenic sedimentary environments, have not been included in this scheme since these are not typical for this climatic zone during Jurassic sedimentation, but for these a separate classification can be created on a similar basis.

The sedimentological characteristics, that describe the reservoirs of humid-climate regions, display a complex hierarchy of the alternating lithogenetictypes that form the sets of facies associations of the main terrigenous environments.

Accordingly, the proposed hierarchy allows creating sedimentological reconstruction within the limits of certain sedimentary systems, represented by typical sets of reservoirs. For each described sedimentary environment are revealed diagnostic criteria, allowing determination of the genesis of ancient sedimentary complexes.

IOP Conf. Series: Earth and Environmental Science **110** (2017) 012024 doi:10.1088/1755-1315/110/1/012024

In the variation in the lowest level of systematics (lithogenetic type), the main characteristic of natural systems, its emergence is reflected. It means that there are no strict unequivocal boundaries between levels of sedimentation -facies, facies associations and sedimentation conditions. There is always a system of transitions between hierarchical levels, as defined by the work of R.C. Selly, H.-E. Reinek and I.B. Singh, H.G. Reading [2, 3, 4].

The proposed systematization of sedimentological objects may be useful for petroleum specialists studying ancient facies sequences where the morphology of the reservoirs is virtually non-preserved in pristine form. However, this systematization does not cover all kinds of natural setting, but can be a starting point for such geological discussions.

References

[1] Chernova OS 2010 The systematic and hierarchy of terrigeneous reservoirs as foundation of palaeosedimentological modeling // *Izvestiya TPUVol 317 Geology, investigation and exploration activities of Siberia* Tomsk pp116-122

[2] Reading HG 1996 *Sedimentary environments: processes, facies and stratigraphy* 3rd ed. Blackwell Oxford 688 p

[3] Reineck HE and Singh IB 1975 *Depositional Sedimentary Environments Springer-Verlag*Berlin Heidelberg 439 p

[4] Selley RC 1996 *Ancient Sedimentary Environments and Their Subsurface Diagnosis* 4th ed. Chapman & Hall London 30 p

Institute of Physics
Dirac House, Temple Back
Bristol BS1 6BE UK

ISSN: 1755-1307
ISBN 978-1-5108-5530-4